Advances in Intelligent Systems and Computing

Volume 238

T0137762

Series Editor

Janusz Kacprzyk, Warsaw, Poland

For further volumes:
http://www.springer.com/series/11156

Jeng-Shyang Pan · Pavel Krömer
Václav Snášel

Editors

Genetic and Evolutionary Computing

Proceedings of the Seventh International
Conference on Genetic and Evolutionary
Computing, ICGEC 2013, August 25–27,
2013 - Prague, Czech Republic

 Springer

Editors
Jeng-Shyang Pan
Department of Electronic Engineering
National Kaohiung University of Applied
   Sciences
Kaohsiung
Taiwan R.O.C.

Václav Snášel
Department of Computer Science
Faculty of Ele. Eng. and Computer Science
VŠB-TUO
Ostrava-Poruba
Czech Republic

Pavel Krömer
Department of Computer Science
Faculty of Ele. Eng. and Computer Science
VŠB-TUO
Ostrava-Poruba
Czech Republic

ISSN 2194-5357   ISSN 2194-5365   (electronic)
ISBN 978-3-319-01795-2   ISBN 978-3-319-01796-9   (eBook)
DOI 10.1007/978-3-319-01796-9
Springer Cham Heidelberg New York Dordrecht London

Library of Congress Control Number: 2013945191

Printed on acid-free paper

Springer is part of Springer Science+Business Media (www.springer.com)

Preface

This volume of Advances in Intelligent Systems and Computing contains accepted papers presented at ICGEC 2013, the 7th International Conference on Genetic and Evolutionary Computing. The conference this year was technically co-sponsored by The Waseda University in Japan, Kaohsiung University of Applied Science in Taiwan, and VŠB-Technical University of Ostrava. ICGEC 2013 was held in Prague, Czech Republic. Prague is one of the most beautiful cities in the world whose magical atmosphere has been shaped over ten centuries. Places of the greatest tourist interest are on the Royal Route running from the Powder Tower through Celetná Street to Old Town Square, then across Charles Bridge through the Lesser Town up to the Hradčany Castle. One should not miss the Jewish Town, and the National Gallery with its fine collection of Czech Gothic art, collection of old European art, and a beautiful collection of French art.

The conference was intended as an international forum for the researchers and professionals in all areas of genetic and evolutionary computing. The main topics of ICGEC 2013 included Intelligent Computing, Evolutionary Computing, Genetic Computing, and Grid Computing.

The organization of the ICGEC 2013 was entirely voluntary. The review process required an enormous effort from the members of the International Technical Program Committee, and we would therefore like to thank all its members for their contribution to the success of this conference. We would like to express our sincere thanks to the invited session organizers, to the host of ICGEC 2013, VŠB – Technical University of Ostrava, and to the publisher, Springer, for their hard work and support in organizing the conference.

August 2013

Jeng-Shyang Pan
Pavel Krömer
Václav Snášel

Preface

This volume of Advances in Intelligent Systems and Computing contains accepted papers presented at ICGEC 2013, the 7th International Conference on Genetic and Evolutionary Computing. The conference this year was technically co-sponsored by The Waseda University in Japan, Kaohsiung University of Applied Science in Taiwan, and VŠB-Technical University of Ostrava. ICGEC 2013 was held in Prague, Czech Republic. Prague is one of the most beautiful cities in the world whose magical atmosphere has been shaped over ten centuries. Places of the greatest tourist interest are on the Royal Route running from the Powder Tower through Celetná Street to Old Town Square, then across Charles Bridge through the Lesser Town up to the Hradčany Castle. One should not miss the Jewish Town, and the National Gallery with its fine collection of Czech Gothic art, collection of old European art, and a beautiful collection of French art.

The conference was intended as an international forum for the researchers and professionals in all areas of genetic and evolutionary computing. The main topics of ICGEC 2013 included Intelligent Computing, Evolutionary Computing, Genetic Computing, and Grid Computing.

The organization of the ICGEC 2013 was entirely voluntary. The review process required an enormous effort from the members of the International Technical Program Committee, and we would therefore like to thank all its members for their contribution to the success of this conference. We would like to express our sincere thanks to the invited session organizers, to the host of ICGEC 2013, VŠB – Technical University of Ostrava, and to the publisher, Springer, for their hard work and support in organizing the conference.

August 2013

Jeng-Shyang Pan
Pavel Krömer
Václav Snášel

Organization

Conference Chair

Václav Snášel   VŠB - Technical University of Ostrava

Program Chair

Yong Xu   Harbin Institute of Technology Shenzhen Graduate School, China

Advisory Chair

Jeng-Shyang Pan   Kaohsiung University of Applied Sciences, Taiwan

Invited Session Chair

Wu-Chih Hu   National Penghu University of Science and Technology, Taiwan

Honorary Chairs

Junzo Watada   Waseda University, Japan
Bin-Yih Liao   Kaohsiung University of Applied Sciences, Taiwan

Publicity Chair

Jaroslav Pokorný   Charles University, Czech Republic

International Program Committee

Akira Asano	Hiroshima University, Japan
Mohsen Askari	University of Technology Sydney, Australia
Dariusz Barbucha	Gdynia Maritime University, Poland
Jonathan Hoyin Chan	King Mongkut's University of Technology Thonburi, Thailand
Chin-Chen Chang	Feng Chia University, Taiwan
Feng-Cheng Chang	Tamkang University, Taiwan
Yueh-Hong Chen	Far East University, Taiwan
Chao-Chun Chen	National Cheng Kung University, Taiwan
Rung-Ching Chen	Chaoyang University of Technology, Taiwan
Shi-Jay Chen	National United University, Taiwan
Pei-Yin Chen	National Cheng Kung University, Taiwan
Tsung-Che Chiang	National Taiwan Normal University, Taiwan
Shu-Chuan Chu	Cheng-Shiu University, Taiwan
Yi-Nung Chung	National Changhua University of Education, Taiwan
Maurice Clerc	Independent Consultant, France
Martine De Cock	Ghent University, Belgium
Jose Alfredo F. Costa	Federal University, Brazil
Zhihua Cui	Taiyuan University of Science and Technology, China
Alexander A. Frolov	Russian Academy of Sciences, Russia
Amparo Fúster-Sabater	Spanish Scientific Research Council, Spain
Petr Gajdoš	VŠB-Technical University of Ostrava, Czech Republic
Xiao-Zhi Gao	Helsinki University of Technology, Finland
Alexander Gelbukh	National Polytechnic Institute, Mexico
Gheorghita Ghinea	Brunel University, UK
Massimo De Gregorio	Istituto di Cibernetica, Italy
Stefanos Gritzalis	University of the Aegean, Greece
Ramin Halavati	Sharif University of Technology, Iran
Enrique Herrera-Viedma	University of Granada, Spain
Jiun-Huei Ho	Cheng Shiu University, Taiwan
Chian C. Ho	National Yunlin University of Science & Technology, Taiwan
Wei-Chiang Hong	Oriental Institute of Technology, Taiwan
Cheng-Hsiung Hsieh	Chaoyang University of Technology, Taiwan
Chien-Chang Hsu	Fu Jen Catholic University, Taiwan
Shu-Hua Hu	Jinwen University of Science and Technology, Taiwan
Yongjian Hu	South China University of Technology, China

Ming-Wen Hu	Tamkang University, Taiwan
Wu-Chih Hu	National Penghu University of Science and Technology, Taiwan
Deng-Yuan Huang	Dayeh University, Taiwan
Tien-Tsai Huang	Lunghwa University of Science and Technology, Taiwan
Hsiang-Cheh Huang	National Kaohsiung University, Taiwan
Yung-Fa Huang	Chaoyang University of Technology, Taiwan
Dusan Husek	Academy of Sciences of the Czech Republic, Czech Republic
Donato Impedovo	University of Bari, Italy
Albert B. Jeng	Jinwen University of Science and Technology, Taiwan
Isabel S. Jesus	Institute of Engineering of Porto, Portugal
Jie Jing	Zhejiang University of Technology, China
Estevam R. Hruschka Jr.	Federal University of Sao Carlos, Brazil
Muhammad Khurram Khan	King Saud University, Kingdom of Saudi Arabia
Mario Koeppen	Kyushu Institute of Technology, Japan
Gabriella Kokai	Fraunhofer Institute for Integrated Circuits, Germany
Miloš Kudělka	VŠB-Technical University of Ostrava, Czech Republic
Hsu-Yang Kung	National Pingtung University of Science and Technology, Taiwan
Yau-Hwang Kuo	National Cheng Kung University, Taiwan
Kun-Huang Kuo	Chienkuo Technology University, Taiwan
Weng Kin Lai	MIMOS Berhad, Malaysia
Jenn-Kaie Lain	National Yunlin University of Science and Technology, Taiwan
Kwok-Yan Lam	Tsinghua University, China
Jouni Lampinen	University of Vaasa, Finland
Sheau-Dong Lang	Central Florida University, USA
Chang-Shing Lee	National University of Tainan, Taiwan
Jung-San Lee	Feng Chia University, Taiwan
Huey-Ming Lee	Chinese Culture University, Taiwan
Shie-Jue Lee	National Sun Yat-Sen University, Taiwan
Jorge E. Núñez Mc Leod	Universidad Nacional' de Cuyo, Argentina
Yue Li	Nankai University, China
Chang-Tsun Li	University of Warwick, UK
Jun-Bao Li	Harbin Institute of Technology, China
Guan-Hsiung Liaw	I-Shou University, Taiwan
Tsung-Chih Lin	Feng-Chia University, Taiwan
Lily Lin	China University of Technology, Taiwan
Yuh-Chung Lin	Tajen University, Taiwan

Chen-Da Tsai	Far East University, Taiwan
Tsung-Han Tsai	National Central University, Taiwan
Eiji Uchino	Yamaguchi University, Japan
Sebastin Ventura	University of Cordoba, Spain
Brijesh Verma	Central Queensland University, Australia
Michael N. Vrahatis	University of Patras, Greece
Shiuh-Jeng Wang	Central Police University, Taiwan
Feng Wang	Kunming University of Science and Technology, China
Ling Wang	Tsinghua University, China
Lidong Wang	National Computer Network Emergency Technical Team/ Coordination Center of China, China
Di Wang	Khalifa University, UAE
Yin Chai Wang	University Malaysia Sarawak, Malaysia
Kwok-Wo Wong	City University of Hong Kong, Hong Kong
Michal Wozniak	Wroclaw University of Technology, Poland
Ruqiang Yan	Southeast University, China
Chyuan-Huei Thomas Yang	Hsuan Chuang University, Taiwan
Chung-Huang Yang	National Kaohsiung Normal University, Taiwan
Sheng-Yuan Yang	St. John's University, Taiwan
Li Yao	University of Manchester, UK
Jui-Feng Yeh	National Chiayi University, Taiwan
Show-Jane Yen	Mining Chuan University, Taiwan
Fa-xin Yu	Zhejiang University, China
Ming Yuchi	Huazhong University of Science and Technology, China
Ivan Zelinka	VŠB-Technical University of Ostrava, Czech Republic
Zhigang Zeng	Wuhan University of Technology, China
Xiao-Jun Zeng	University of Manchester, UK
Zhiyong Zhang	Henan University of Science and Technology, China
Yong Zhang	Shenzhen University, China
Xinpeng Zhang	Shanghai University, China
Yongping Zhang	Hisilicon Technologies Co., Ltd, China

Sponsoring Institutions

VŠB - Technical University of Ostrava, Czech Republic
Waseda University, Japan
Kaohsiung University of Applied Science,Taiwan

Contents

Advances in Genetic and Evolutionary Computing

Multimedia Innovative Computing

Innovative Intelligent Internet Technology

Multimedia Social Networks and Soft Computing

Pattern Recognition and Multimedia Signal Processing

High Speed Computation and Applications in Information Security Systems

Forecast Models of Partial Differential Equations Using Polynomial Networks

Ladislav Zjavka

VŠB-Technical University of Ostrava, IT4innovations Ostrava, Czech Republic
ladislav.zjavka@vsb.cz

Abstract. Unknown data relations can describe lots of complex systems through partial differential equation solutions of a multi-parametric function approximation. Common neural network techniques of pattern classification or function approximation problems in general are based on whole-pattern similarity relationships of trained and tested data samples. They apply input variables of only absolute interval values, which may cause problems by far various training and testing data ranges. Differential polynomial neural network is a new type of neural network developed by the author, which constructs and substitutes an unknown general sum partial differential equation, defining a system model of dependent variables. It generates a total sum of fractional polynomial terms defining partial relative derivative dependent changes of some combinations of input variables. This type of regression is based only on trained generalized data relations. The character of relative data allows processing a wider range of test interval values than defined by the training set. The characteristics of differential equation solutions also in general facilitate a greater variety of model forms than allow standard soft computing methods.

Keywords: polynomial neural network, partial differential equation composition, sum relative derivative term, multi-parametric function approximation.

1   Introduction

Differential equation solutions allow define models for a variety of pattern recognition [10] and primarily function approximation problems, applying genetic programming (GP) techniques [3] or an artificial neural network (ANN) construction [9]. A common ANN operating principle is based on entire similarity relationships of new presented input patterns with the trained ones. It does not allow for eventual forthright data relations of variables, which might define a generalized model. Common soft-computing techniques utilize only absolute interval values of input variables, which are not able to describe a wider range of applied data [1]. The generalization from the training data set may be difficult or problematic if the model has not been trained with inputs around the range covered testing data [2]. If training data involve relations, which may become stronger or weaker character, the network

J.-S. Pan et al. (eds.), *Genetic and Evolutionary Computing*,
Advances in Intelligent Systems and Computing 238,
DOI: 10.1007/978-3-319-01796-9_1, © Springer International Publishing Switzerland 2014

model could generalize it into wide-range valid values. Differential polynomial neural network (D-PNN) is a new neural network type, which creates and resolves an unknown partial differential equation (DE) following a data description of a multi-parametric function approximation. A general DE is substituted producing sum of fractional polynomial derivative terms, forming a system model of dependent variables. In contrast with the ANN functionality, each neuron can direct take part in the network total output calculation, which is formed by the sum of active neuron outputs. The study tried to create a neural network, which function approximation is based on any dependent data relations. ANN solutions do not provide model specifications in the form of a math description. The model appears to the users as a "black box". D-PNN combines the neural network functionalities with some math techniques of differential equation solutions. Its models are the boundary of neural network and exact computational techniques.

$$y = a_0 + \sum_{i=1}^{m} a_i x_i + \sum_{i=1}^{m} \sum_{j=1}^{m} a_{ij} x_i x_j + \sum_{i=1}^{m} \sum_{j=1}^{m} \sum_{k=1}^{m} a_{ijk} x_i x_j x_k + ... \qquad (1)$$

m – number of variables $X(x_1, x_2, ... , x_m)$ $A(a_1, a_2, ... , a_m), ...$ - vectors of parameters

D-PNN's block skeleton is formed by the GMDH (Group Method of Data Handling) polynomial neural network, which was created by a Ukrainian scientist Aleksey Ivakhnenko in 1968, when the back-propagation technique was not known yet [4]. General connection between input and output variables is possible to express by the Volterra functional series, a discrete analogue of which is Kolmogorov-Gabor polynomial (1). This polynomial can approximate any stationary random sequence of observations and can be computed by either adaptive methods or system of Gaussian normal equations [6]. GMDH decomposes the complexity of a process into many simpler relationships each described by low order polynomials (2) for every pair of the input values. Typical GMDH network maps a vector input x to a scalar output y, which is an estimate of the true function $f(x) = y^t$.

$$y = a_0 + a_1 x_i + a_2 x_j + a_3 x_i x_j + a_4 x_i^2 + a_5 x_j^2 \qquad (2)$$

2   General Partial Differential Equation Composition

The basic idea of the D-PNN is to compose and substitute a general sum partial differential equation (3), which is not known in advance and can describe a system of dependent variables, with a generated sum of fractional relative multi-parametric polynomial derivative terms (5).

$$a + \sum_{i=1}^{n} b_i \frac{\partial u}{\partial x_i} + \sum_{i=1}^{n} \sum_{j=1}^{n} c_{ij} \frac{\partial^2 u}{\partial x_i \partial x_j} + ... = 0 \qquad u = \sum_{k=1}^{\infty} u_k \qquad (3),(4)$$

$u = f(x_1, x_2, ... , x_n)$ – searched function of all input variables
$a, B(b_1, b_2, ..., b_n), C(c_{11}, c_{12}, ...)$ – polynomial parameters

Partial DE terms are formed according to the adapted method of integral analogues, which is a part of the similarity model analysis. It replaces mathematical operators and symbols of a DE by ratio of corresponding values. Derivatives are replaced by their integral analogues, i.e. derivative operators are removed and simultaneously with all operators are replaced by similarly or proportion signs in equations to form dimensionless groups of variables [5].

$$u_i = \frac{\left(a_0 + a_1 x_1 + a_2 x_2 + a_3 x_1 x_2 + a_4 x_1^2 + a_5 x_2^2 + ...\right)^{m/n}}{b_0 + b_1 x_1 + ...} = \frac{\partial^m f(x_1,...,x_n)}{\partial x_1 \partial x_2 ... \partial x_m} \tag{5}$$

n – combination degree of a complete polynomial of n-variables
m – combination degree of denominator variables

The fractional polynomials (5) define partial derivative relations of n-input variables. The numerator of a DE term (5) is a polynomial of all n-input variables and partly defines an unknown function u of eq. (4). The denominator is a derivative part, arose from the partial derivation of the complete n-variable polynomial in respect to competent variable(s). The root function of numerator takes the polynomial into competent combination degree to get the dimensionless values [5].

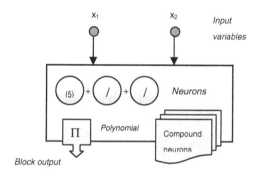

Fig. 1. D-PNN block of basic and compound neurons

Blocks of the D-PNN (Fig.1.) consist of derivative neurons, one for each fractional polynomial derivative combination, so each neuron is considered a summation DE term (4). Each block contains a single output polynomial (2), without derivative part. Neurons do not affect the block output but participate direct in the total network output sum calculation of a DE composition. Each block has 1 and neuron 2 vectors of adjustable parameters a, resp. a, b.

$$F\left(x_1, x_2, u, \frac{\partial u}{\partial x_1}, \frac{\partial u}{\partial x_2}, \frac{\partial^2 u}{\partial x_1^2}, \frac{\partial^2 u}{\partial x_1 \partial x_2}, \frac{\partial^2 u}{\partial x_2^2}\right) = 0 \tag{6}$$

where $F(x_1, x_2, u, p, q, r, s, t)$ is a function of 8 variables

In the case of 2 input variables the 2nd odder partial DE can be expressed in the form of eq. (6), which involve all derivative terms of variables applied by the GMDH polynomial (2). D-PNN processes these 2-combination square polynomials of blocks and neurons, which form competent DE terms of eq. (5). Each block so include 5 basic neurons of derivatives x_1, x_2, x_1x_2, x_1^2, x_2^2 of the 2nd order partial DE (6), which is most often used to model physical or natural systems.

3   Differential Polynomial Neural Network

Multi-layered networks forms composite polynomial functions (Fig.2.). Compound terms (CT), i.e. derivatives in respect to variables of previous layers, are calculated according to the composite function partial derivation rules (7)(8). They are formed by products of partial derivatives of external and internal functions.

$$F(x_1, x_2, \ldots , x_n) = f(y_1, y_2, \ldots , y_m) = f(\phi_1(X),\ \phi_2(X),\ldots,\ \phi_m(X)) \qquad (7)$$

$$\frac{\partial F}{\partial x_k} = \sum_{i=1}^{m} \frac{\partial f(y_1, y_2,\ldots,y_m)}{\partial y_i} \cdot \frac{\partial \phi_i(X)}{\partial x_k} \qquad k=1, \ldots , n \qquad (8)$$

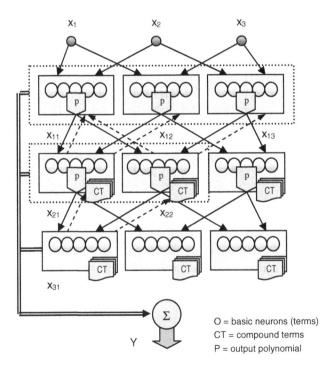

Fig. 2. 3-variable multi-layered backward D-PNN with 2-variable combination blocks

Thus blocks of the 2nd and following hidden layers are additionally extended with compound terms (neurons), which form composite derivatives utilizing outputs and inputs of back connected previous layer blocks. The 1st block of the last (3rd) hidden layer (Fig.2.) forms neurons e.g. (9)(10)(11) [10].

$$y_1 = \frac{\partial f(x_{21}, x_{22})}{\partial x_{21}} = w_1 \frac{\left(a_0 + a_1 x_{21} + a_2 x_{22} + a_3 x_{21} x_{22} + a_4 x_{21}^2 + a_5 x_{22}^2\right)^{1/2}}{\frac{3}{2} \cdot (b_0 + b_1 x_{21})} \tag{9}$$

$$y_2 = \frac{\partial f(x_{21}, x_{22})}{\partial x_{11}} = w_2 \frac{(a_0 + a_1 x_{21} + a_2 x_{22} + a_3 x_{21} x_{22} + a_4 x_{21}^2 + a_5 x_{22}^2)^{1/2}}{\frac{3}{2} \cdot x_{22}} \cdot \frac{(x_{21})^{1/2}}{\frac{3}{2} \cdot (b_0 + b_1 x_{11})} \tag{10}$$

$$y_3 = \frac{\partial f(x_{21}, x_{22})}{\partial x_1} = w_3 \frac{(a_0 + a_1 x_{21} + a_2 x_{22} + a_3 x_{21} x_{22} + a_4 x_{21}^2 + a_5 x_{22}^2)^{1/2}}{\frac{3}{2} \cdot x_{22}} \cdot \frac{(x_{21})^{1/2}}{\frac{3}{2} \cdot x_{12}} \cdot \frac{(x_{11})^{1/2}}{\frac{3}{2} \cdot (b_0 + b_1 x_1)} \tag{11}$$

The square (12) and combination (13) derivative terms are also calculated according to the composite function derivation rules.

$$y_4 = \frac{\partial^2 f(x_{21}, x_{22})}{\partial x_{11}^2} = w_4 \frac{(a_0 + a_1 x_{21} + a_2 x_{22} + a_3 x_{21} x_{22} + a_4 x_{21}^2 + a_5 x_{22}^2)^{1/2}}{1.5 \cdot x_{22}} \cdot \frac{x_{21}}{3.8 \cdot (b_0 + b_1 x_{11} + b_2 x_{11}^2)} \tag{12}$$

$$y_5 = \frac{\partial^2 f(x_{21}, x_{22})}{\partial x_1 \partial x_{12}} = w_5 \frac{(a_0 + a_1 x_{21} + a_2 x_{22} + a_3 x_{21} x_{22} + a_4 x_{21}^2 + a_5 x_{22}^2)^{1/2}}{1.5 \cdot x_{22}} \cdot \frac{x_{21}}{3.3 \cdot (b_0 + b_1 x_{11} + b_2 x_{12} + b_3 x_{11} x_{12})} \tag{13}$$

The best-fit neuron selection is the initial phase of the DE composition, which may apply a proper genetic algorithm (GA). Parameters of polynomials might be adjusted by means of difference evolution algorithm (EA), supplied with sufficient random mutations. The parameter optimization is performed simultaneously with the GA term combination search, which may result in a quantity of local or global error solutions. The number of network hidden layers coincides with a total amount of input variables. There would be welcome to apply an adequate gradient descent method too, which parameter updates result from partial derivatives of polynomial DE terms in respect with the single parameters [7].

$$Y = \frac{\sum_{i=1}^{k} y_i}{k} \qquad k = \text{amount of active neurons} \tag{14}$$

Only some of all potential combination DE terms (neurons) may participate in the DE composition, in despite of they have an adjustable term weight (w_i). D-PNN's total output Y is the sum of all active neuron outputs, divided by their amount k (14). The root mean square error (RMSE) method (15) was applied for the polynomial parameter optimization and neuron combination selection. D-PNN is trained only with a small set of input-output data samples likewise the GMDH algorithm does [6].

$$E = \sqrt{\frac{\sum_{i=1}^{M} \left(y^d - y_i\right)^2}{M}} \to \min \tag{15}$$

4   Test Experiments

The presented 3-variable multi-layered D-PNN (Fig.2.) can be tested to approximate non-linear multi-parametric functions. The D-PNN and ANN models were trained with 24 data samples, randomly generated by benchmark functions from the interval $<10,400>$. The ANN approximation ability falls rapidly outside of training range, while the D-PNN's alternate errors grow just slowly (Fig.3. and Fig.4.). Experiments with other benchmarks (e.g. $x_1+x_2^2+x_3^3$) result in similar outcome graphs. The ANN with 2-hidden layers of neurons applied the sigmoidal activation function and standard back-propagation algorithm. The parameter and weight adjustment of both methods appeared heavy time-consuming and have not succeed any experiment.

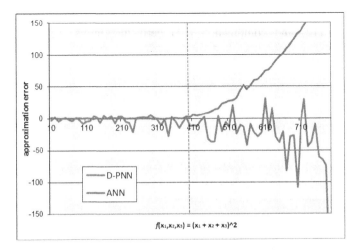

Fig. 3. Comparison of the $f(x_1, x_2, x_3)=(x_1+x_2+x_3)^2$ function approximation

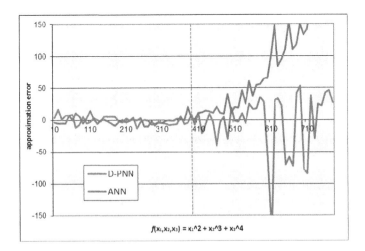

Fig. 4. Comparison of the $f(x_1, x_2, x_3)= x_1^2+x_2^3+x_3^4$ function approximation

5   Real Data Multi-parametric Models

A real data multi-parametric function can be represented by the relative humidity model which inputs are 3 weather variables of wind speed, temperature and sea level pressure of a 1-site locality. The test model of real meteorological data can roughly estimate the time and amount of precipitations. The relative humidity values increase at night hours (with temperature decrease), upswing or day grows can indicate precipitations (Fig.5a-d). The comparisons were done with 1-layer recurrent neural network (RNN), which applies as inputs also its neuron outputs from a previous time estimate. D-PNN applies only 3 current state variables, which disadvantages it, as it does not allow for time sequences. Both networks were trained with previous day hourly data series (24 or 48 hours, i.e. data samples) free on-line available [11].

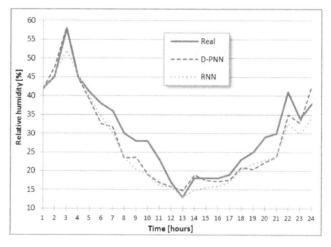

Fig. 5a. $RMSE_{D\text{-}PNN} = 4.16$, $RMSE_{RNN} = 4.72$

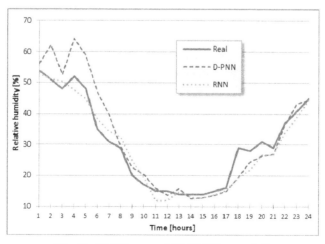

Fig. 5b. $RMSE_{D\text{-}PNN} = 5.75$, $RMSE_{RNN} = 3.48$

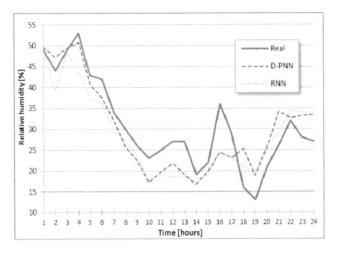

Fig. 5c. $RMSE_{D\text{-}PNN} = 5.33$, $RMSE_{RNN} = 5.70$

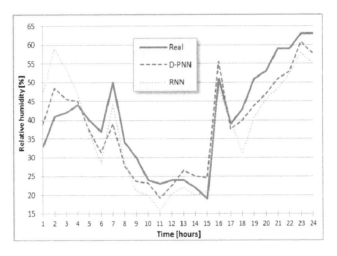

Fig. 5d. $RMSE_{D\text{-}PNN} = 5.27$, $RMSE_{RNN} = 7.99$

The humidity value estimation model (Fig.5.) is only a search test of unknown real meteorological data function relations. The model could be formed with reference to other weather variables forecasts, as meteorological predictions of this very complex dynamic system are sophisticated and not any time faithful, using simple neural network models requiring as a rule high amount of input variables. This method could try to improve an official "Aladin" forecast model provided by Czech hydro-meteorological institute. The model comprises 2-day chart prognoses of temperature, humidity, wind speed, pressure and cloudiness in a selected locality [12]. Thus D-PNN can be trained with real data observations of previous 1 or 2 days to define a true multi-parametric function relation, exactly actual for this time interval. After it can form a new revised 24-hour estimation of some variable (e.g. relative humidity), applying the previous day trained real model of data relations and input variables of the "Aladin" predictions. In the case of an unexpected weather change from a day to day the model will be not equally true, however this trend is not very frequent. The estimation will also depend on prediction accuracies of other "Aladin" model variables, which form the D-PNN's input vector.

6   Time-Series Predictions

The simple neural network models applying 3 state weather variables of 1 site locality are able to predict their values simultaneously in this very complex system (Fig.6a-c.). The 3 meteorological variables (wind speed, dew point, see level pressure) and their 3-time series of hour stamps form 9 variables of the input vectors totally [11]. However D-PNN applies only 3 hidden layers of blocks, i.e. 3 inter-connected networks of Fig.2, which disallows it to define all possible data relations.

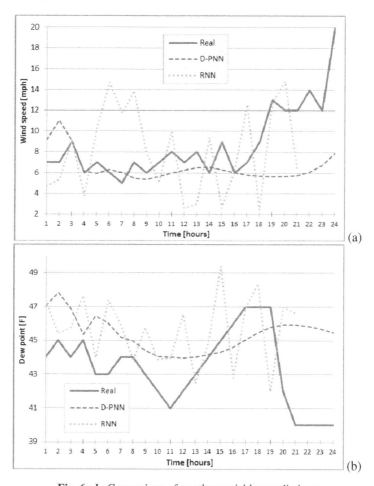

Fig. 6a-b. Comparison of weather variables predictions

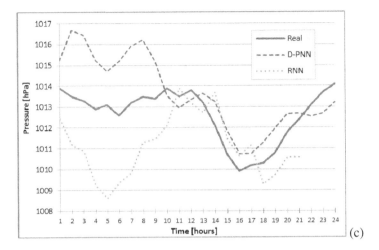

Fig. 6c. Comparison of weather variables predictions

The D-PNN and RNN (Fig. 7) predict values of the 3-variables at a next time state, which form the 3 latest time inputs of a following prediction. Dew point is the non-linear function of temperature and relative humidity. There was need of doing more time-consuming experiments, which were not always succeeded and also not valid on all tested data as other factors can fair influence the very complex weather system [8].

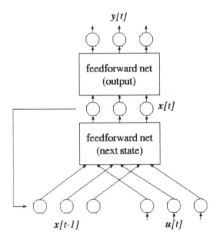

Fig. 7. Recurrent neural network

7 Conclusion

D-PNN is a new neural network type, which function approximation and identification is based on generalized data relations. Its relative data processing is contrary to the common soft-computing method functionality (e.g. ANN, GP), which applications are subjected to a fixed interval of absolute values. This handicap disallows it to use far various training and testing data range values, which may involve real data applications. Regarding to test experiments the D-PNN's non-linear regression can cover a wider interval of input values. It forms and resolves an unknown general DE with a composition of sum fractional derivative terms, defining a system model of dependent variables. The inaccuracies of presented experiments can result from applied incomplete rough training and selective methods, requiring improvements. The D-PNN's operating principle differs by far from other common neural network techniques.

Acknowledgement. This work has been elaborated in the framework of the IT4Innovations Centre of Excellence project, reg. no. CZ.1.05/1.1.00/02.0070 supported by Operational Programme 'Research and Development for Innovations' funded by Structural Funds of the European Union and by the Ministry of Industry and Trade of the Czech Republic, under the grant no. FR-TI1/420, and by SGS, VŠB – Technical University of Ostrava, Czech Republic, under the grant No. SP2012/58.

References

1. Dutot, A.L., Rynkiewicz, J., Steiner, F.E., Rude, J.: A 24-h forecast of ozone peaks and exceedance levels using neural classifiers and weather predictions. Environmental Modelling & Software 22 (2007)
2. Giles, C.L.: Noisy Time Series Prediction using Recurrent Neural Networks and Grammatical Inference. Machine Learning 44, 161–183 (2001)
3. Iba, H.: Inference of differential equation models by genetic programming. Information Sciences 178(23), 4453–4468 (2008)
4. Hall, T., Brooks, H.E., Doswell, C.A.: Precipitation Forecasting Using a Neural Network. Weather and Forecasting 14 (1998)
5. Kuneš, J., Vavroch, O., Franta, V.: Essentials of modeling. SNTL Praha (1989) (in Czech)
6. Nikolaev, N.Y., Iba, H.: Adaptive Learning of Polynomial Networks. Springer, New York (2006)
7. Nikolaev, N.Y., Iba, H.: Polynomial harmonic GMDH learning networks for time series modelling. Neural Networks 16, 1527–1540 (2003)
8. Shrivastava, G., Karmakar, S., Kowar, M.K.: Application of Artificial Neural Networks in Weather Forecasting. International Journal of Computer Applications 51(18) (2012)
9. Tsoulos, I., Gavrilis, D., Glavas, E.: Solving differential equations with constructed neural networks. Neurocomputing 72, 2385–2391 (2009)
10. Zjavka, L.: Recognition of Generalized Patterns by a Differential Polynomial Neural Network. Engineering, Technology & Applied Science Research 2(1) (2012)
11. National Climatic Data Center of National Oceanic and Atmospheric Administration (NOAA), http://cdo.ncdc.noaa.gov/qclcd_ascii/, http://cdo.ncdc.noaa.gov/cdo/3505dat.txt
12. "Aladin" forecast model of Czech hydro-meteorological institute, http://www.chmi.cz/files/portal/docs/meteo/ov/aladin/results/public/meteogramy/meteogram_page_portal/m.html

PSO-2S Optimization Algorithm for Brain MRI Segmentation

Abbas El Dor[1], Julien Lepagnot[2], Amir Nakib[1], and Patrick Siarry[1]

[1] Université de Paris-Est Créteil, Laboratoire LISSI, E.A. 3956
122 rue Paul Armangot, 94400 Vitry-sur-Seine, France
siarry@u-pec.fr
[2] Université de Haute-Alsace, Laboratoire LMIA, E.A. 3993
4 rue des Frères Lumière, 68093 Mulhouse, France

Abstract. In image processing, finding the optimal threshold(s) for an image with a multimodal histogram can be done by solving a Gaussian curve fitting problem, *i.e.* fitting a sum of Gaussian probability density functions to the image histogram. This problem can be expressed as a continuous nonlinear optimization problem. The goal of this paper is to show the relevance of using a recently proposed variant of the Particle Swarm Optimization (PSO) algorithm, called PSO-2S, to solve this image thresholding problem. PSO-2S is a multi-swarm PSO algorithm using charged particles in a partitioned search space for continuous optimization problems. The performances of PSO-2S are compared with those of SPSO-07 (*Standard Particle Swarm Optimization in its 2007 version*), using reference images, *i.e.* using test images commonly used in the literature on image segmentation, and test images generated from brain MRI simulations. The experimental results show that PSO-2S produces better results than SPSO-07 and improves significantly the stability of the segmentation method.

1 Introduction

Digital image processing has attracted a growing interest, due to its practical relevance in many fields of research and in industrial and medical applications. Image segmentation is typically used to locate objects and boundaries in images. It is one of the main components of several image analysis systems, thus it received a great deal of attention. Several surveys and comparative papers are available in the literature [13,10,14,7]. Image thresholding is one of the most popular segmentation approaches. It makes use of the image histogram to partition the images into several meaningful groups of pixels. In automatic image thresholding methods, the segmentation problem can be formulated as a continuous nonlinear optimization problem. Hence, the use of a metaheuristic is a relevant choice to solve it efficiently.

In this paper, we propose to use a recently proposed algorithm [5], called PSO-2S, which is a new variant of particle swarm optimization (PSO) [8]. PSO is inspired by social behavior simulations of bird flocking. It has already been

J.-S. Pan et al. (eds.), *Genetic and Evolutionary Computing,*
Advances in Intelligent Systems and Computing 238,
DOI: 10.1007/978-3-319-01796-9_2, © Springer International Publishing Switzerland 2014

applied successfully to image processing problems [6,9]. This algorithm optimizes a problem by iteratively improving a candidate solution with regard to a given measure of quality. PSO-2S is a multi-swarm PSO algorithm based on several initializations in different zones of the search space, using charged particles. This algorithm uses two kinds of swarms, one main and several auxiliary swarms. The best particles of the auxiliary ones generate the main swarm. More precisely, the auxiliary swarms are initialized several times in different zones. An electrostatic repulsion heuristic is then applied in each zone to increase the diversity of the particles. Each auxiliary swarm performs several generations based on standard PSO algorithm to provide the best solution in its related zone. The provided solutions are then used as the main swarm.

This paper is structured as follows: Section 2 presents an overview of the standard particle swarm optimization and its new variant PSO-2S. Section 3 is dedicated to the presentation of the image thresholding method. The image segmentation criterion is given in Section 4. Experimental protocol and parameter setting are presented in Section 5. Experimental results are discussed in Section 6. The work in this paper is concluded in section 7.

2   Presentation of the PSO-2S Algorithm

2.1   Review of the Standard PSO

The particle swarm optimization (PSO) [8] is inspired originally by the social and cognitive behavior existing in the bird flocking. The algorithm is initialized with a population of particles randomly distributed in the search space, and each particle is assigned a randomized velocity. Each particle represents a potential solution to the problem.

In this paper, the swarm size is denoted by s, and the search space is n-dimensional. In general, the particles have three attributes: the current position $X_i = (x_{i,1}, x_{i,2}, ..., x_{i,n})$, the current velocity vector $V_i = (v_{i,1}, v_{i,2}, ..., v_{i,n})$ and the past best position $Pbest_i = (p_{i,1}, p_{i,2}, ..., p_{i,n})$. The best position found in the neighborhood of the particle i is denoted by $Gbest_i = (g_1, g_2, ..., g_n)$. These attributes are used to update iteratively the state of each particle in the swarm. The objective function to be minimized is denoted by f. The velocity vector V_i of each particle is updated using the best position it visited so far and the overall best position visited by its neighbors. Then, the position of each particle is updated using its updated velocity per iteration. At each step, the velocity of each particle and its new position are updated as follows:

$$v_{i,j}(t+1) = wv_{i,j}(t)+c_1r_{1_{i,j}}(t)\,[pbest_{i,j}(t)-x_{i,j}(t)]+c_2r_{2_{i,j}}(t)\,[gbest_{i,j}(t)-x_{i,j}(t)] \tag{1}$$

$$x_{i,j}(t+1) = x_{i,j}(t) + v_{i,j}(t+1) \tag{2}$$

where w is called inertia weight, c_1, c_2 are the learning factors and r_1, r_2 are two random numbers selected uniformly in the range $[0, 1]$.

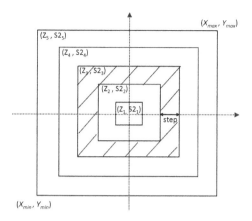

Fig. 1. Partitioning of the search space

2.2   PSO Improved Variant: PSO-2S

An improved variant of the original PSO algorithm, called PSO-2S, was proposed by El Dor et al. [5]. This variant consists of using three main ideas: the first is to use two kinds of swarms: a main swarm, denoted by S1, and s auxiliary ones, denoted by $S2_i$, where $1 \leq i \leq s$. The second idea is to partition the search space into several zones in which the auxiliary swarms are initialized (the number of zones is equal to the number of auxiliary swarms s). The last idea is to use the concept of the electrostatic repulsion heuristic to diversify the particles for each auxiliary swarm in each zone.

To construct S1, the auxiliary swarms $S2_i$ evolve several times in different areas, and then each best particle for each $S2_i$ is saved and considered as a new particle of S1. To do so, the population of each auxiliary swarm is initialized randomly in different zones (each $S2_i$ is initialized in its corresponding zone i). After each of these initializations, $nb_{generation}$ displacements of particles, for each $S2_i$, are performed in the same way as standard PSO. Then the best solution found by each auxiliary swarm, named $gbest_i$, is added to S1. The number of initializations of $S2_i$ is equal to the number of particles in S1.

As mentioned above, the second idea is to partition the search space $[min_d, max_d]^D$ into several zones (max_{zone} zones). Then, one calculates the $center_d$ and the $step_d$ of each dimension separately, according to (3) and (4). In the case of using an uniform (square) search space, the $step_d$ are similar for all dimensions.

$$center_d = (max_d + min_d)/2 \qquad (3)$$

$$step_d = (max_d - min_d)/2 \times max_{zone} \qquad (4)$$

where max_{zone} is a fixed value, and d is the current dimension ($1 \leq d \leq D$).

This process is illustrated in Figure 1, where the ith swarm $S2_i$ and its attributed zone Z_i are denoted by $(Z_i, S2_i)$.

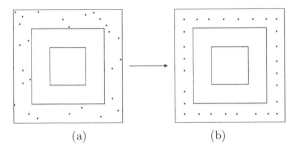

(a)   (b)

Fig. 2. Repulsion process: (a) $(Z_3, S2_3)$ before repulsion, (b) $(Z_3, S2_3)$ after repulsion

The sizes of the zones of the partitioned search space are different ($Z_1 < Z_2 < \ldots < Z_{max_{zone}}$). Therefore, the number of particles in S2$_i$, denoted by $S2_{isize}$, depends on its corresponding zone size. Indeed, a small zone takes less particles and the number of particles increases when the zone becomes larger. The size of each auxiliary swarm is calculated as follows:

$$S2_{isize} = num_{zone} \times nb_{particle} \qquad (5)$$

where $num_{zone} = 1, 2, \ldots, max_{zone}$ is the current zone number and $nb_{particle}$ is a fixed value.

After the initializations of the auxiliary swarms in different zones $(Z_i, S2_i)$, an electrostatic repulsion heuristic is applied to diversify the particles and to widely cover the search space [4]. This technique is used in an agent-based optimization algorithm for dynamic environments [11]. Therefore, this procedure is applied in each zone separately, hence each particle is considered as an electron. Then a force of $1/r^2$ is applied, on the particles of each zone, until the maximum displacement of a particle during an iteration becomes lower than a given threshold ϵ (where r is the distance between two particles, ϵ is typically equal to 10^{-4}). At each iteration of this procedure, the particles are projected in the middle of the current zone, before a new application of the repulsion heuristic. Figure 2 presents an example of the repulsion applied to $(Z_3, S2_3)$.

3   The Problem at Hand

The segmentation problem has received a great deal of attention, thus any attempt to survey the literature would be too space-consuming. The most popular segmentation methods may be found in [15]. In this work, image segmentation is performed using the thresholding approach. Image thresholding is a supervised segmentation method, *i.e.* the number of regions (classes of pixels) and their properties are known in advance by the user. The segmentation is done by determining, for each pixel, the class whose properties are the closest to those observed for that pixel. The thresholding technique is based on the assumption that different regions of the image can be distinguished by their gray levels.

It makes use of the histogram $h(j)$ of the processed image, *i.e.* the observed probability of gray level j. It can be defined as follows:

$$h(j) = \frac{g\,(j)}{\sum_{i=0}^{L-1} g\,(i)} \qquad (6)$$

where $g(j)$ denotes the occurrence of gray-level $j \in \{0, 1, \dots, L-1\}$ in the image.

Thresholding the image into N classes is to find the $N-1$ thresholds that will partition the histogram into N zones.

The main contribution of the work we present here is to show the significance of using PSO-2S for MR image segmentation. The performances of PSO-2S are first compared with those of SPSO-07 (*Standard Particle Swarm Optimization* in its 2007 version) [2], using reference images, commonly used in the literature on image segmentation. Then, the performances of both algorithms are compared using images from a database generated by brain MRI simulation [1,3]. This database, called *BrainWeb*, provides images for which an "optimal" segmentation is known. Indeed, the BrainWeb MRI simulations are based on a predefined anatomical model of the brain. The images generated by these simulations can then be used to validate a segmentation method, or to compare the performance of different methods.

4   Image Segmentation Criterion

Before using this criterion we must fit the histogram of the image to be segmented to a sum of Gaussian probability density functions (pdf's). This procedure is named Gaussian curve fitting, more details about it are given below. The pdf model must be fitted to the image histogram, typically by using the maximum likelihood or mean-squared error approach, in order to find the optimal threshold(s). For the multimodal histogram $h(i)$ of an image, where i is the gray level, we fit $h(i)$ to a sum of d probability density functions [12]. The case where the Gaussian pdf's are used is defined by:

$$p\,(x) = \sum_{i=1}^{d} P_i \exp\left[-\frac{(x-\mu_i)^2}{\sigma_i^2} \right] \qquad (7)$$

where P_i is the amplitude of Gaussian pdf on μ_i, μ_i is the mean and σ_i^2 is the variance of mode i, and d is the number of Gaussians used to approximate the original histogram and corresponds to the number of segmentation classes.

Our goal is to find a vector of parameters, Θ, that minimizes the fitting error J, given by the following expression:

$$J(\Theta) = \sum_{i=0}^{L-1} |h(i) - p(\Theta, i)|^2 \qquad (8)$$

where $h(i)$ is the measured histogram. Here, J is the objective function to be minimized with respect to Θ, a set of parameters defining the Gaussian pdf's and the probabilities, given by $\Theta = \{P_i, \mu_i, \sigma_i; \ i = 1, 2, \cdots, d\}$. After fitting the multimodal histogram, the optimal threshold could be determined by minimizing the overall probability of error, for two adjacent Gaussian pdf's, given by:

$$e(T_i) = P_i \int_{-\infty}^{T_i} p_i(x)\,dx + P_{i+1} \int_{T_i+1}^{\infty} p_{i+1}(x)\,dx \qquad (9)$$

with respect to the threshold T_i, where $p_i(x)$ is the ith pdf and $i = 1, \ldots, d-1$. Then the overall probability to minimize is:

$$E(T) = \sum_{i=1}^{d-1} e(T_i) \qquad (10)$$

where T is the vector of thresholds: $0 < T_1 < T_2 < ... < T_{(d-1)} < L - 1$. In our case L is equal to 256.

5   Experimental Protocol and Parameter Setting

To compare the performance of PSO-2S and SPSO-07, the criterion (8) is minimized for each test image in Figure 3 (a). The stagnation criterion used is satisfied if no significant improvement (greater than $1E-10$) in the current best solution is observed during $1E+4$ successive evaluations of the objective function. In addition, the maximum number of evaluations allowed is set to 300000.

In this figure, LENA and BRIDGE are reference images used for the validation of segmentation methods in the literature. The images MRITS and MRICS are obtained from BrainWeb [1] and correspond to transverse and coronal sections of a brain, respectively. The parameters used for the MRI simulation are a T1-weighted sequence, a slice thickness of 1mm, a Gaussian noise of 3% calculated relative to the brightest tissue, and a 20% level of intensity non-uniformity (radio frequency bias).

The values of the PSO and SPSO-07 parameters used for the segmentation problem are defined below:

- PSO-2S using 30 zones and $\frac{p^4}{7000} + 10$ particles in each zone, where p is the zone number. The parameter K used to generate the neighborhood of the particles is set to $K = 3$. The parameter $Nb_{generation}$ is set to 15 ;
- SPSO-07 (*Standard Particle Swarm Optimization* in its 2007 version) [2] using $10+2\sqrt{D}$ particles (the formula recommended by the authors of SPSO-07), where D is the dimension of the problem. The parameter K is set to $K = 3$.

6   Experimental Results and Discussion

In this section, the experimental results obtained with PSO-2S and SPSO-07 are presented. The segmentation results are shown in Figure 3. In this figure, the

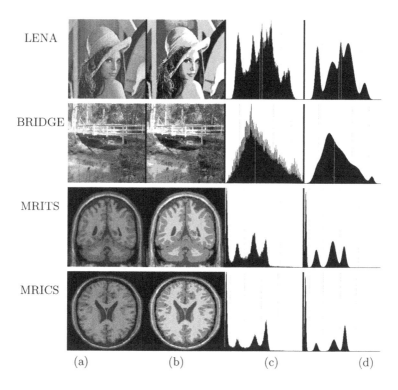

LENA

BRIDGE

MRITS

MRICS

(a)   (b)   (c)   (d)

Fig. 3. Illustration of the segmentation process. (a) Original images. (b) Segmented images using thresholds in Table II. (c) Original histograms. (d) Approximated histograms.

original images and their histograms are illustrated in (a) and (c), respectively. Approximated histograms are presented in (d), and segmented images (using 5 classes) are shown in (b). For each test image, one can see that the approximation of its histogram, illustrated in detail for LENA in Figure 4, leads to a good image segmentation.

The histogram approximation results, for each test image, are presented in Table 1. In this table, the parameters of each of the five Gaussian pdf's used to approximate the histogram of each image are given. The parameters of the ith pdf of an image are denoted by P_i, μ_i and σ_i. Threshold values between the different classes of pixels, calculated for each image using its approximated histogram, are given in Table 2.

For each test image, the number of evaluations performed by each algorithm, averaged over 100 runs, is given in Table 3. The success rate (the percentage of acceptable solutions found among the ones of the 100 runs, $i.e.$ the percentage of solutions with an objective function value lower or equal to 5.14E−4, 5.09E−4, 7.51E−4, 7.99E−4 for LENA, BRIDGE, MRITS, MRICS, respectively), and the average approximation error (the average value of the objective function for the

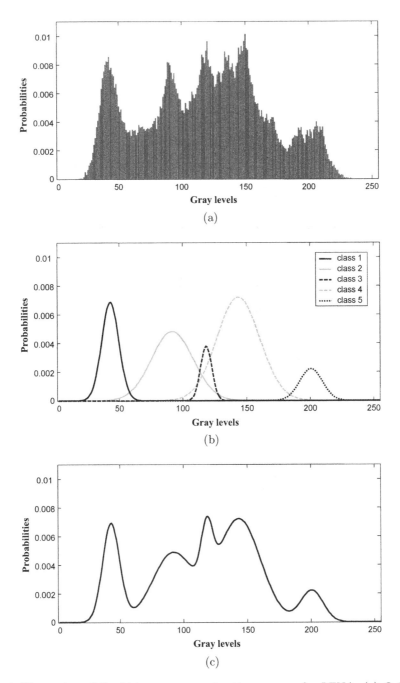

Fig. 4. Illustration of the histogram approximation process for LENA. (a) Original histogram. (b) Gaussian pdf's for each class of pixels. (c) Sum of the Gaussian pdf's (approximated histogram).

Table 1. Parameters of the Gaussian pdf's used to approximate the histogram of each test image

Image	μ_1	P_1	σ_1	μ_2	P_2	σ_2	μ_3	P_3	σ_3	μ_4	P_4	σ_4	μ_5	P_5	σ_5
LENA	41.62	0.17	9.87	90.57	0.28	23.14	117.21	0.06	6.33	142.54	0.43	23.82	199.56	0.07	12.64
BRIDGE	41.13	0.06	15.36	76.75	0.39	25.97	119.25	0.38	34.73	173.38	0.14	28.43	225.53	0.02	9.09
MRITS	4.71	0.28	3.78	40.79	0.17	10.05	94.50	0.36	14.26	131.55	0.19	9.17	225.60	0.00	252.92
MRICS	4.66	0.41	3.66	41.76	0.09	7.64	99.57	0.21	11.92	135.53	0.29	7.46	192.73	0.00	239.82

Table 2. Threshold values for each test image

Image	Thresholds
LENA	57 112 120 183
BRIDGE	46 98 156 215
MRITS	15 62 117 184
MRICS	17 64 121 177

Table 3. Average number of evaluations for the segmentation of an image, approximation error and success rate obtained for each algorithm, for each test image

Image	Algorithm	Evaluations	Approximation error	Success rate
LENA	PSO2S	119721.7 ± 64844.3	5.44E−4 ± 3.06E−5	41 %
	SPSO-07	67057.1 ± 64316.9	5.52E−4 ± 3.08E−5	25 %
BRIDGE	PSO2S	241537.8 ± 70707.8	5.27E−4 ± 9.54E−5	48 %
	SPSO-07	125524.4 ± 63277.9	5.16E−4 ± 6.01E−6	24 %
MRITS	PSO2S	81080.9 ± 66569.8	7.69E−4 ± 1.26E−4	95 %
	SPSO-07	44212.0 ± 57321.5	8.79E−4 ± 2.73E−4	77 %
MRICS	PSO2S	72922.5 ± 35894.4	8.68E−4 ± 1.18E−4	70 %
	SPSO-07	28185.4 ± 16188.4	9.46E−4 ± 2.71E−4	54 %

best solution found) of an image histogram are also given in this table, for 100 runs of an algorithm.

In this table, we see that PSO-2S requires more evaluations than SPSO-07 to converge to an acceptable solution. However, its success rate is significantly higher than the one of SPSO-07 for all images, according to the Fisher's exact test with a 95% confidence level. Indeed, PSO-2S is designed to prevent premature convergence of PSO algorithm. Hence, it significantly improves the stability of the segmentation method. It shows the significance of using PSO-2S for this class of problems.

7   Conclusion

In this paper, we present an image segmentation method using the thresholding approach to identify several classes of pixels in standard and medical images. This method includes an optimization step in which we integrated our PSO-2S algorithm. We also tested the method using the algorithm SPSO-07.

Segmentation results obtained on several test images, commonly used in the literature in image processing and on synthetic images obtained from simulations of brain MRI, are satisfactory. We show that using PSO-2S provides greater stability for this segmentation method, compared with SPSO-07. It shows the relevance of using PSO-2S for this type of problems. Our work in progress consists in the improvement of the segmentation criterion in order to enhance the segmentation quality and accelerate the optimization process.

References

1. BrainWeb: Simulated Brain Database,
   http://brainweb.bic.mni.mcgill.ca/brainweb (2012)
2. Clerc, M., et al.: The Particle Swarm Central website (2012),
   http://www.particleswarm.info
3. Collins, D.L., Zijdenbos, A.P., Kollokian, V., Sled, J.G., Kabani, N.J., Holmes, C.J., Evans, A.C.: Design and construction of a realistic digital brain phantom. IEEE Transactions on Medical Imaging 17(3), 463–468 (1998)
4. Conway, J., Sloane, N.: Sphere Packings, Lattices and Groups. Springer (1999)
5. El Dor, A., Clerc, M., Siarry, P.: A multi-swarm PSO using charged particles in a partitioned search space for continuous optimization. Computational Optimization and Applications 53(1), 271–295 (2012)
6. Feng, H.-M., Horng, J.-H., Jou, S.-M.: Bacterial Foraging Particle Swarm Optimization Algorithm Based Fuzzy-VQ Compression Systems. Journal of Information Hiding and Multimedia Signal Processing 3(3), 227–239 (2012)
7. Gonzalez, R.C., Woods, R.E.: Digital Image Processing, 3rd edn. Prentice-Hall, Inc., Upper Saddle River (2006)
8. Kennedy, J., Eberhart, R.C.: Particle swarm optimization. In: The IEEE International Conference on Neural Networks IV, Perth, Australia, November 27-December 1, pp. 1942–1948 (1995)
9. Kwok, N.M., Wang, D., Ha, Q.P., Fang, G., Chen, S.Y.: Locally-Equalized Image Contrast Enhancement Using PSO-Tuned Sectorized Equalization. In: Chatterjee, A., Siarry, P. (eds.) Computational Intelligence in Image Processing, pp. 21–36. Springer (2013)
10. Lee, S.U., Chung, S.Y., Park, R.H.: A comparative performance study of several global thresholding techniques for segmentation. Computer Vision, Graphics, and Image Processing 52(2), 171–190 (1990)
11. Lepagnot, J., Nakib, A., Oulhadj, H., Siarry, P.: A new multiagent algorithm for dynamic continuous optimization. International Journal of Applied Metaheuristic Computing 1(1), 16–38 (2010)
12. Nakib, A., Oulhadj, H., Siarry, P.: Non-supervised image segmentation based on multiobjective optimization. Pattern Recognition Letters 29(2), 161–172 (2008)
13. Pitas, I.: Digital Image Processing Algorithms and Applications. John Wiley & Sons (2000)
14. Sahoo, P.K., Soltani, S., Wong, A.K.C., Chen, Y.C.: A survey of thresholding techniques. Comput. Vision Graph. Image Process. 41(2), 233–260 (1988)
15. Sezgin, M., Sankur, B.: Survey over image thresholding techniques and quantitative performance evaluation. Journal of Electronic Imaging 13, 146–165 (2004)

Task Scheduling in Grid Computing Environments

Yi-Syuan Jiang and Wei-Mei Chen*

Department of Electronic and Computer Engineering,
National Taiwan University of Science and Technology,
Taipei 106, Taiwan
{M9902144,wmchen}@mail.ntust.edu.tw

Abstract. A grid computing environment is a parallel and distributed system that brings together various computing capacities to solve large computation problems. Task scheduling is a critical issue for grid computing, which maps tasks onto a parallel and distributed system for achieving good performance in terms of minimizing the overall execution time. This paper presents a genetic algorithm to solve this problem for improving the existing genetic algorithm with two main ideas: a new initialization strategy is introduced to generate the first population of chromosomes and the good characteristics of found solutions are preserved for new generations. Our proposed algorithm is implemented and evaluated using a set of well-known applications in our specific-defined system environment. The experimental results show that the proposed algorithm outperforms other algorithms within several parameter settings.

1 Introduction

In past few years, grid computing systems and applications become popular [3], due to a rapid development of many-core. A grid computing environment is a parallel and distributed system that brings together various computing capacities to solve large computation problems. In grid environments, task scheduling, which plays an important role, divides a larger job into smaller tasks and maps tasks onto a parallel and distributed system [1,6]. The goal of a task scheduling is typically to schedule all the tasks on a given number of available processors so as to minimize the overall length of time required to execute the whole program.

A parallel and distributed computing system may be homogeneous [10] or heterogeneous systems [7,11,13]. A homogeneous system means that the processors are the same performance in processing capabilities. On the other hand, heterogeneous systems have different processing capabilities in the target system. In general, the processors are connected by an interconnection network, which is either fully-connected [11,13] or partially-connected [2]. In the fully-connected network every processor can communicate with each other, whereas data can be transferred to some specified processors in a partially-connected network. Besides, the task duplication issue [10] was also discussed to reduce the communication time by duplicating some tasks on more than one processor to eliminate communication cost. To avoid increasing energy consumption,

---

* This work is partially supported by National Science Council under the Grant NSC 101-2221-E-011-039-MY2.

here we consider the target system which is the fully-connected heterogeneous systems without task duplication.

The genetic algorithm (GA), first proposed by Holland [5], provides a popular solution for application problems [4,9]. GAs have been shown that outperforms several algorithms in the task scheduling problem, which simply define the search space to be the solution space in which each point is denoted by a number string, called a chromosome. Based on these solutions, three operators which are selection, crossover, and mutation, are employed to transform a population of chromosomes to better solutions iteratively. In order to keep the good features from the previous generation, the crossover operator exchanges the information from two chromosomes chosen randomly, and the mutation operator alters one bit of a chromosome.

In this paper, we proposed a genetic algorithm for task scheduling on a grid computing system, called TSGA. In general, GA approaches directly initialize the first population by some uniform random process. TSGA develops a new initialization policy, which divides the search space into specific patterns in order to accelerate the convergence of solutions. To solve the task scheduling problem, a chromosome usually contains a mapping part and an order part to indicate the corresponding computer and the executing order. In the standard GA, when crossover and mutation operators are applied, both of the mapping part and the order part will be changed, which brings that the parents' characteristics cannot be kept in the next generation. Inspired by the idea of eugenics, TSGA presents new operators for crossover and mutation to preserve good features from the previous generation.

The remainder of the paper is organized as follow. In the next section, we provide the problem definition. The proposed genetic scheduling algorithm is presented in Section 3. We describe our experimental results in Section 4. Finally, conclusions are drawn in Section 5.

2   Problem Definition

Task scheduling is mapping smaller tasks to multiprocessors. Tasks with data precedence are modeled by a Directed Acyclic Graph (DAG) [13]. The main idea of DAG scheduling is minimizing the makespan which is the overall execution time for all tasks.

2.1   DAG Modeling

A DAG $G = (V, E)$ is depicted in Fig. 1(a), where V is a set of N nodes and E is a set of M directed edges. For the problem of task scheduling, V represents the set of tasks and each task contains a sequence of instructions that should be completed in a particular order. Let $w_{i,j}$ be the computation time to finish a particular task $t_i \in V$ on the processor P_j, detailed in Fig. 1(b). Each edge $e_{i,j} \in E$ in the DAG indicates the precedence constraint that task t_i should complete its execution before task t_j starts. Let $c_{i,j}$ denote the communication cost needed to transport the data between task t_i and task t_j, which is the weight on an edge $e_{i,j}$. If t_i and t_j is assigned to the same processor, the communication cost $c_{i,j}$ is zero.

The source node of an edge is called a predecessor of that node. Similarly, the destination node emerged from a node is called a successor of that node. In Fig. 1(a), t_1 is the predecessor of t_2, t_3, t_4, and t_5. On the other hand, t_2, t_3, t_4, and t_5 are the successor of t_1. In a graph, a node with no parent is called an entry node, and a node with no child is called an exit node. If a node t_i is scheduled to a processor P_j, the start-time and the finish-time of t_i are denoted by $ST(t_i, P_j)$ and $FT(t_i, P_j)$, respectively.

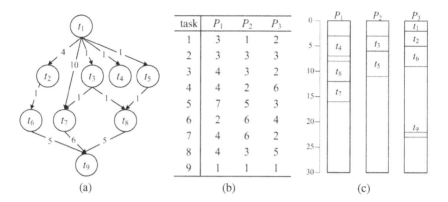

task	P_1	P_2	P_3
1	3	1	2
2	3	3	3
3	4	3	2
4	4	2	6
5	7	5	3
6	2	6	4
7	4	6	2
8	4	3	5
9	1	1	1

(a)   (b)   (c)

Fig. 1. An example of (a) DAG, (b) the computation cost matrix, and (c) an example of scheduling

2.2 Makespan

After all tasks are scheduled onto parallel processors, considering a particular task t_i on the processor P_j, the start-time $ST(t_i, P_j)$ can be defined as

$$ST(t_i, P_j) = \max\{RT_j, DAT(t_i, P_j)\},$$

where $DAT(t_i, P_j)$ is the data arrival time of task t_i at the processor P_j, which is the time when all the needed data have been transmitted. On the other hand, $DAT(t_i, P_j)$ is defined as

$$DAT(t_i, P_j) = \max_{t_k \in pred(t_i)} \{(FT(t_k, P_j) + c_{k,i})\},$$

where $pred(t_i)$ denotes the set of immediate predecessor tasks of the task t_i. Since

$$FT(t_k, P_j) = w_{k,j} + ST(t_k, P_j)$$

and

$$RT_j = \max_{t_k \in exe(P_j)} \{FT(t_k, P_j)\},$$

where $exe(P_j)$ is the set containing tasks which executes on the processor P_j, the overall schedule length of the entire program is the largest finish time among all tasks and can be expressed as

$$makespan = \max_{t_i \in V}\{FT(t_i, P_j)\}.$$

Fig. 1(c) demonstrates a scheduling for the graph described in Fig. 1(a). The makespan of this scheduling is 23.

3   Proposed Method

In this section, we introduce TSGA algorithm in detail, including the encoded and decoded representations and five important operators.

3.1   The Representation of Solutions

The representation of a chromosome is given in Fig. 2, which is divided into a mapping part (S_M) and an order part (S_O). We use integer arrays to store S_M and S_O and the size of arrays is equal to the number of tasks. If $S_M[i]$ is j and $S_O[i]$ is k, it means that a task t_k is executed on the processor P_j.

According to the chromosome represented in Fig. 2, the solution of a DAG in Fig. 1(a) can be scheduled in Fig. 1(c). First, we assign tasks into the mapping processor according to the index of S_M. Tasks t_4, t_7, and t_8 are scheduled on processor P_1. Tasks t_3, and t_5 are executed on processor P_2. Tasks t_1, t_2, t_6, and t_9 are assigned to the processor P_3. Following the order in S_O, we schedule t_4, t_7, and t_8 in the order of t_4, t_8, t_7 in P_1. For P_2, t_3 is executed before t_5. Tasks t_1, t_2, t_6, and t_9 are taken in the order of t_1, t_2, t_6, t_9 in P_3. Finally, we should count the wait time for communicating, if two dependent tasks are scheduled on a different processor.

TSGA defines the fitness function in order to measure the quality of solutions. The purpose of the scheduling problem is minimizing the makespan. Thus, the fitness function is defined as the makespan.

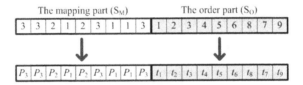

Fig. 2. A representation of chromosome

3.2   TSGA

An algorithmic flowchart of TSGA is given in Fig. 3. If the number of generation is not smaller than the maximum generation G, TSGA will output the best solution.

Initialization. As shown in Algorithm 1, TSGA initializes the first population that consists of encoded chromosomes. Each chromosome is composed of the processor assignment and the execution order. Instead of using the random strategy to give the processor assignment, we devise a new method dividing the search space into specific patterns equally. The search space is divided into $\log_2 n$ subspaces, where n is the number of processors. Since the best scheduling solution may occupy either few processors or most processors in different cases, we give some patterns with a different number of processors in order to explore the solution space in different aspects.

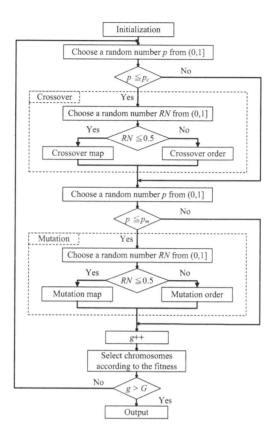

Fig. 3. TSGA flowchart

Crossover Map Operator. As shown in Fig. 4(a), the crossover map operator is used for changing the mapping processor of two chromosomes. The crossover map operator chooses two chromosomes S and T from the population and an integer I between 1 and N randomly. TSGA keeps the processor assignment which is located on the left of I. For the processors on the right of I, we exchange the processors of S and T which are assigned to execute the same task. The processor assignments of tasks t_5, t_6, and t_9 are exchanging directly, since those tasks are occupied in the same place in both chromosomes. On the other hand, tasks t_7 and t_8 are located at different places in these two chromosomes, so they are scheduled to the processor in which they are assigned to another chromosome, and TSGA exchanges the processor assignments. The chromosome S'' and T'' are generated by the crossover map operator in SGA.

Crossover Order Operator. The crossover order operator is practiced to the second part of the chromosome. In Fig. 4(b), after choosing two chromosomes S and T, TSGA chooses a crossover point I between 1 and N. Then, Step 1 copies the left portion of I in S and T to the new chromosome S' and T', respectively. In order to complete the right

---

Algorithm 1. Initialization_operator(C, n, DAG)

Require: C: population set; n: the number of processors
Ensure: C
1: $d \leftarrow \log_2 n$ {Divide chromosomes into d groups}
2: **for** group =1 to d **do**
3:      **for** $i = 1$ to population_size$/d$ **do**
4:          $S_M \leftarrow$ choose a processor for each task from $\{1, 2, \ldots, 2^d\}$
5:          $S_O \leftarrow$ an executed order according to a topological ordering
6:          $S \leftarrow S_M \oplus S_O$
7:          $C \leftarrow C \bigcup \{S\}$
8:      **end for**
9: **end for**
10: **return** C

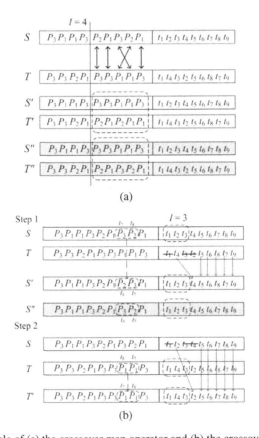

Fig. 4. An example of (a) the crossover map operator and (b) the crossover order operator

segment of I, the crossover order operator carries tasks into S' according to the order in T. Step 2 uses the same rules in step 1 to generate the second offspring T'. Note that the processor assignment should be adjusting to corresponding to the original processor assignment. The chromosome S'' is created by the crossover order operator in SGA.

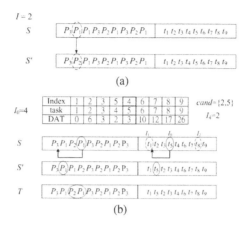

(a)

(b)

Fig. 5. An example of (a) the mutation map operator and (b) the mutation order operator

Table 1. Parameter settings used in the experiment

Population size	Generation size	Crossover rate	Mutation rate	Selection operator
400	1000	0.8	0.2	Binariesry tournament

Mutation Map Operator. The mutation map operator is used for the mapping section of the chromosome, shown in Fig. 5(a). Like the crossover operator, choosing a random number I between 1 to N is essential. Then, the mutation map operator changes the processor in which the chosen task $S_O[I]$ is executed to another processor chosen randomly.

Mutation Order Operator. The mutation order operator is applied to the second part of the chromosome. We select an index I_0 from 1 to N randomly, which decides a task $S_O[I_0]$ to will be mutated. We create an empty set *cand* which would contain indexes of tasks that can be inserted. I_1 and I_2 are the index of the first related task of task $S_O[I_0]$ on the left and right of I_0, respectively. For the index i between $I_1 + 1$ and I_0, if DAT of $S_O[i]$ is larger than DAT of $S_O[I_0]$, task $S_O[I_0]$ should be executed before task $S_O[i]$, since task $S_O[I_0]$ can be executed early. For the index i between I_0 and $I_2 - 1$, if DAT of $S_O[i]$ is smaller than DAT of $S_O[I_0]$, task $S_O[i]$ should be executed before task $S_O[I_0]$, since task $S_O[i]$ can be executed early. We add tasks which meets those conditions to the set *cand*. If *cand* is empty, we fill *cand* with indexes between $I_1 + 1$ and $I_2 - 1$. We select an index I_4 from *cand* randomly, and move task $S_O[I_0]$ to the location I_4. Finally, we adjust the processor assignment to correspond with the original processor assignment, for the same reason mentioned in the crossover order operator. Fig. 5(b) demonstrates the mutation order operator.

4   Experimental Results

This paper considers five well-known applications, Gauss-Jordan elimination (GJ) [14], the fast Fourier transformation (FFT) [11], Robot control, Sparse matrix solver, and

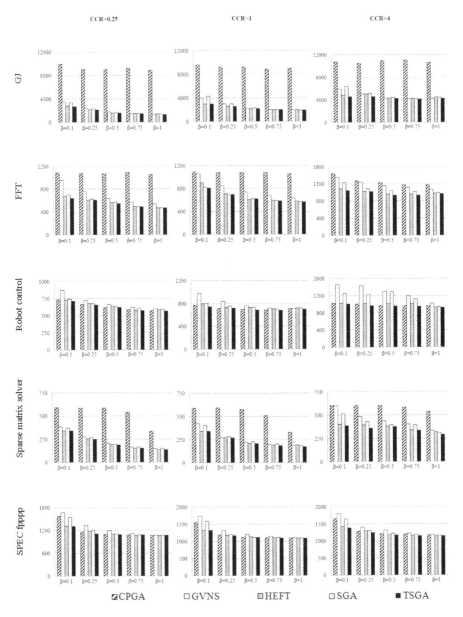

Fig. 6. Experimental results with varying parameters

SPEC fpppp are regared as the benchmark in the problem. Robot control, Sparse matrix solver, and SPEC fpppp are taken from a Standard Task Graph (STG) archive [15]. GJ, FFT, Robot control, Sparse matrix solver, and SPEC fpppp have 300, 223, 88, 96, and 334 tasks, and 552, 382, 131, 67, and 1145 edges, respectively. Each application has various features depending on different settings as below.

1) The number of processors (P):
   $$SET_P = \{4, 8, 16\}$$
2) The range percentage of computation costs on processors (β):
   $$SET_\beta = \{0.1, 0.25, 0.5, 0.75, 1.0\}$$
3) The communication to computation ratio (CCR):
   $$SET_{CCR} = \{0.25, 0.5, 1.0, 2.0, 4.0\}$$

To evaluate our proposed algorithms, we have implemented them using an AMD FX(tm)-8120 eight-core processor (3.10 GHz) using C++ language. Generally speaking, the excellent solution of various problems would be generated by various parameters for a specific algorithm. However, we use the same parameter values listed in Table 1, to show the performance in terms of makespan in this paper.

The performance of TSGA is compared with four algorithms, CPGA [10], SGA, GVNS [13], and HEFT [11]. For each data configuration of five DAGs, the average of makespan obtained over 10 runs is computed for CPGA, SGA, GVNS, and TSGA. On the other hand, HEFT is run only once, since it is a deterministic algorithm.

The experimental results of five DAGs with P fixed to 16 and varying β value are given in Fig. 6, which presents histograms in a table way. Column 1 indicates that a specific application was tested. The parameter CCR is recorded in the row 1, if the CCR value is changing. Every application, shown in each row, was tested with different CCR value and varying β value. Each column contains five applications with fixed CCR value. For instance, the result of testing FFT with $CCR = 4$ and varying β value is given in the row 3 and column 4, and so on. Note that, the vertical coordinate shows the makespan.

5   Conclusion

In this paper, we presented a genetic algorithm for task scheduling, called TSGA, to solve the problem of task scheduling on parallel and distributed computing systems by improving the standard genetic algorithm by increasing the convergence speed and preserving the good features of previous generation. To demonstrate the TSGA performance, we introduced new genetic operators in TSGA. The initialization operator divides the search space into specific patterns so that we can save much time to explore the whole search space. The crossover map operator and the mutation map operator help us to find more suitable processor assignments and the crossover order operator and the mutation order operator provide us with more efficient execution order. For some cases with high CCR values, TSGA schedules tasks to occupy fewer processors to reduce the communication time. For some cases with small β values, TSGA assigns tasks to occupy the processors which have higher processing capability to minimize the execution time. The experimental results show that the proposed algorithm outperforms other algorithms within several parameter settings.

References

1. Culler, D., Singh, J., Gupta, A.: Parallel Computer Architecture: A Hardware/Software Approach. Morgan Kaufmann Publisher, San Francisco (1998)
2. Choudhury, P., Chakrabarti, P.P., Kumar, R.: Online Scheduling of Dynamic Task Graphs with Communication and Contention for Multiprocessors. IEEE Trans. on Parallel and Distributed Systems 23(1), 126–133 (2012)
3. Falzon, G., Li, M.: Enhancing Genetic Algorithms for Dependent Job Scheduling in Grid Computing Environments. Journal of Supercomputing 62(1), 290–314 (2012)
4. Han, Q., Yu, L., Zheng, W., Cheng, N., Niu, X.: A Novel QKD Network Routing Algorithm Based on Optical-Path-Switching. Journal of Information Hiding and Multimedia Signal Processing 5(1), 13–19 (2014)
5. Holland, J.H.: Adaptation in Natural and Artificial Systems. University of Michigan Press, Ann Arbor (1975)
6. Hwang, K.: Advanced Computer Architecture: Parallelism, Scalability, Programmability. McGraw-Hill, Inc., New York (1993)
7. Kwok, Y.K., Ahmad, I.: Efficient Scheduling of Arbitrary Task Graphs to Multiprocessors Using a Parallel Genetic Algorithm. Journal of Parallel and Distributed Computing 47(1), 58 (1997)
8. Kwok, Y.K., Ahmad, I.: Static Scheduling Algorithms for Allocating Directed Task Graphs to Multiprocessors. ACM Computing Surveys 31(4), 406–471 (1999)
9. Loukhaoukha, K.: On the Security of Digital Watermarking Scheme Based on Singular Value Decomposition and Tiny Genetic Algorithm. Journal of Information Hiding and Multimedia Signal Processing 3(2), 135–141 (2012)
10. Omara, F.A., Arafa, M.M.: Genetic algorithms for task scheduling problem. Journal of Parallel and Distributed Computing 70(1), 13–22 (2010)
11. Topcuoglu, H., Hariri, S., Wu, M.-Y.: Performance-effective and Low-complexity Task Scheduling for Heterogeneous Computing. IEEE Trans. on Parallel and Distributed Systems 13(3), 260–274 (2002)
12. Wu, A.S., Yu, H., Jin, S., Lin, K., Schiavone, G.: An Incremental Genetic Algorithm Approach to Multiprocessor Scheduling. IEEE Trans. on Parallel and Distributed Systems 15(9), 824–834 (2004)
13. Wen, Y., Xu, H., Yang, J.: A Heuristic-based Hybrid Genetic-variable Neighborhood Search Algorithm for Task Scheduling in Heterogeneous Multiprocessor System. Information Sciences 181(3), 567–581 (2011)
14. Yu, H.: Optimizing Task Schedules using an Artificial Immune System Approach. In: Proceedings of the 10th Annual Conference on Genetic and Evolutionary Computation, pp. 151–158 (2008)
15. http://www.Kasahara.Elec.Waseda.ac.jp/schedule/

A Swarm Random Walk Algorithm for Global Continuous Optimization

Najwa Altwaijry and Mohamed El Bachir Menai

Department of Computer Science, College of Computer and Information Sciences,
King Saud University, P.O. Box 51178 Riyadh 11453, Saudi Arabia
{ntwaijry,menai}@ksu.edu.sa

Abstract. Many real–world problems are modeled as global continuous optimization problems with a nonlinear objective function. Stochastic methods are used to solve these problems approximately, when solving them exactly is impractical. In this class of methods, swarm intelligence (SI) presents metaheuristics that exploit a population of interacting agents able to self–organize, such as ant colony optimization (ACO), particle swarm optimization (PSO), and artificial bee colony (ABC). This paper presents a new SI-based method for solving continuous optimization problems. The new algorithm, called Swarm Random Walk (SwarmRW), is based on a random walk of a swarm of potential solutions. SwarmRW is validated on test functions and compared to PSO and ABC. Results show improved performance on most of the test functions.

Keywords: Continuous Optimization, Swarm Intelligence, Metaheuristic.

1 Introduction

A continuous optimization problem is defined by a given objective function. Optimization algorithms search the solution space of such problems in order to find a solution, possibly subject to constraints [6]. An unconstrained continuous optimization problem is defined as:

$$
\begin{aligned}
minimize \quad & f(\mathbf{x}), \ \mathbf{x} = (x_1, x_2, \ldots, x_{n_x}) \\
subject\ to \quad & x_j \in \mathbb{R}, \quad j = 1, \ldots, n_x
\end{aligned}
\tag{1}
$$

where $\mathbf{x} \in \mathbb{R}^n$, $\mathbb{R}^n$ is the feasible domain, and $\mathbb{R}$ is the domain of the variable x_j. The search terminates when the optimum is found, or when a number of iterations has been exceeded, or if progress is insufficient.

Many algorithms exist to solve continuous optimization problems, such as the Genetic Algorithm [11], Simulated Annealing [16], and Tabu Search [10]. Swarm Intelligence (SI) is a major paradigm that includes many of these algorithms. In species that live in groups without a leader, individuals have no universal knowledge of the environment or the behavior of the group as a whole. These species

J.-S. Pan et al. (eds.), *Genetic and Evolutionary Computing*,
Advances in Intelligent Systems and Computing 238,
DOI: 10.1007/978-3-319-01796-9_4, © Springer International Publishing Switzerland 2014

depend on local interactions between individuals of the swarm to share information that enables them to move and survive in their environment. Examples from nature include a school of fish, or a flock of birds.

Studies of social organisms have resulted in a number of metaheuristics for SI. A swarm generally refers to a collection of interacting agents or individuals. Each individual is relatively simple; local interactions with neighboring individuals and the environment result in collective emergent swarm behavior that can be used to solve complex problems. Examples of SI based metaheuristics abound: Particle Swarm Optimization (PSO) [15], introduced by Eberhart and Kennedy, simulates the social behaviour of bird flocking. PSO can optimize multidimensional nonlinear functions, and has been applied to several real-world problems, e.g. [9,21,19,7]. Ant Colony Optimization (ACO) [4] can be thought of as a swarm whose individual agents are ants. ACO is used to solve a wide range of problems (e.g. [8,20,1]). The Artificial Bee Colony algorithm (ABC) [13] is based on the foraging behavior of bees while searching for food, and has been used to solve many problems (e.g. [12,18,3]). Many other swarm–based metaheuristics exist, and most derive their power from mimicking natural phenomena.

This paper aims to formulate a new algorithm based on trial and error behavior exhibited during the learning process of the swarm. The rest of this paper is organized as follows: Section 2 presents some relevant related work. Section 3 presents our proposed algorithm, SwarmRW. Next, Section 4 presents results on the chosen benchmark functions. Finally, Section 5 concludes this paper.

2   Related Work

The basic PSO [15] maintains a swarm of potential solutions called particles. Particles move in the search space by adding a velocity vector to the current particle position vector. There are basic variations to the original PSO that improve the quality of solutions and convergence speed. Velocity clamping was introduced to control particle divergence, especially of those particles that are far from the global best position. If the particle's velocity exceeds a certain threshold, it is set to the maximum velocity allowed. The inertia weight w controls the momentum of the particle, and is applied to the previous velocity. For $w < 1$, particles decelerate until their velocities are zero, where larger values of w encourage exploration, and small values encourage exploitation. In addition to the basic PSO, two other variants are used for comparison purposes in this study: PSO–TVAC and HPSO–TVAC [21]. PSO with Time Varying Acceleration Coefficients (PSO–TVAC) varies the acceleration coefficients and the intertia weight. The cognitive component is reduced over time, while the social component is increased over time, to facilitate global search in the early stages of the optimization process, and convergence towards the global optimium at the end. c_1 is varied from 2.5 to 0.5, and c_2 is varied from 0.5 to 2.5. The inertia weight is varied from $w = 0.9$ to $w = 0.4$. HPSO–TVAC "Self-Organizing Heirarchical Particle Swarm Optimizer with TVAC" keeps the previous velocity term $\mathbf{v}_i(t)$ constant at 0, and whenever particles stagnate in the search space (i.e. $\mathbf{v}_i(t + 1) = 0$), their velocities are reinitialized as $\mathbf{v_i} \sim (-rV, rV)$, where r is randomly generated number in the range $[0,1]$, and V is a time-varying reinitialization velocity that decays from the maximum velocity allowed V_{max} to $0.1 * V_{max}$. ABC [13] is based on the foraging behavior of bees while searching for food. ABC has been widely studied and used to solve a variety of problems [14].

3 The Proposed Algorithm

Trial and error [22] is a basic method for solving problems. Trial and error attempts to solve a problem through repeated, varied methods that are continued until the problem is successfully solved, or until the agent decides to stop. For example, a dog might learn how to open a gate through trial and error, using a series of movement approximations, with no insight by the dog. A particular movement that helps the dog achieve its goal is learned and adopted, and moves that are detrimental to its goal are abandoned. Trial and error is unsystematic and does not require insight. Learning is promoted by positive results. Trial and error is a heuristic method for problem solving; in the field of computer science, it is called generate and test. Many metaheuristics apply the basic idea of generate and test, such as the Genetic Algorithm [11] and Simulated Annealing [16].

3.1 Proposed Method

SwarmRW also applies the basic trial and error method to solve optimization problems. A swarm of potential solutions is randomly generated. Each individual then generates a new solution. However, in contrast to the simple generate and test methodology, individuals in the swarm cooperate —to a certain extent— to generate the next solution for a particular individual. Individuals "learn" from each other the best values, or positions, and change themselves accordingly. This cooperative behavior is illustrated in the fact that an individual will "imitate" its neighbor in a single trait (dimension) at a time, and if this improves the position of the individual, it will be accepted, otherwise, it will be rejected.

3.2 Definition

For an optimization problem with many dimensions d, a number of potential solutions $\mathbf{x}_i = (x_{i1}, \ldots, x_{id})$, $1 \leq i \leq n$, are randomly generated in a swarm of size n. A change to $\mathbf{x}_i$ is made in a single dimension j, and greedy selection is performed: if this move results in a better fitness value, the change is accepted. If, however, the move decreases fitness, the move is rejected. The change made to x_{ij} is done through swarm cooperation. A random solution $\mathbf{x}_k$, $1 \leq k \leq n$ and $k \neq i$, is selected, and solution $\mathbf{x}_i$ imitates solution $\mathbf{x}_k$ in dimension j.

Compare the method described above to PSO. PSO creates a population of solutions, and attempts to solve the problem by adding velocity values to all dimensions j, $1 \leq j \leq max\ dimensions$, simultaneously. For a small number of dimensions, PSO can converge to the optimum, but as the number of dimensions increases, solution quality could decrease given limited execution time. Intuitively, changing the solution in a single dimension at a time allows the search process to better judge if the proposed change is beneficial.

Based on this observation, a new algorithm is proposed, where a solution is changed in only one dimension j at a time t, using a simple formula:

$$x_l^{t+1} = \begin{cases} x_l^t + \rho * (x_l^t - y_l^t) & l = j, \\ x_l^t & l \neq j. \end{cases} \tag{2}$$

where j is a randomly selected dimension, and $\mathbf{y}$ is a randomly selected solution, distinct from $\mathbf{x}$. The parameter ρ is a randomly generated number that describes how much $\mathbf{x}$ is attracted or repelled by $\mathbf{y}$. This process is a random walk performed by the swarm.

The SwarmRW method is similar to how employed bees in the ABC algorithm produce neighboring solutions. However, the main differences between SwarmRW and ABC are that no onlooker bees are employed to probabilistically improve the best solutions produced by employed bees. In addition, solutions are never abandoned and scout bees are not used to replace abandoned solutions with new solutions in the search space. See Algorithm 1.

Algorithm 1. SwarmRW Pseudo Code

Input: S, max, ρ
Output: Best Solution
Initialise: S: swarm size, max: maximum runs, ρ: parameter;

begin
    Initialize swarm randomly;
    for $iter = 0 \to max - 1$ **do**
        for $i = 0 \to S - 1$ **do**
            Randomly select k, $0 \leq k < S$, $k \neq i$, and set $\mathbf{y} = \mathbf{x}_k$;
            Randomly select dimension j;
            Calculate x_{ij}^{t+1} by eq. (2), and set $\bar{\mathbf{x}}_i = (x_{i1}^t, \ldots, x_{ij}^{t+1}, \ldots, x_{id}^t)$;
            if $fitness(\mathbf{x}_i) < fitness(\bar{\mathbf{x}}_i)$ **then**
                $\mathbf{x}_i = \bar{\mathbf{x}}_i$;
            end
        end
    end
    return *Best Solution*
end

---

4   Experimental Results

The benchmark functions selected for testing [6], their ranges of search and initialization, and maximum velocities allowed for PSO are shown in Table 1. The

Sphere function is the first function of De Jong's test set [2], a unimodel function with one global minimum. The Rosenbrock function, also called Rosenbrock's valley or Rosenbrock's banana function, is a non-convex unimodel function where the global minimum is inside a flat, long, and narrow valley. Optimization algorithms can easily find the valley, but find it difficult to converge to the global minimum. The Rastrigin function is a non-convex function with a large number of local minima, making it difficult to locate the global minimum. The Griewank function and Schaffer's F6 function both have a large number of local minima. Schwefel's function is a multimodal function with a global minimum that is distant from the next best local minimum. Consequently, search algorithms can potentially converge in the wrong direction. Ackley's function is a widely used moderately complex multimodal test function. The Weierstrass function is a continuous function that is differentiable only on a set of points. The Katsuura function is a multi-modal, non-separable, asymmetrical, continuous everywhere yet differentiable nowhere function. The Lunacek bi-Rastrigin function is a hybrid function consisting of a Rastrigin and a double-sphere part with two funnels. It is a multi-modal, non-separable, asymmetrical, continuous everywhere yet differentiable nowhere function. f_{10} is shifted, while f_{11} is shifted and rotated. Katsuura and Lunacek are rotated and shifted as in the CEC2013 benchmark suite [17]. These functions are used extensively in the literature, and as such, were chosen to illustrate the performance of SwarmRW. SwarmRW is compared with PSO, PSO–TVAC, HPSO–TVAC [21], and ABC. We evaluate the different algorithms according to their precision and execution time, where all are designed to break early if the function minimum is found.

Swarm size is set to 40, except for ABC, where it is set to 80 as ABC divides the swarm into employed and onlooker bees. Using the same swarm size would result in 20 solutions being evaluated instead of 40, as in the other algorithms. PSO parameters are set as follows: $c_1 = c_2 = 1.4962$ as recommended in [5], velocity clamping is employed with maximum velocity set as shown in Table 1, and a time-varying inertia weight $c = 0.9$ to $c = 0.4$. PSO–TVAC and HPSO–TVAC parameters are set as specified in [21]. SwarmRW's parameter ρ is sampled from a uniform distribution in the range $[-1.5, 1.5]$ for all experiments. Results are shown in Tables (2, 3), the average value is shown on top, the standard deviation is shown in parenthesis, and the execution time on the bottom. Gmax is the maximum number of generations allowed, and Dim is the number of dimensions. For ABC, the number of generations is set to half that of PSO and its variants and SwarmRW, as ABC combines the employed and onlooker bee phases in one loop (generation). Setting the same number of generations would allow some solutions (the best solutions) to be optimized at least twice as many times as allowed for by PSO and its variants and SwarmRW. Results for 50 and 100 dimensions are unreported for HPSO–TVAC. The stopping condition in [21] is set to 0.01. For the rest of the algorithms, it is set to 0.0. Averages, standard deviations and times are reported for 100 trials, except for HPSO–TVAC, where they are reported for 50 trials. Experiments were performed on a 2.6 GHz Intel Core i7 machine with 16GB of RAM. PSO, PSO–TVAC, HPSO–TVAC, ABC,

Table 1. Benchmark Functions: Function name, Function Definition, Range of Search, Initialization Range, Maximum Velocity, and Global Minimum

Function	Definition	Search Range	Init. Range	V_{max}	$f(\mathbf{x}^*)$
Sphere	$f_1(\mathbf{x}) = \sum\limits_{i=1}^{n} x_i^2$	$(-100, 100)^n$	$(50, 100)^n$	100	0
Rosenbrock	$f_2(\mathbf{x}) = \sum\limits_{i=1}^{n-1} [100(x_i^2 - x_{i+1})^2 + (x_i - 1)^2]$	$(-100, 100)^n$	$(15, 30)^n$	100	0
Rastrigin	$f_3(\mathbf{x}) = \sum\limits_{i=1}^{n} [x_i^2 - 10\cos(2\pi x_i) + 10]$	$(-10, 10)^n$	$(2.56, 5.12)^n$	10	0
Griewank	$f_4(\mathbf{x}) = \frac{1}{4000} \sum\limits_{i=1}^{n} x_i^2 - \prod\limits_{i=1}^{n} \cos(\frac{x_i}{\sqrt{i}}) + 1$	$(-600, 600)^n$	$(300, 600)^n$	600	0
Schaffer's f_6	$f_5(\mathbf{x}) = 0.5 - \frac{\left(\sin\sqrt{x^2+y^2}\right)^2 - 0.5}{(1.0+0.001(x^2+y^2))^2}$	$(-100, 100)^2$	$(15, 30)^2$	100	0
Schwefel	$f_6(\mathbf{x}) = 418.9829 * n - \sum\limits_{i=1}^{n} x_i \sin\sqrt{|x_i|}$	$(-500, 500)^n$	$(-500, 500)^n$	500	0
Ackley	$f_7(\mathbf{x}) = -20\exp\left(-0.2\sqrt{\frac{1}{n}\sum\limits_{i=1}^{n} x_i^2}\right)$ $- \exp\left(\frac{1}{n}\sum\limits_{i=1}^{n}\cos 2\pi x_i\right) + 20 + e$	$(-30, 30)^n$	$(-30, 30)^n$	30	0
Weierstrass	$f_8(\mathbf{x}) = \sum\limits_{i=1}^{n}\left(\sum\limits_{k=0}^{k_{max}}[a^k\cos(2\pi b^k(x_i+0.5))]\right)$ $-n\sum\limits_{k=0}^{k_{max}}[a^k\cos(\pi b^k)], a=0.5, b=3, k_{max}=20$	$(-0.5, 0.5)^n$	$(-0.5, 0.2)^n$	0.5	0
Katsuura	$f_9(\mathbf{x}) = \frac{10}{n^2}\prod\limits_{i=1}^{n}\left(1+i+\sum\limits_{j=1}^{32}\frac{|2^j z_i - \text{round}(2^j z_i)|}{2^j}\right)^{\frac{10}{n^{1.2}}}$ $-\frac{10}{n^2}+200, z = \mathbf{M}_2 \wedge^{100}(\mathbf{M}_1\frac{5(x-o)}{100})$	$(-100, 100)^n$	$(-100, 100)^n$	100	200
Lunacek	$f_{10}(\mathbf{x}) = \min\left(\sum\limits_{i=1}^{n}(\hat{x}_i - \mu_0)^2, dn + s\sum\limits_{i=1}^{n}(\hat{x}_i - \mu_1)^2\right)$ $+10\left(n - \sum\limits_{i=1}^{n}\cos(2\pi\hat{z}_i)\right) + 300, s = 1 - \frac{1}{2\sqrt{n+20}-8.2}$ $d=1, \mu_0 = 2.5, \mu_1 = -\sqrt{\frac{\mu_0^2 - d}{s}}, y = \frac{10(x-o)}{100}$ $\hat{x}_i = 2\text{sign}(x_i^*)y_i + \mu_0, z = \wedge^{100}(\hat{x} - \mu_0)$	$(-100, 100)^n$	$(-100, 100)^n$	100	300
Lunacek	$f_{11}(\mathbf{x}) = \min\left(\sum\limits_{i=1}^{n}(\hat{x}_i - \mu_0)^2, dn + s\sum\limits_{i=1}^{n}(\hat{x}_i - \mu_1)^2\right)$ $+10\left(n - \sum\limits_{i=1}^{n}\cos(2\pi\hat{z}_i)\right) + 400, s, d, \mu_0, \mu_1, y, \hat{x}_i$ as in $f_{10}, z = \mathbf{M}_2 \wedge^{100}(\mathbf{M}_1(\hat{x} - \mu_0))$	$(-100, 100)^n$	$(-100, 100)^n$	100	400

Table 2. Average results, standard deviation and time (m:s) for the different methods on $f_1 - f_5$ functions

Function	Dim.	Gmax	PSO	PSO TVAC	HPSO TVAC	ABC	SwarmRW
f_1	30	3000	1.21e−7 (4.32e−7) 0:19	1.50e−4 (5.77e−4) 0:19	0.01 (n/a) n/a	7.88e−16 (1.43e−16) 0:02	**7.65e−22** (6.40e−22) 0:01
	50	5000	0.026 (0.11) 0:56	0.322 (0.47) 0:59	n/a (n/a) n/a	1.74e−15 (3.06e−16) 0:03	**3.22e−21** (2.10e−21) 0:02
	100	6000	342.94 (498.38) 2:21	358.06 (382.22) 2:28	n/a (n/a) n/a	3.24e−8 (1.15e−7) 0:06	**2.70e−9** (1.38e−9) 0:05
f_2	30	5000	59.19 (55.95) 0:56	187.49 (347.38) 0:58	13.67 (11.0) n/a	10.43 (6.96) 0:24	**4.53** (4.19) 0:24
	50	6000	289.71 (395.94) 1:56	738.91 (903.11) 2:00	n/a (n/a) n/a	11.63 (6.81) 0:47	**6.31** (5.14) 0:47
	100	7000	5.98e+5 (2.20e+6) 4:39	2.13e+5 (4.02e+6) 4:46	n/a (n/a) n/a	37.01 (17.33) 1:46	**22.36** (8.06) 1:51
f_3	30	5000	68.56 (20.37) 1:06	57.08 (17.68) 1:08	0.044 (0.196) n/a	1.56e−13 (6.84e−13) 0:33	**0.0** (0.0) 0:23
	50	6000	203.23 (43.96) 2:19	183.05 (39.75) 2:22	n/a (n/a) n/a	0.05 (0.21) 1:07	**1.64e−13** (9.56e−13) 1:03
	100	7000	767.41 (108.09) 5:37	731.24 (91.91) 5:45	n/a (n/a) n/a	13.67 (3.66) 2:44	**5.52** (1.29) 2:40
f_4	30	5000	6.31e−14 (2.80e−13) 1:11	4.89e−8 (1.96e−7) 1:12	0.01 (0.0035) n/a	7.59e−16 (1.82e−16) 0:43	**2.22e−16** (3.85e−17) 0:42
	50	6000	3.72e−5 (1.16e−4) 2:24	0.002 (0.003) 2:32	n/a (n/a) n/a	1.77e−15 (3.74e−16) 1:24	**4.77e−16** (5.98e−17) 1:24
	100	7000	2.46 (5.09) 5:56	2.82 (3.16) 6:40	n/a (n/a) n/a	2.01e−11 (1.62e−11) 3:15	**4.44e−14** (2.71e−14) 3:14
f_5	2	1000	0.0025 (1.11e−17) 0:00	0.0025 (0.0) 0:00	0.01 (0.007) n/a	0.0025 (1.50e−8) 0:00	0.0025 (2.05e−9) 0:00

Table 3. Average results, standard deviation and time (m:s) for the different methods on $f_6 - f_{11}$ functions

Function	Dim.	Gmax	PSO	PSO TVAC	ABC	SwarmRW
f_6	30	5000	2420.91 (390.10) 1:00	2381.68 (503.33) 1:03	2.24e−6 (1.67e−5) 0:33	**5.71e−8** (9.09e−13) 0:29
	50	6000	5513.21 (654.85) 2:08	6491.22 (975.18) 2:10	63.82 (74.78) 1:04	**2.73** (16.84) 0:57
	100	7000	15168.92 (986.78) 5:04	18289.87 (1269.45) 5:07	1313.21 (249.40) 2:27	**1019.01** (189.45) 2:16
f_7	30	5000	1.21 (0.69) 0:54	1.24 (0.62) 0:55	5.32e−14 (6.80e−15) 0:24	**3.28e−14** (3.71e−15) 0:23
	50	6000	2.69 (0.64) 1:53	2.73 (0.53) 2:00	1.05e−11 (4.05e−12) 0:46	**1.37e−13** (2.35e−14) 0:45
	100	7000	6.75 (0.95) 4:44	6.74 (0.97) 4:54	7.50e−6 (3.57e−6) 1:51	**1.98e−6** (5.78e−7) 1:46
f_8	30	5000	1.91 (1.13) 41:28	1.06 (0.83) 41:21	7.11e−16 (2.36e−15) 42:15	**0.0** (0.0) 27:56
	50	6000	10.65 (2.79) 79:20	8.51 (2.39) 79:23	1.06e−12 (1.08e−12) 79:29	**1.28e−14** (1.13e−14) 79:12
	100	7000	48.89 (6.09) 177:51	48.37 (5.03) 177:45	9.61e−4 (2.08e−4) 179:37	**3.48e−4** (5.03e−5) 179:32
f_9	30	5000	202.54 (0.30) 60:46	202.29 (0.35) 60:49	201.67 (0.22) 60:56	**201.18** (0.22) 61:13
	50	6000	203.51 (0.34) 123:22	203.41 (0.41) 123:43	202.39 (0.24) 122:10	**201.75** (0.23) 121:29
	100	7000	204.26 (0.25) 289:18	204.11 (0.34) 289:20	203.09 (0.20) 288:4	**202.38** (0.24) 286:43
f_{10}	10	5000	314.05 (2.94) 0:52	313.66 (1.65) 0:53	307.92 (2.66) 0:46	**304.29** (4.49) 0:44
	20	6000	344.03 (7.34) 2:16	340.17 (6.19) 2:17	320.12 (1.49) 1:54	**316.80** (6.69) 1:54
	30	7000	391.47 (15.94) 4:03	377.04 (12.39) 4:04	330.52 (0.04) 3:22	**329.83** (3.63) 3:22
f_{11}	10	5000	421.73 (7.11) 0:55	**417.38** (4.17) 0:55	436.83 (4.17) 0:50	430.16 (4.87) 0:46
	20	6000	469.11 (19.97) 2:33	**456.00** (11.00) 2:33	554.79 (15.30) 2:13	527.49 (17.44) 2:10
	30	7000	531.36 (34.85) 4:41	**505.57** (22.31) 4:41	743.56 (23.55) 4:08	680.14 (27.93) 4:01

and SwarmRW were tested on functions f_1 to f_5. The remaining functions were used to test PSO, PSO-TVAC, ABC, and SwarmRW. Results on functions f_1 to f_5 show that SwarmRW outperforms PSO, PSO–TVAC and HPSO–TVAC on all benchmark functions except Schaffer's F6, which has only two dimensions. It also outperforms ABC, however, their performance is comparable. ABC and SwarmRW are faster than PSO and its variants, except on Schaffer's F6 function, and their execution time is again comparable. The performance gap between SwarmRW and PSO and its variants increases with an increase in the number of dimensions for these functions. For the Sphere function, SwarmRW achieves the best results in all dimensions, and as the number of dimensions increases, the gap between it and PSO and its variants increases. This trend is seen in all test functions except for Schaffer"s F6 function. ABC and SwarmRW perform similarly. For the Rosenbrock function, SwarmRW outperforms all other algorithms for all dimensions, however, it is again comparable with ABC. On the Rastrigin function, SwarmRW locates the global minimum at 30 dimensions with a standard deviation of 0.0, and outperforms the others in 50 and 100 dimensions. The gap with ABC in 50 dimensions is quite large, however, similar performance is noted in 100 dimensions. On the Griewank function, SwarmRW outperforms all others in all tested dimensions, marginally outperforming ABC. Finally, on Schaffer's F6 function, all algorithms perform comparably.

Results on functions f_6, f_7, f_8, f_9, and f_{10} show that SwarmRW outperforms PSO and PSO–TVAC. It also outperforms ABC in all tested dimensions on the Schwefel function; the largest gap is observed in 30 dimensions, however, the performance gap is decreased as dimensions increase. On the Ackley function, SwarmRW and ABC are comparable, and both outperform PSO and PSO–TVAC. On the Weierstrass function, SwarmRW locates the minimum with a standard deviation of 0 for 30 dimensions, and outperforms PSO and PSO–TVAC. Its performance is close to that of ABC. On the Katsuura function, all algorithms perform similarly. On Lunacek bi-Rastrigin without rotation, SwarmRW outperforms all other algorithms, however, it is comparable with ABC. On Lunacek bi-Rastrigin with rotation, PSO–TVAC is the best performer followed by PSO and SwarmRW, with ABC showing the worst performance. This might be due to dependencies among variables, which SwarmRW is unaware of. A niching strategy might be beneficial for SwarmRW on this and similar functions.

Overall, SwarmRW has shown successful performance on test functions. It outperforms ABC on all instances, is quite simple to implement and makes use of a single parameter, easing the process of parameter selection.

5   Conclusion and Future Work

This paper presented SwarmRW, a new algorithm for continuous unconstrained optimization problems. It was tested on eleven well-known benchmark functions, and performed favorably compared with PSO, PSO–TVAC, HPSO–TVAC, and ABC, with regards to precision as well as execution time. It makes use of a single parameter, ρ, to decide the step size in each dimension. Future work includes finding the best parameter values for different functions, extensive testing of SwarmRW on other benchmark functions, as well as investigating other versions of SwarmRW for other problems such as TSP and graph coloring problems.

References

1. Blum, C., Vallès, M., Blesa, M.: An ant colony optimization algorithm for DNA sequencing by hybridization. Computers & Operations Research 35(11), 3620–3635 (2008)
2. De Jong, K.A.: Analysis of the behavior of a class of genetic adaptive systems. PhD thesis, University of Michigan, MI, USA (1975)
3. de Oliveira, I., Schirru, R.: Swarm intelligence of artificial bees applied to in-core fuel management optimization. Annals of Nuclear Energy 38(5), 1039–1045 (2011)
4. Dorigo, M.: Optimization, Learning and Natural Algorithms. PhD thesis, Dipartimento di Elettronica, Politecnico di Milano, Milan, Italy (1992) (in Italian)
5. Eberhart, R.C., Shi, Y.: Comparing inertia weights and constriction factors in particle swarm optimization. In: Proceedings of the IEEE 2000 Congress on Evolutionary Computation, vol. 1, pp. 84–88 (2000)
6. Engelbrecht, A.: Fundamentals of computational swarm intelligence, vol. 1. Wiley, London (2005)
7. Feng, H.M., Chen, C.Y., Ye, F.: Evolutionary fuzzy particle swarm optimization vector quantization learning scheme in image compression. Expert Systems with Applications 32(1), 213–222 (2007)
8. Fuellerer, G., Doerner, K., Hartl, R., Iori, M.: Ant colony optimization for the two-dimensional loading vehicle routing problem. Computers & Operations Research 36(3), 655–673 (2009)
9. Fukuyama, Y., Yoshida, H.: A particle swarm optimization for reactive power and voltage control in electric power systems. In: Proceedings of the 2001 Congress on Evolutionary Computation, vol. 1, pp. 87–93 (2001)
10. Glover, F.: Tabu search – part i. ORSA Journal on Computing 1(3), 190–206 (1989)
11. Holland, J.: Adaptation in natural and artificial systems, vol. 1(97), p. 5. University of michigan press, Ann Arbor (1975)
12. Kang, F., Li, J., Xu, Q.: Hybrid simplex artificial bee colony algorithm and its application in material dynamic parameter back analysis of concrete dams. Journal of Hydraulic Engineering 6, 014 (2009)
13. Karaboga, D.: An idea based on honey bee swarm for numerical optimization. Technical report tr06, Erciyes University Press, Erciyes (2005)
14. Karaboga, D., Gorkemli, B., Ozturk, C., Karaboga, N.: A comprehensive survey: Artificial Bee Colony (ABC) algorithm and applications. Artificial Intelligence Review, 1–37 (2012)
15. Kennedy, J., Eberhart, R.: Particle swarm optimization. In: Proceedings of the IEEE International Conference on Neural Networks, vol. 4, pp. 1942–1948 (November/December 1995)
16. Kirkpatrick, S., Gelatt, C., Vecchi, M.: Optimization by simulated annealing. Science 220(4598), 671–680 (1983)
17. Liang, J.J., Qu, B.Y., Suganthan, P.N., Hernández-Daz, A.G.: Problem definitions and evaluation criteria for the CEC 2013 special session on real-parameter optimization (January 2013)

18. Omkar, S., Senthilnath, J.: Artificial bee colony for classification of acoustic emission signal source. International Journal of Aerospace Innovations 1(3), 129–143 (2009)
19. Omran, M., Engelbrecht, A., Salman, A.: Particle swarm optimization method for image clustering. International Journal of Pattern Recognition and Artificial Intelligence 19(03), 297–321 (2005)
20. Rajendran, C., Ziegler, H.: Ant-colony algorithms for permutation flowshop scheduling to minimize makespan/total flowtime of jobs. European Journal of Operational Research 155(2), 426–438 (2004)
21. Ratnaweera, A., Halgamuge, S., Watson, H.: Self-organizing hierarchical particle swarm optimizer with time-varying acceleration coefficients. IEEE Transactions on Evolutionary Computation 8(3), 240–255 (2004)
22. Thorpe, W., Thorpe, W.: The origins and rise of ethology: The science of the natural behaviour of animals. Heinemann Educational Books (1979)

A Sampling-PSO-K-means Algorithm for Document Clustering

Nadjet Kamel[1,2], Imane Ouchen[2], and Karim Baali[2]

[1] Computer Science Department, Faculty of Sciences, UFAS
Setif, Algeria
[2] LRIA, Computer Science Department, USTHB
Algiers, Algeria
nkamel@usthb.dz, imane.ouchen@gmail.com, karimbaali@hotmail.com

Abstract. Clustering is grouping objects into clusters such that objects within the same cluster are similar and objects of different clusters are dissimilar. Several clustering algorithms have been proposed in the literature, and they are used in several areas: security, marketing, documentation, social networks etc. The K-means algorithm is one of the best clustering algorithms. It is very efficient but its performance is very sensitive to the initialization of clusters. Several solutions have been proposed to address this problem. In this paper we propose a hybrid algorithm for document web clustering. The proposed algorithm is based on K-means, PSO and Sampling algorithms. It is evaluated on four datasets and the results are compared to those obtained by the algorithms: K-means, PSO, Sampling+K-means, and PSO+K-means. The results show that the proposed algorithm generates the most compact clusters.

Keywords: Clustering algorithms, PSO, K-means, Sampling.

1 Introduction

Clustering is grouping objects into groups such that objects within a same group are similar and objects into different groups are dissimilar. The main problem of clustering is to obtain optimal grouping. This issue arises in many scientific applications, such as biology, education, genetics, criminology, etc… Many approaches [1], [2], [3], [4], [5], [6], [7] have been developed in this regard, but the K-means algorithm [8] is the most popular for its simplicity and efficiency. The main problem of K-means algorithm is its sensitivity to the initial cluster centroids. This influences seriously its optimal solution which may be local rather than global. Several solutions were proposed to overcome these shortcomings. In [9] the authors proposed the Global K-means algorithm which is an incremental approach that dynamically adds one centroid at a time, followed by the processing of the K-means algorithm until the convergence. In this algorithm, the centroids are chosen one by one in the following way: the first centroid is chosen to be the centroid of all the data. Other centroids are chosen in the data set where every data is a candidate to become a centroid. The latter centroid will be tested with the rest of the

J.-S. Pan et al. (eds.), *Genetic and Evolutionary Computing*,
Advances in Intelligent Systems and Computing 238,
DOI: 10.1007/978-3-319-01796-9_5, © Springer International Publishing Switzerland 2014

data set. The best candidate is the one who minimizes the objective function. In [10] the authors proposed to use the hierarchical clustering to choose the initial centroids. This approach proposes to use a hierarchical clustering at a first step, then we compute the centroids of each resulting cluster, and at the end we use the K-means algorithm with the obtained results. In [11] the authors propose to use bootstrap to determine the initial centroids. It consists of dividing the data set into a set of samples and then we use the K-means algorithm on each sample. Each cluster of the different groups produces a set of candidate centroids to be used for the initialization of the K-means algorithm. Another approach proposed in [12], uses MaxMin method for the initialization. Many Hybridizations approaches are also used to face the problem of the initialization of K-means. These approaches use metaheuristics such genetic algorithms [13], Particle Swarm Optimization (PSO) [14,21], and Ant Colony Organization (ACO) [15,22], to choose the initial centroids.

In this paper we propose a clustering algorithm that hybridizes PSO, K-means and stamping techniques to improve the clustering results. It is based on the PSO-K-means algorithm presented in [14] and the initialization by sampling presented in [11]. The proposed algorithm is used to cluster a set of documents collection. The results show that the proposed algorithm produce more compact clusters than PSO, K-means, and PSO-K-means algorithms.

The remaining of the paper is organized as follows: the next section presents the PSO-K-means algorithm on which our proposition is based. The section 3 presents our proposition. The results of our algorithm are presented in section 4. Finally, section 5 presents our conclusion and future works.

2   PSO-K-means Algorithm

In this section, we present an algorithm that hybridizes the PSO and the K-means algorithm. The PSO algorithm is used to generate initial centroids for K-means. This algorithm is proposed in [14] and it was proposed for document clustering.

The Particle Swarm Optimization (PSO) algorithm is a stochastic optimization technique based on population. It can find an optimal solution, or near the optimal of a quantitative or qualitative problem. [16, 17,18].

The hybrid algorithm PSO+K-means includes two modules: the PSO module and the K-means module. Firstly, the PSO module is executed to determine the centroids of the clusters. The algorithm PSO is used at first step to find the neighboring of the optimal solution thanks to a global search. Then the resulting centroids are used by the K-means module in order to refine and generate the optimal solution.

2.1   PSO Module

A particle in the swarm represents a possible solution for the clustering of a collection of documents. A swarm represents a set of possible solutions.

As the problem of clustering of documents is multi-dimensional, the PSO method must be adapted to this problem. To do this, a matrix (Nt,k) is used, where Nt is the number of terms in the collection, and k is the number of clusters. So, each particle is represented as follows:

$$X_i=(C_1, C_2, ..., C_i, ..., C_k)$$

Where C_i represents the ith centroid vectord, and k represents the number of clusters.

At each iteration, the particle changes the position of each of its centroids by taking into account its experience and the neighbors' experience. The distance between the centroid and a document of the collection is used as *fitness* to evaluate the obtained solution of each particle. The value of the *fitness* is measured by the following formula:

$$f = \frac{\sum_{i=1}^{k}\left\{\frac{\sum_{j=1}^{p_i}d(o_i,m_{ij})}{p_i}\right\}}{k} \tag{1}$$

Where m_{ij} represents the jth document in the ith cluster; o_i is the centroid of the ith cluster; $d(o_i,m_{ij})$ is the distance between the document m_{ij} and the centroid o_i; p_i is the number of documents in the cluster c_i; and k is the number of clusters.

The PSO module can be defined by the following steps:

(1) Initially, each particle chooses randomly k vectors of documents from the collection as cluster centroids.
(2) For each particle:
    (a) Assign each document vector of the collection to the nearest cluster (minimal distance).
    (b) Compute the fitness value.
    (c) Change the position and the speed of each particle.
(3) Repeat the step (2) until one of the following stopping criteria is reached:
    (a) The maximum number of iterations is exceeded
    or
    (b) The average change in centroid vectors is less than a predefined value.

2.2   K-means Module

The K-means module uses as initial centroids those resulting from the PSO algorithm, and executes the remaining of the K-means algorithm by computing the optimal centroids. The algorithm is given through the following steps:

(1) Use the centroids vectors resulting from PSO algorithm.

(2) Assign each document vector of the collection to the nearest cluster (minimal distance).

(3) Compute each centroid by using the following formula:

$$C_i = \frac{1}{n_j} \sum_{\forall d_j \in S_j} d_j$$

Where d_j is the document vector of the cluster S_j; C_i is the centroid ; n_j is the number of documents in the cluster S_j.

(4) Repeat the steps (2) and (3) until convergence.

In the PSO+K-means algorithm, the global search of PSO and the rapidity of convergence of K-means are combined. The algorithm PSO is used at first step to find the neighboring of the optimal solution thanks to a global search. The obtained result is then used for the K-means algorithm, by using a local search to generate the final result.

3   Sample-PSO-K-means Algorithm: Our Contribution

In this section we present our proposed algorithm: Sample-PSO-K-means. As outlined above, the hybridization of PSO and K-means algorithm avoids the problem of the initialization of initial centroids, and combines a local and a global optimum search. On the other hand, and according to the experiment and the analysis presented in [14], the PSO approach needs a big number of iterations to insure a good convergence. Larger the database is, more particles are needed, and higher will be the number of iterations.

Refining the algorithm of the initialization by stamping [11] is also a solution for the initialization problem of K-means. It consists of stamping a collection of data in a set of sub-groups, and then using the K-means algorithm on each group. The next step consists of applying K-means with the different solutions, but in this case, it is applied on the whole data sets. Finally the best found solution is chosen. This approach is called also, K-means refinement clustering algorithm, because it applies the algorithm twice. In this approach the sub-groups collaborate to have best results.

The sampling approach is efficient when the size of the data set is large. For this reason, we propose to use this approach with PSO. The main idea is that the sub-samples collaborate by taking the principal of the particles in the swarm and the PSO algorithm can have a good initial configuration and reduce the number of particles to be created. The data set is divided into a known number of sub-samples. The same number represents the size of the PSO swarm (number of the particles).

Our algorithm is summarized as follows:

1. Choose S sub-samples in the initial collection (this number represents the number of PSO particles).

2. Apply the K-means on each sub-sample using random initial centroids. We obtain for each sub-sample a set of centroids.

3. Apply PSO having S particles on the howl data using the centroids resulting from the previous step.
4. Apply the K-means on all data using as initial configuration the results obtained by PSO.

4   Implementation

As application, we choose to cluster web documents. We do this in two steps: the first step preprocesses the web documents. The second step encodes the web documents and implements the algorithm.

4.1   Web Document Preprocessing

Before clustering the data, a preprocessing is necessary. This step is done through two steps: the first one prepare the data (cleaning, lemmatization, and tokenizing), and the second one creates the index file containing all the important information in each document of the collection. The most used method for the index file is TF*IDF.

4.2   Representation of the Different Components

We have three components: documents, centroids and clusters. Particularly, for the PSO optimization, the components are particles and the swarm to which are assigned. For the initialization by sample method, we have to define a structure for a sub-sampling.

A document is represented as a vector of size m, where m is the number of terms in the collection. We note $D_i=(w_{i1},w_{i2},w_{i3},...,w_{im})$ the ith document vector of the collection, where, w_{ij} represents the weight of the jth term in the document D_i, with $i \in 1,2,...,n$ and $j \in 1,2,3,...,m$. n is the number of documents in the collection.

Each cluster is represented by a vector. Its size is unknown, because it can't be determined in advance. We note $C_i=(D_{i1},D_{i2},D_{i3},...)$ the ith cluster, with $i \in 1,2,3,...,k$. k is the number clusters.

A particle is a vector of k centroids. A centroid is represented by a document vector. We note $P_i=(C_{i1},C_{i2},C_{i3},...,C_{ik})$ the ith particle, with $i \in 1,2,3,...,N$. N is the number of particles (Swarm size).

A swarm is a vector of N particles. We note $S=(P_1,P_2,P_3,...,P_N)$ with $i \in 1,2,3,...,N$

A sub-stamp is a sub-group in the collection. It can be represented as a vector of M documents. M is the size of the sub-stamp. We note $E_i=(D_{i1},D_{i2},D_{i3},...,D_{iM})$ the ith sub-stamp with $i \in 1,2,3,...,N$

4.3   Evaluation of the Solutions

To evaluate the solutions, we use the formula defined by the equation (1). This function is used as a fitness function for the PSO algorithm and to evaluate the different results. It measures the average distance between the documents of a cluster and its centroid. The smaller this value is, the more the cluster is compact. So, it can be used to evaluate the quality of the clusters.

To calculate the similarity between two documents m_p and m_j, we often use a distance measure. The most used measures are based on the Minkowski formula given by:

$$D_n(m_p, m_j) = \left(\sum_{i=1}^{d_m} |m_{i,p} - m_{i,j}|^n \right)^{1/n}$$

For $n=2$, the formula defines the Euclidian distance. For our work we use the normalized Euclidian distance. The normalized Euclidian distance between the documents m_p and m_j is given as follows:

$$d(m_p, m_j) = \sqrt{\sum_{k=1}^{d_m} (m_{pk} - m_{ik})^2 / d_m}$$

Where m_p and m_j are two document vectors; d_m is the size of the vectors; m_{pk} and m_{jk} represents respectively the weight values of the kth term in the collection for the documents m_p and m_j.

5   Experimentation and Results

In this section we give the details of the implementation of our algorithm and its results.

5.1   Experimentation

We implemented the K-means algorithm, the initialization by stamping approach, the PSO, the PSO + K-means algorithm, and finally our algorithm.

To evaluate these algorithms, we used four datasets [7]. Since this document collection is large (11 000 documents with 12 621 201 terms), and The results of some researches [19,20], say that it is preferred to use the data with low dimensionality for the PSO + K-means algorithm, we used the sub-categories separately. For example, we used the algorithm for the category with the subject « Banking and Finance » to find the clusters: Commercial Banks, Building societies and Insurance Agencies. We selected the documents with low size to reduce the number of terms. For each algorithm, we use the Euclidian distance as a similarity measure.

5.2   Parameters

After several executions, we noticed that K-means is stable after 20 iterations for the most datasets. The PSO algorithm needs generally 100 iterations to reach a stable solution. Our results show that hybridization of PSO and K-means returns the best solution than the previous algorithms, and it stabilizes after 50 iterations.

We have two kinds of parameters. The parameters that are fixed in the program and they do not change (see table 1.), and the parameters we change at each test (see table 2.). The choice of the fixed parameters is based on the results of [18]. In order to explore the large data space, we choose to use 50 particles for the PSO algorithm and 25 particles where PSO is used (PSO, PSO + K-means, Stamping + PSO + K-means). Since the number of stamps is related to the number of particles, we used 25 sub-stamps.

Table 1. Fixed PSO Parameters

Parameter	Value
Inertia factor	0.3
Confident coefficient at its best position.	0.72
Confident coefficient at its neighboring	1.49

Table 2. Parameters of the Stamping+PSO+K-means algorithm

Parameter	Value
Number of clusters	5
Number of iteration in PSO	5-50
Number of iteration in K-means	10-25
Number of particles	10-25
Dimension of the problem	500-10.000
Size of a stamp	26-65
Number of documents to classify	137-655

5.3   Results

The fitness function (formula (1)) evaluates the different generated solutions. The best solution is the one minimizing the value of this function. The Table 3. shows the results obtained by using the algorithms K-means, PSO, Sampling+K-means, PSO+K-means and Sampling+PSO+K-means. The values represent the average of the fitness values for 10 simulations performed separately. The data sets: Dataset 1, Dataset 2, Dataset 3, Dataset 3 and Dataset 4 represent respectively the document collections: "Banking and Finance", "Programming Languages", "Science" and "Sport".

The results show that the Sampling+PSO+K-means algorithm generates the lowest values of the fitness function. This means that the clusters generated by this algorithm are the most compact.

Table 3. Performances of the implemented algorithms (fitness values)

	K-means	PSO	Sampling + K-means	PSO+ K-means	Sampling+ pso+K-means
Dataset 1	5.093	6.982	4.098	3.211	2.911
Dataset 2	4.788	4.871	4.021	2.922	3.109
Dataset 3	7.245	8.632	6.883	5.447	4.398
Dataset 4	4 9.09	10.093	7.902	4.63	4.877

The figure 1 illustrates the convergence of the different algorithms on the dataset 1. In this figure, we notice that the algorithm K-means converges quickly but prematurely. Its fitness value grows from 11.45 to 8.26 after only 10 iterations, and after that, it doesn't change. For PSO and PSO+K-means, we notice that the two curves have the same trajectories until the 25th iteration because the two algorithms do the same work (they use PSO). After the 25th iteration, the trajectories change. PSO converges slowly. In contrast, the PSO+K-means algorithm converges quickly. For the Sampling + K-means, we notice that the curve is identical to the K-means one. The difference is the initial point which is 9.34. Finally, we can deduce that the Sampling+PSO+K-means algorithm is the fusion of the two curves of Sampling +K-means and PSO+K-means. It is the algorithm which gives the better results.

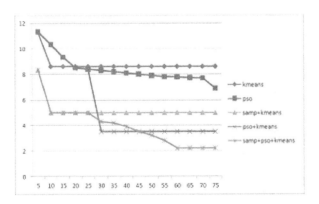

Fig. 1. Convergence graph of the clustering algorithms: K-means, PSO, PSO+K-means, Sampling+K-means, Sampling+PSO+K-means)

6   Conclusion

The use of hybrid algorithms to improve the performance of the clustering is not a new idea. At our knowledge, there is no gain using the stamping with PSO and K-means for the initialization. From our experimentations, the hybrid algorithm PSO+K-means returns better result than the classical K-means algorithm.

In contrast, we notice that PSO needs an important number of iterations and particles because its initialization is random. In the case of a large collection of data, the complexity of the problem is empiric and the use of PSO+K-means algorithm for clustering needs large memory, and high execution time. This makes it not an interesting solution.

We have used the stamping with PSO and not directly with K-means. Experimentations achieved in [14] shows that the hybrid algorithm PSO+K-means does not improve the results. Using K-means first, generates a local optimum. Using the obtained result as the initial value of one of the particles of the swarm, the others will be chosen randomly. This particle has the best fitness of the group. This will help the convergence of the other particles, but the considered particle does not converge even it is the best solution. In our algorithm, the stamping approach returns a number of solutions equal to the number of particles. So, all the particles will have a good initial configuration and they can all cooperate to converge to a global optimum. The tests we did show that our proposition returns the best results than PSO+K-means algorithm. In fact, with a more large data collection, the results will be more visible.

The proposed algorithm gives good initialization to the K-means algorithm. It improves the performance of the hybrid algorithm PSO+K-means and makes it usable for large data sets.

As future work, we tend to use parallelism for the stamping, by executing the K-means algorithm on the different stamps at the same time, and for the computation of the fitness value for each particle.

References

1. Agrawal, R., Gehrke, J., Gunopulos, D., Raghavan, P.: Automatic subspace clustering of high dimensional data for data mining applications (1999)
2. Nagesh, H., Goil, S., Choudhary, A.: Efficient and scalable subspace clustering for every large data sets (1999)
3. Sheikholeslami, G., Chatterjee, S., Zhang, A.: Wavecluster: A multi-resolution clustering approach for very large spatial databases. In: Proc. 24th Int. Conf. Very Large Data Bases, VLDB, New York City, USA, pp. 428–439 (1998)
4. Kaufman, L., Rousseeuw, P.J.: Finding groups in data. In: An Introduction to Cluster Analysis. John Wiley & Sons (1990)
5. Sneath, P.H.A., Sokal, R.R.: Numerical Taxonomy. The Principles and Practice of Numerical Classification. W. H. Freeman and Company, San Francisco (1973)
6. Vazirani. Algorithmes d'approximation, V. Collection IRIS. Springer (2006)
7. TREC. Text Retrieval Conference (1999), http://trec.nist.gov
8. MacQueen, J.B.: Some Methods for classification and Analysis of Multivariate Observations. In: Proceedings of 5th Berkeley Symposium on Mathematical Statistics and Probability, vol. 1, pp. 281–297. University of California Press, Berkeley (1967)
9. Likas, A., Vlassis, M., Verbeek, J.: The global k-means clustering algorithm. Pattern Recognition 36, 451–461 (2003)
10. Milligan, G.W.: The validation of four ultrametric clustering algorithms. Pattern Recognition 12, 41–50 (1980)

11. Bradley, P.S., Fayyad, U.M.: Refining initial points for K-Means clustering. In: Proc. 15th International Conf. on Machine Learning, pp. 91–99. Morgan Kaufmann, San Francisco (1998)
12. Mirkin, B.: Clustering for data mining: A data recovery approach. Chapman and Hall, London (2005)
13. Kwedlo, W., Iwanowicz, P.: Using Genetic Algorithm for Selection of Initial Cluster Centers for the K-Means Method. In: Rutkowski, L., Scherer, R., Tadeusiewicz, R., Zadeh, L.A., Zurada, J.M. (eds.) ICAISC 2010, Part II. LNCS, vol. 6114, pp. 165–172. Springer, Heidelberg (2010)
14. Xiaohui, C., Potok, T.E.: Document Clustering Analysis Based on Hybrid PSO+K-means Algorithm. Applied Software Engineering Research Group, Computational Sciences and Engineering Division, Oak Ridge National Laboratory, Oak Ridge, TN 37831- 6085, USA (2005)
15. Saatchi, S., Hung, C.-C.: Hybridization of the Ant Colony Optimization with the K-Means Algorithm for Clustering. In: Kalviainen, H., Parkkinen, J., Kaarna, A. (eds.) SCIA 2005. LNCS, vol. 3540, pp. 511–520. Springer, Heidelberg (2005)
16. Carlisle, A., Dozier, G.: An Off-The- Shelf PSO. In: Proceedings of the 2001 Workshop on Particle Swarm Optimization, Indianapolis, IN, pp. 1–6 (2001)
17. Kennedy, J., Eberhart, R.C., Shi, Y.: Swarm Intelligence. Morgan Kaufmann, New York (2001)
18. Shi, Y., Eberhart, R.C.: Parameter selection in particle swarm optimization. In: Porto, V.W., Waagen, D. (eds.) EP 1998. LNCS, vol. 1447, pp. 591–600. Springer, Heidelberg (1998)
19. Omran, M., Salman, A., Engelbrecht, A.P.: Image classification using particle swarm optimization. In: Proceedings of the 4th Asia-Pacific Conference on Simulated Evolution and Learning 2002 (SEAL 2002), Singapore, pp. 370–374 (2002)
20. Van, D.M., Engelbrecht, A.P.: Data clustering using particle swarm optimization. In: Proceedings of IEEE Congress on Evolutionary Computation 2003 (CEC 2003), Canbella, Australia, pp. 215–220 (2003)
21. Alireza, A., Hamidreza, M.: Combining PSO and k-means to Enhance Data Clustering. In: International Symposium on Telecommunication, vol. 1 and 2, pp. 688–691 (2008)
22. Taher, N., Babak, A.: An efficient hybrid approach based on PSO, ACO and k-means for cluster analysis. Applied Soft Computing 10(1), 183–197 (2010)

A New Algorithm for Data Clustering Based on Cuckoo Search Optimization

Ishak Boushaki Saida[1], Kamel Nadjet[2], and Bendjeghaba Omar[3]

[1] University M'hamed Bougara of Boumerdes (UMBB) and LRIA (USTHB), Algeria
saida_2005_compte@yahoo.fr
[2] University Farhat Abbes of Setif (UFAS) and LRIA (USTHB), Algeria
nkamel@usthb.dz
[3] LREEI, University M'hamed Bougara of Boumerdes (UMBB), Algeria
benomar75@yahoo.fr

Abstract. This paper presents a new algorithm for data clustering based on the cuckoo search optimization. Cuckoo search is generic and robust for many optimization problems and it has attractive features like easy implementation, stable convergence characteristic and good computational efficiency. The performance of the proposed algorithm was assessed on four different dataset from the UCI Machine Learning Repository and compared with well known and recent algorithms: K-means, particle swarm optimization, gravitational search algorithm, the big bang–big crunch algorithm and the black hole algorithm. The experimental results improve the power of the new method to achieve the best values for three data sets.

Keywords: Data Clustering, Cuckoo Search, Metaheuristic, Optimization.

1 Introduction

Clustering is an unsupervised classification technique of data mining [1] [2]. It divides a set of data into groups or clusters based on the similarity between the data objects, such that similar objects fall in the same cluster and different objects in different clusters.

Clustering is used in several applications like document clustering [3], image segmentation [4] and pattern recognition [5]. For each application we have to select and extract a set of features to represent the data objects and also we have to define measuring proximity between these data objects.

Many clustering methods have been proposed. They are classified into several major algorithms: hierarchical clustering, partitioning clustering, density based clustering and graph based clustering.

One of the most popular and famous partitioning algorithm is K-means because its efficiency and simplicity [6]. Unfortunately, the K-means algorithm suffers from two problems: It needs to define the number of clusters before starting and in addition, its performance depends strongly on the initial centroids and may get trapped in local optimal solutions.

J.-S. Pan et al. (eds.), *Genetic and Evolutionary Computing*,
Advances in Intelligent Systems and Computing 238,
DOI: 10.1007/978-3-319-01796-9_6, © Springer International Publishing Switzerland 2014

Recently, nature inspired approaches have received increased attention from researchers dealing with data clustering problems [7].

To avoid the inconvenience of K-means, we propose in this paper to use a new metaheuristic approach. It is mainly based on the cuckoo search (CS) algorithm which was proposed by Xin-She Yang and Suash Deb in 2009 [8] [9]. Cuckoo search is based on the interesting breeding behaviour such as brood parasitism of certain species of cuckoos and typical characteristics of Lévy flights. The CS is generic and robust for many optimization problems [10] [11]. It is a population based and this algorithm overcomes the problem of local optimum to global one.

The efficiency of the proposed algorithm is tested on four different data sets issued from literature [12] and the obtained results are compared with some recent well known algorithms reported in [13].

The remaining of this paper is organized as follows: In section 2, related works is presented. In section 3, we present cluster analysis. In section 4, we describe the basics of cuckoo search algorithm. The proposed approach for data clustering is explained in section 5. Numerical experimentation and comparisons are provided in Section 6. Finally, conclusions and our future work are drawn in Section 7.

2   Related Works

Several metaheuristic were developed to overcome the disadvantage of K-means. Most of them are evolutional and population based. For instance the genetic algorithm is evolutionary population optimization based; it uses natural genetics and evolution: selection, mutation, and crossover [14]. It is still suffers from the difficulty of coding modelling. Also, the operation of crossover and mutation are too expensive. More over it needs much parameter to handle. The ant colony algorithm is another metaheuristic inspired from the behaviour of the real ants to find the shortest path between a food source and their nest [15] [16]. Particle Swarm Optimization (PSO) incorporates swarming behaviours observed in flocks of birds, schools of fish, or swarms of bees, and even human social behaviour, from which the idea is emerged [17][18]. Like the genetic algorithm, it needs much parameter to manipulate. A data clustering algorithm based on the gravitational search algorithm (GSA) was proposed in [19] [20]. It is based on the law of gravity and the notion of mass interactions. The Big Bang–Big Crunch (BB–BC) algorithm was also applied for resolving the problem of clustering [21]. It is an   optimization method that is based on one of the theories of the evolution of the universe namely the Big Bang and Big Crunch theory. Another heuristic algorithm namely the black hole algorithm was defined to resolve the problem of clustering, which is inspired from the black hole phenomenon [13].

The new algorithm proposed in this paper is based on the cuckoo search optimization. In this metaheuristic no much parameters is used. We need only to define the worse nests probability which does not really affect in the results of clustering. More over, the research of the optimal solution is done by a mathematical function. In each generation we select the best solution and the next generation is calculated by the cuckoo search function using the best solution. Thereby, we always convert to the optimal solution.

3   Cluster Analysis

The main goal of the clustering process is to group the most similar objects in the same cluster or group. Each object is defined by a set of attributes or measurements.

To determine the similar objects, we use the measure of similarity between them. Several similarity measures are defined in the literature. In this paper we use the Euclidean distance to calculate the similarity between the objects. It is the most popular metric done by the formula (1):

$$distance(o_i, o_j) = \left(\sum_{p=1}^{m} |o_{ip} - o_{jp}|^{\frac{1}{2}} \right)^2 \tag{1}$$

Where: m is the number of attributes and o_{ip} is the value of the attribute number p of the object number i (o_i).

The result of a clustering algorithm must be evaluated and validated. This is done by using validity indexes. They are classified into internal and external one [22]. In this paper we use the sum intra cluster (SSE) which is an internal validity index, and the error rate which is an external validity index. These indexes are defined by the formula (2) and (3).

$$SSE = \sum_{i=1}^{k} \sum_{j=1}^{n} W_{ij} * \sqrt{\sum_{p=1}^{m} (o_{jp} - c_{ip})^2} \tag{2}$$

Where: $W_{ij} = 1$ if the object is in the cluster and 0 otherwise. k is the number of clusters, n is the number of objects, m is the number of attributes and c_{ip} is the value of the attribute number p of the centroid of the cluster number i (c_i).

$$ER = \frac{number\ of\ misplaced\ objects}{total\ of\ objects\ whithin\ dataset} * 100 \tag{3}$$

4   Basics of Cuckoo Search Algorithm

The Cuckoo search (CS) is a new metaheuristic optimisation algorithm, proposed by Xin-She Yang and Suash Deb [8] [9]. The algorithm is based on the obligate brood parasitic behaviour of some cuckoo species in combination with the Lévy flight behaviour of some birds and fruit flies. In fact, the algorithm has three particular idealized rules [8]:

1. Each cuckoo lays one egg at a time, and dumps its egg in randomly chosen nest;
2. The best nests with high quality of eggs will carry over to the next generations;

3. The number of available host nests is fixed and the egg laid by a cuckoo is discovered by the host bird with a probability $pa \in [0, 1]$. In this case, the host bird can either throw the egg away or abandon the nest, and build a completely new nest. For simplicity, this last assumption can be simulated by the fraction (pa) of the n worse nests that are replaced by new random nests.

Based on these three rules, the basic steps of the Cuckoo Search (CS) can be summarized by the pseudo code shown in Figure 1.

Cuckoo Search via Lévy Flights
begin
Objective function f(x), x = $(x_1, ..., x_d)^T$
Generate initial population of n host nests $x_i (i = 1, 2, ..., n)$
while *(t <MaxGeneration) or (stop criterion)*
Get a cuckoo randomly by Lévy flights
Evaluate its quality/fitness F_i
Choose a nest among n (say, j) randomly
if *($F_i > F_j$),*
Replace j by the new solution;
end
A fraction (p_a) of worse nests are abandoned and new ones are built;
Keep the best solutions (or nests with quality solutions);
Rank the solutions and find the current best
end while
Post process results and visualization
End

Fig. 1. Pseudo code of the standard Cuckoo Search (CS) [8]

5   Clustering with Cuckoo Search

For solving the data clustering problem, the standard cuckoo search algorithm is adapted to reach the centroids of the clusters. For doing this, we suppose that we have n objects, and each object is defined by m attributes. In this work, the main goal of the CS is to find k centroids of clusters which minimize the formula (2). Knowing that the problem is multi-dimensional, the data set must be represented by a matrix (n, m), such as the row i corresponds to the object number.

In cuckoo search mechanism, the solutions are the nests and each nest is represented by a matrix with k rows and m columns, where, the matrix rows are the centroids of clusters.

We propose a cuckoo search algorithm for the data clustering throught the following steps:

1. Generate randomly the initial population of *nb_nest* host nests;
2. Calculate the fitness of these solutions and find the best solution;
   While (t < MaxGeneration) or (stop criterion);
3. Generate *nb_nest new* solutions with the cuckoo search;
4. Calculate the fitness   of the new solutions;
5. Compare the new solutions with the old solutions, if the new solution is better than the old one, replace the old solution by the new one ;
6. Generate a fraction (p_a) of new solutions to replace the worse nests;
7. Compare these solutions with the old solutions. If the new solution is better than the old solution, replace the old solution by the new one;
8. Find the best solution;
   End while;
9. Print the best nest and fitness;

6 Implementation and Results

In order to test the validity and the efficiency of the proposed approach, we have selected four data sets from the literature [12]. We have used an internal and an external quality measure in order to evaluate and compare this method with the other ones cited in [13].

For the Internal quality measure, we consider the sum of intra-cluster distances represented by the formula (2). The goal is to minimize this function called fitness function. For the External quality measure, we calculate the error rate (ER), which represents the percentage of misplaced objects as given in formula (3). It is the same one used in [13], thereby we can compare the performance of the CS algorithm to the most recent algorithms reported in [13]: K-means, particle swarm optimization (PSO), gravitational search algorithm (GSA) the big bang–big crunch algorithm (BB-BC) and the black hole algorithm (BH).

We should note that, for all datasets the population size and the probability of worse nests were set to 20 and 0.25 respectively.

6.1 Iris Dataset

The Iris dataset contains 150 objects with four attributes. They are unscrewed into 3 classes of 50 instances, where each class represents a type of iris plant. The best obtained solution using the cuckoo search for the iris data is given in Table 1, where the row i represents the value of all the attributes of the centroid of the cluster number i. The variation graph of the fitness function according to the number of generations is represented in Figure 2.

Table 1. The best solution for Iris data by CS

Center 1	5.9347	2.7979	4.4179	1.4171
Center 2	6.7336	3.0664	5.6301	2.1055
Center 3	5.0130	3.4040	1.4710	0.2358

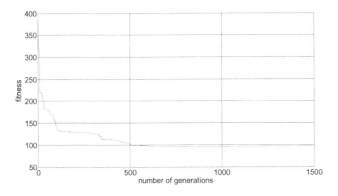

Fig. 2. Fitness function for Iris dataset

6.2   Wine Dataset

The Wine dataset describes the quality of wine from physicochemical properties. There are 178 instances with 13 features grouped into 3 classes.

The best obtained solution using the cuckoo search for the wine data is given in Table 2, and the variation graph of the fitness function according to the number of generations is represented in Figure 3.

Table 2. The best solution for Wine data by CS

Center 1	Center 2	Center 3
13.69227	12.48561	12.80205
1.82475	2.29206	2.52757
2.52063	2.42019	2.43197
16.89081	21.29222	19.61010
105.29640	92.54816	98.89915
2.81080	2.04322	2.10036
3.15359	1.73150	1.45264
0.30168	0.36522	0.45469
2.02348	1.42479	1.41032
5.73206	4.43186	5.75083
1.08947	1.02753	0.86257
3.13876	2.41258	2.19260
1137.21760	463.65681	687.03916

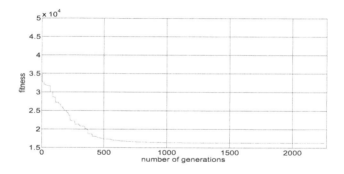

Fig. 3. Fitness function for Wine dataset

6.3   Cancer Dataset

The Cancer dataset represents the Wisconsin breast cancer databases. The dataset contains 683 instances with 9 features. Each instance has one of two possible classes: benign or malignant.

The best obtained solution using the cuckoo search for the cancer data is shown in Table 3, and the variation graph of the fitness function according to the number of generations is represented in Figure 4.

Table 3. The best solution for cancer data by CS

| Center 1 | 2.88848 | 1.12802 | 1.20058 | 1.16359 | 1.99256 | 1.11893 | 2.00507 | 1.10059 | 1.03144 |
| Center 2 | 7.11641 | 6.64365 | 6.62561 | 5.61432 | 5.24276 | 8.10514 | 6.07841 | 6.02254 | 2.32808 |

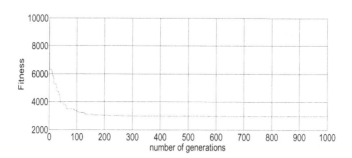

Fig. 4. Fitness function for Cancer dataset

6.4   Vowel Dataset

The Vowel dataset consists of 871 instances. Each point are represented by 3 features. These instances are grouped into 6 classes.

The best obtained solution using the cuckoo search for the Vowel data is given in Table 4, and the variation graph of the fitness function versus to the number of generations is represented in Figure 5.

Table 4. The best solution for Vowel data by CS

Center 1	360.48780	2290.36553	2976.77797
Center 2	437.21449	993.57547	2658.09454
Center 3	407.75777	1011.99989	2310.59089
Center 4	507.54437	1839.75796	2555.73930
Center 5	374.96411	2149.78838	2678.23073
Center 6	622.99723	1308.62969	2332.83028

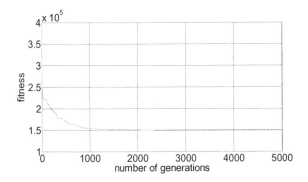

Fig. 5. Fitness function for Vowel dataset

For the considered dataset, the best value of fitness functions given by the cuckoo search are compared with those obtained by the different algorithms: K-means, particle swarm optimization (PSO), gravitational search algorithm (GSA), big bang–big crunch algorithm (BB-BC) and the black hole algorithm (BH). The comparison results are shown in Table 5.

From Table 5, it is obvious that the Cuckoo Search algorithm can reach very important results. In the case of Iris data, the value of the best fitness function obtained by the Cuckoo Search is 96.65564, which is better than all other ones. Also, for the Wine data the value of the best fitness function given by the Cuckoo Search is 16292.24388, which is significantly better than the all the other ones. As we can see in table 5, the value of the best fitness function achieved by the Cuckoo Search for the Cancer data is 2964.38839, which is the best one. However, the value of the best fitness function found by the Cuckoo Search for Vowel data is 148990.15884, which is the best one after the black hole algorithm (BH).

Table 6 compares the error rate obtained by the clustering with the CS on the four dataset with different clustering algorithms (K-means, PSO, GSA, BB-BC and BH). As shown in the Table 6, CS gives a minimum error rate for all the datasets except the Vowel data.

Table 5. Best fitness obtained by different algorithms on Iris, Wine, Cancer and Vowel dataset

Approach	Iris	Wine	Cancer	Vowel
K_means	97.32592	16,555.67942	2986.96134	149 394.80398
PSO	96.87935	16,304.48576	2974.48092	152 461.56473
GSA	96.68794	16,313.87620	2965.76394	151 317.56392
BB-BC	96.67648	16,298.67356	2964.38753	149 038.51683
BH	96.65589	16,293.41995	2964.38878	148 985.61373
CS	**96.65564**	**16292.24388**	**2964.38839**	**148990.15884**

Table 6. The error rate (ER) of clustering algorithm on the different dataset

Dataset	K-means	PSO	GSA	BB-BC	BH	CS
Iris	13,42	10,06	10,04	10,05	10,02	**10,01**
Wine	31,14	28,79	29,15	28,52	28,47	**27,07**
Cancer	4,39	3,79	3,74	3,70	3,70	**3,52**
Vowel	43,57	42,39	42,26	41,89	41,65	**42,45**

7   Conclusion

In this paper, we have presented a new approach for solving the data clustering problem. The approach is principally based on the cuckoo search algorithm. The proposed algorithm is applied to four different data sets. Simulation experiments show that the proposed approach gives better results compared to some other more frequently used clustering approaches. The cuckoo search algorithm is useful for solving the data clustering problem. Moreover it is easy to implement and it manipulates a few parameters. In order to improve the obtained results and as a future work, we plan to hybridize the proposed approach with other algorithms.

References

1. Jain, K., Murthy, M.N., Flynn, P.J.: Data Clustering: A Review. ACM Computing Surveys 31(3), 264–323 (1999)
2. Xu, R., Wunsch, D.C.: Clustering, 2nd edn., pp. 1–13. IEEE Press, John Wiley and Sons, Inc. (2009)
3. Verma, H., Kandpal, E., Pandey, B., Dhar, J.: A Novel Document Clustering Algorithm Using Squared Distance Optimization Through Genetic Algorithms. (IJCSE) International Journal on Computer Science and Engineering 02(5), 1875–1879 (2010)
4. Hsuan-Ming, F., Ji-Hwei, H., Shiang-Min, J.: Bacterial Foraging Particle Swarm Optimization Algorithm Based Fuzzy-VQ Compression Systems. Journal of Information Hiding and Multimedia Signal Processing 3(3), 227–239 (2012)
5. Wong, K.C., Li, G.C.L.: Simultaneous Pattern and Data Clustering for Pattern Cluster Analysis. IEEE Transaction on Knowledge and Data Engineering Los Angeles 20, 911–923 (2008)

6. Jain, A.K.: Data clustering: 50 Years beyond K-means. Pattern Recognition Letters 31, 651–666

7. Colanzi, T.E., Assunção, W.K.K.G., Pozo, A.T.R., Vendramin, A.C.B.K., Pereira, D.A.B., Zorzo, C.A., de Paula Filho, P.L.: Application of Bio-inspired Metaheuristics in the Data Clustering Problem. Clei Electronic Journal 14(3) (2011)

8. Yang, X.-S., Deb, S.: Cuckoo Search via Levy Flights. In: Proc. of World Congress on Nature and Biologically Inspired Computing (NaBIC 2009), India, pp. 210–214. IEEE Publications, USA (2009)

9. Yang, X.-S., Deb, S.: Engineering Optimisation by Cuckoo Search. International Journal of Mathematical Modelling and Numerical Optimisation 1(4-30), 330–343 (2010)

10. Jothi, R., Vigneshwaran, A.: An Optimal Job Scheduling in Grid Using Cuckoo Algorithm. International Journal of Computer Science and Telecommunications 3(2), 65–69 (2012)

11. Noghrehabadi, A., Ghalambaz, M., Ghalambaz, M., Vosough, A.: A hybrid Power Series – Cuckoo Search Optimization Algorithmto Electrostatic Deflection of Micro Fixed-fixed Actuators. International Journal of Multidisciplinary Sciences and Engineering 2(4), 22–26 (2011)

12. Merz, C.J., Blake, C.L.: UCI Repository of Machine Learning Databases, http://www.ics.uci.edu/-mlearn/MLRepository.html

13. Hatamlou, A.: Black hole: A New Heuristic Optimization Approach for Data Clustering. Information Sciences 222, 175–184 (2013)

14. Auffarth, B.: Clustering by a Genetic Algorithm with Biased Mutation Operator. In: IEEE Congress Evolutionary Computation (CEC), pp. 1–8 (July 2010)

15. Shelokar, P.S., Jayaraman, V.K., Kulkarni, B.D.: An Ant Colony Approach for Clustering. Analytica Chimica Acta 509, 187–195 (2004)

16. Liu, X., Fu, H.: An Effective Clustering Algorithm with Ant Colony. Journal of Computers 5(4), 598–605 (2010)

17. Premalatha, K., Natarajan, A.M.: A New Approach for Data Clustering Based on PSO with Local Search. Computer and Information Science 1(4), 139–145 (2008)

18. Chuang, L.-Y., Lin, Y.-D., Yang, C.-H.: An Improved Particle Swarm Optimization for Data Clustering. In: Proceedings of the International Multiconference of Engineers and Computer Scientists Hong Kong, vol. I, pp. 440–445 (March 2012)

19. Hatamlou, A., Abdullah, S., Nezamabadi-pour, H.: Application of Gravitational Search Algorithm on Data Clustering. In: Yao, J., Ramanna, S., Wang, G., Suraj, Z. (eds.) RSKT 2011. LNCS, vol. 6954, pp. 337–346. Springer, Heidelberg (2011)

20. Hatamlou, A., Abdullah, S., Nezamabadi-pour, H.: A Combined Approach for Clustering Based on K-means and Gravitational Search Algorithms. Swarm and Evolutionary Computation 6, 47–52 (2012)

21. Hatamlou, A., Abdullah, S., Hatamlou, M.: Data Clustering Using Big Bang–Big Crunch Algorithm. In: Pichappan, P., Ahmadi, H., Ariwa, E. (eds.) INCT 2011. CCIS, vol. 241, pp. 383–388. Springer, Heidelberg (2011)

22. Rendón, E., Abundez, I., Arizmendi, A., M Quiroz, E.: Internal Versus External Cluster Validation Indexes. International Journal 5(1), 27–34 (2011)

Object Detection Using Scale Invariant Feature Transform

Thao Nguyen[1], Eun-Ae Park[1], Jiho Han[1],
Dong-Chul Park[1], and Soo-Young Min[2]

[1] Dept. of Electronics Engineering, Myong Ji University, Korea
[2] SOC Platform Research Division, Korea Electronics Tech. Inst., Korea
parkd@mju.ac.kr

Abstract. An object detection scheme using the Scale Invariant Feature Transform (SIFT) is proposed in this paper. The SIFT extracts distinctive invariant features from images and it is a useful tool for matching between different views of an object. This paper proposes how the SIFT can be used for an object detection problem, especially human detection problem. The Support Vector Machine (SVM) is adopted as the classifier in the proposed scheme. Experiments on INRIA Perdestrian dataset are performed. Preliminary results show that the proposed SIFT-SVM scheme yields promising performance in terms of detection accuracy.

1 Introduction

Human detection problem has been an important issue in computer vision with applications involved with locations and movements of target human. The increase in the use of digital cameras for various applications including security purposes requires for an automatic method which yields efficient detection of human objects. Various approaches have proposed for this problem[1–3]. Approaches to the human detection problem can be largely classified into two categories. The first category of methods tries to detect parts of the human body and then combine them to detect an entire human[1] while the other category of methods considers statistical characteristics of given image data that utilize a set of low level features within a target window to find if the target window contains a human[2]. The method proposed in this paper can be considered as a method in the latter category. However, our main emphasis are given to how the SIFT can be used for the feature extraction in human detection problem.

The SIFT proposed by David Lowe is one of the most famous algorithms in computer vision [4–6]. The SIFT has been widely used for its advantageous features including its invariance to scale, orientation, and view point. The SIFT has been successfully applied to several computer vision problems including image matching, stereo matching, and motion tracking[3, 7].

Typically, a human detection system consists of two main parts: selection of a proper feature extraction method and design of an efficient classifier. Much effort on the human detection problem has been given to adopting various combinations of these two parts. One of the most widely used methods is HOG-SVM(Support Vector Machine)[1]. HOG feature uses the spatial distribution of

J.-S. Pan et al. (eds.), *Genetic and Evolutionary Computing*,
Advances in Intelligent Systems and Computing 238,
DOI: 10.1007/978-3-319-01796-9_7, © Springer International Publishing Switzerland 2014

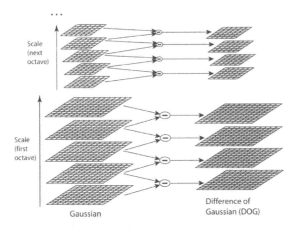

Fig. 1. *DoG space*

edge orientations and its dimension is usually very high enough to cause computational problems. In this paper, a simple feature extraction method using SIFT to improve the training time is proposed. Since SVM is known as a powerful classifier, we can expect an acceptable performance in the proposed human detection scheme.

The content of this paper is organized as follows: Section 2 summarizes the SIFT and describes how the SIFT can be used as a feature extraction method for the proposed human detection scheme. A brief review on SVM is given in Section 3. Experiments and results on a set of data are presented in Section 4. The final concluding remarks on this paper is given in Section 5.

2   Feature Extraction Method

2.1   Scale Invariant Feature Transform

The SIFT has been used for extracting highly distinctive invariant features from image data. The SIFT can provide features invariant to scale, rotation, illumination and viewpoint[4]. By adopting the features extracted by adopting the SIFT, reliable matching of the same object among different images can be obtained[7]. When the SIFT is applied to images, the followings are the major computation processes to generate the set of image features:

Scale-space Extrema Detection. The first stage in the SIFT searches over all scales and image locations. It is implemented efficiently by using a DoG (Difference of Gaussian method) shown in Fig. 1.

Keypoint Localization. At each potential location, a detailed model is fit to determine location and scale. Keypoints are finally passed through a contrast and edge test to keep the stable one.

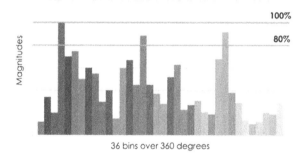

Fig. 2. Histogram of dominant orientations

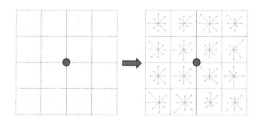

Fig. 3. Keypoint descriptor

Orientation Assignment. One or more dominant orientations are assigned to each keypoint in order to perform the rotation invariance as explained in Fig. 2.

Keypoint Descriptor. A descriptor is created to represent each keypoint based on the orientation assigned in the previous stage. The descriptor based on the histogram of gradient in the image is calculated as shown in Fig. 3. Finally, the descriptor is processed to reduce changes in illumination. Once the keypoint descriptors are retrieved, the descriptors can be used as a feature for data for different problems. More detailed information on SIFT computation can be found in [1].

2.2   Feature Extraction Using SIFT for Human Detection Problem

Many different feature extraction methods including HOG, HSV, LBP, and ULBP have been proposed for object detection problems [1, 8, 9]. In this paper, we propose an extraction method using the SIFT algorithm which is extremely strong in image matching and object tracking problems. The SIFT, however, is rarely used in human detection problem because the SIFT catches only details around the keypoints, not the human shape. Note that the number of keypoints in SIFT is not a static number. After careful observations on keypoint locations in each category (which is negative for backgrounds and positive for human objects), we find that keypoint positions can be used for human detection.

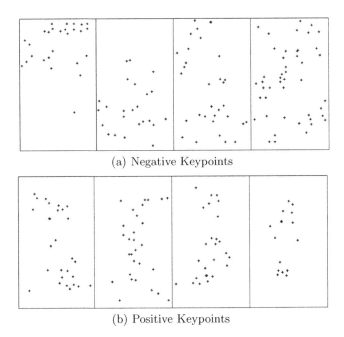

(a) Negative Keypoints

(b) Positive Keypoints

Fig. 4. Keypoints in Negative and Positive images

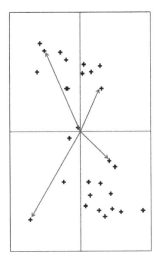

Fig. 5. Histogram computation

Fig. 4-(a) and Fig. 4-(b) are the keypoint locations for negative and positive images, respectively.

As can be seen from Fig. 4, the distribution of keypoints in positive images mostly has distinctive differences from the distribution in negative images while the keypoints in positive images always spread out in the whole spatial image space. Based on this observation, we proposed a method to compute the feature descriptor for each image as shown in Fig. 5. The extracted features are computed based on the center point of the picture. The orientation histograms are computed based on the relative angle between each keypoint and the center point. The accumulative magnitude is computed by the Euclidean distance from each keypoint to the center point. The number of histogram bins used for our experiments is 8. Furthermore, the number of keypoints is stable for both negative and positive images. In our experiments, the numbers of keypoints for negative and positive images are set to 105 and 150, respectively. This method also yields a lower dimensional feature space but still discriminative enough for our classification task.

3   Support Vector Machine

SVM is first introduced by Boser, Guyon and Vapnik [10]. In machine learning, SVM is a supervised learning algorithm with associated learning algorithms that analyze data and recognize patterns, used for classification and regression analysis. The basic SVM takes a set of input data and predicts, for each given input, which of two possible classes forms the output, making it a non-probabilistic binary linear classifier. In the case of linear SVM, the SVM algorithm will try to find the function which minimizes an objective function that combines the training error and the complexity as follows:

$$\frac{1}{m}\sum_{i=1}^{m} l(f(x_i, \alpha), y_i) + ||w||^2 \tag{1}$$

where $X = \{x_1, x_2, .., x_m\}$ is the input dataset and y_i is the binary values, which only has value of 1 or -1 and w is the margin. It was said that we only need to minimize $||w||^2$, subject to $y_i(wx_i + b) \geq 1$. The classification task is then performed by checking this decision function: $f(x) = wx + b$. More details on SVM can be found in[10].

4   Experiments and Results

INRIA is a data set which has been used for testing in many human detection systems. 1,000 images are randomly selected for positive training examples, together with their left-right reflections (2,000 images total). A fixed set of 10,000 patches randomly sampled from the 1,000 negative images yields the negative training data. Fig. 6 and Fig 7 show examples of positive and negative training images, respectively.

For a training data set including positive and negative training data, SIFT is applied on each image. Then these SIFT keypoint locations are processed to obtain the histogram of orientation as described in section 2.

Fig. 6. Positive training images

Fig. 7. Negative training images

(a) Negative Keypoints   (b) Positive Keypoints

Fig. 8. Experimental results on INRIA dataset

Typically, the system searches human by using a scanning window with various sizes over the whole image. For each window size and location, a feature extraction process such as HOG is perform and then passed to a classifier to determine if it is a human object or not. The most advantageous feature of the proposed method is that we only need to perform SIFT only once for each test image since SIFT is scale invariant. Hence, for each window size and position, we do not need to perform SIFT more than once. We only collect all keypoints inside the window and compute the histogram, then pass it to SVM for classification task. Moreover, the number of keypoints in positive image is usually quite higher than the number in negative images. It is stable with 150 keypoints per image while negative has only about 105 keypoints per image. Hence, before computing the histogram, we can pre-check and eliminate window which does not have enough number of keypoints. By doing so, scanning process will be more effective and can save us a lot of computing time as well as complexity in computation. In experiments on the cropped positive and negative images, we perform SIFT with SVM. The results show that the detection accuracy with the proposed SIFT-SVM detection scheme is 91.56 % on average. The average training time for the proposed detection scheme on our PC (Intel Core Quad 2.33GHz, 4GB of RAM) is 2 hour 37 minutes 26 seconds. The detection accuracy and the proposed extraction method are acceptable for the system that requires real time performance or needs to update the training database quickly. However, the training time for SVM is quite a burden for practical uses. Fig. 8 shows some results on testing example.

5   Conclusions

In this paper, we propose an feature extraction method using SIFT and a human object detection scheme by adopting the SVM. In the proposed scheme, the SIFT produces keypoints from images first. Based on the observation on the distinctive difference in the distribution of keypoints in positive images and negative images, a computing method of feature descriptor for each image is proposed in this paper. The orientation histograms are computed based on the relative angle between each keypoint and the center point and this histogram is used as the input to the SVM-based classifier. Preliminary results from experiments show that the detection accuracy obtained from the proposed detection scheme is extremely high on a set of INRIA database. However, the training time for SVM is somewhat too much. Topics in future research include a way how to alleviate the computational burden in training SVM.

Acknowledgement. This work was supported by National Research Foundation of Korea Grant funded by the Korean Government (2010-0009655) and by the IT R&D program of The MKE/KEIT (10040191, The development of Automotive Synchronous Ethernet combined IVN/OVN and Safety control system for 1Gbps class).

References

1. Dalal, N., Triggs, B.: Histograms of oriented gradients for human detection. In: Proc. of CVPR, vol. 1, pp. 886–893 (2005)
2. Felzenszwalb, P., McAllester, D., Ramanan, D.: A discriminatively trained, multiscale, deformable part model. In: Proc. of CVPR, vol. 1, pp. 1–8 (2008)

3. Zhou, H., Yuan, Y., Shi, Y.: Object tracking using SIFT features and mean shift. Journal on Computer Vision and Image Understanding 113(3), 345–352 (2009)
4. Lowe, D.: Distinctive Image Features from Scale-Invariant Keypoints. International Journal of Computer Vision 60(2), 91–110 (2004)
5. Lowe, D.: Object Recognition from Local Scale-Invariant Features. In: Proc. of IEEE Int. Conf. Computer Vision, vol. 2, pp. 1150–1157 (1999)
6. Brown, M., Lowe, D.: Invariant Features from Interest Point Groups. In: Proc. of British Machine Vision Conference, vol. 1, pp. 656–665 (2002)
7. Brown, M., Lowe, D.: Recognising panoramas. In: Proc. of IEEE Int. Conf. Computer Vision, vol. 2, pp. 1218–1225 (2003)
8. Ahonen, T., Hadid, A., Pietikainen, M.: Face Description with Local Binary Patterns. IEEE Trans. on Pattern Analysis and Machine Intelligence 28(12), 2037–2041 (2006)
9. Novak, C.: Anatomy of a color histogram. In: Proc. of CVPR, vol. 1, pp. 599–605 (1992)
10. Cortes, C., Vapnik, V.: Support-Vector Networks. Journal on Machine Learning 20(3), 273–297 (1995)

Nearest Feature Line and Extended Nearest Feature Line with Half Face

Qingxiang Feng, Lijun Yan, Tien-Szu Pan, and Jeng-Shyang Pan

Shenzhen Graduate School,
Harbin Institute of Technology,
Shenzhen, China
jengshyangpan@gmail.com

Abstract. In this paper, nearest feature line with half face (NFLhalfface) classifier and extended nearest feature line with half face (ENFLhalfface) classifier are proposed for face recognition. Nearest feature line (NFL) classifier and extended nearest feature line (ENFL) classifier both constitute the feature line by a pair of the samples belonging to the same class. Being different from them, NFLhalfface and ENFLhalfface constitute the feature line by left half face and right half face of a prototype face, which reduce the computational complexity. At the same time, the experimental results on AR face database and Yale face database show that they are superior to nearest neighbour (NN), NFL, ENFL, nearest feature mid-point (NFM) and shortest feature line segment (SFLS).

Keywords: Nearest Feature Line, Extended nearest Feature Line, Nearest Neighbor, Face Recognition.

1 Introduction

The procedure of face recognition contains two steps. The first step is feature extraction. For instance PCA [1], LDA [2], ICA [3] and other methods [4-6]. The second step is classification. One of the most popular classifiers is the nearest neighbor (NN) classifier [7]. However, the performance of NN is limited by the available prototypes in each class. To overcome this drawback, nearest feature line (NFL) [8] was proposed by Stan Z. Li. NFL was originally used in face recognition, and later began to be used in many other applications.

NFL attempts to enhance the representational capacity of a sample set of limited size by using the lines through each pair of the samples belonging to the same class. NFL shows good performance in many applications, including face recognition [9-12], audio retrieval [13], speaker identification [14], image classification [15], object recognition [16] and pattern classification [17]. The authors of NFL explain that the feature line can give information about the possible linear variants of the corresponding two samples very well.

J.-S. Pan et al. (eds.), *Genetic and Evolutionary Computing,*
Advances in Intelligent Systems and Computing 238,
DOI: 10.1007/978-3-319-01796-9_8, © Springer International Publishing Switzerland 2014

Though successful in improving the classification ability, there are still some drawbacks in NFL that limit their further application in practice, which can be summarized as two main points. Firstly, NFL will have a large computation complexity problem when there are many samples in each class. Secondly, NFL may fail when the prototypes in NFL are far away from the query sample, which is called as extrapolation inaccuracy of NFL.

To solve the above problems, extended nearest feature line [18] (ENFL), nearest feature mid-point [19] (NFM), shortest feature line segment [20] (SFLS) and nearest feature centre [21] (NFC) are proposed. They gains better performance in some situation. However, they are not so good in other situation.

In this paper, two classifiers, called nearest feature line with half face (NFLhalfface) classifier and extended nearest feature line with half face (ENFLhalfface) classifier, are proposed for face recognition. A large number of experiments are executed on Yale face database and AR face database. Detailed comparison result is given.

The rest of this article is organized as follows. In Section 2, we introduced the background. In section 3, we describe the new classifiers. In section 4, the analysis of the new classifiers is introduced. In the fifth quarter we compare the NN, NFL, ENFL, NFM, SFLS, NFLhalfface and ENFLhalfface by the experiment on Yale face database and AR face database. Finally, a brief summary is given.

2   Background

In this section, we will introduce nearest feature line classifier, extended nearest feature line classifier, nearest feature centre classifier and shortest feature line segment classifier. Suppose that $Y = \{ y_i^c, c = 1, 2, \cdots, M, i = 1, 2, \cdots, N_c \} \subset R^D$ denote the prototype set, where y_i^c is the ith prototype belonging to c-class, M is the number of class, and N_c is the number of prototypes belonging to the c-class.

2.1   Nearest Feature Line

The core of NFL is the feature line metric. As is shown in Fig. 1, the NFL classifier doesn't compute the distance of query sample y and y_i^c; doesn't calculate the distance of y and y_j^c, while NFL classifier calculates the feature line distance between query sample y and the feature line $\overline{y_i^c y_j^c}$. The feature line distance between point y and feature line $\overline{y_i^c y_j^c}$ is defined as:

$$d(y, \overline{y_i^c y_j^c}) = \| y - y_p^{ij,c} \| \tag{1}$$

where $y_p^{ij,c}$ is the projection point of y on the feature line $\overline{y_i^c\,y_j^c}$, $\|.\|$ means the L2-norm.

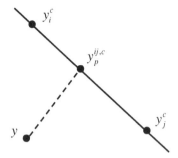

Fig. 1. The metric of NFL

2.2   Extended Nearest Feature Line

Borrowing the concept of feature line spaces from the NFL method, the extended nearest feature line (ENFL) is proposed in 2004. However, the distance metric of ENFL is different from the feature line distance of NFL.

ENFL does not calculate the distance between the query sample and the feature line. Instead, ENFL calculates the product of the distances between query sample and two prototype samples. Then the result is divided by the distance between the two prototype samples. As shown in Fig. 2. The new distance metric of ENFL is described as

$$d_{ENFL}(y,\overline{y_i^c\,y_j^c}) = \frac{\|\,y - y_i^c\,\| \times \|\,y - y_j^c\,\|}{\|\,y_i^c - y_j^c\,\|} \tag{2}$$

The distance between the pair of prototype samples can strengthen the effect when the distance between them is large.

Fig. 2. The metric of ENFL

2.3   Shortest Feature Line Segment

Instead of calculating the distance between the query sample and the feature line, SFLS tries to find the shortest feature line segment which satisfies the given geometric relation constraints together with the query sample. As shown in Fig. 3. The pair of samples in the sample class constitute a feature line segment. If the query sample is inside or on the hyper sphere centered at the midpoint of the feature line segment, the corresponding feature line segment will be tagged and the distance metric of SFLS can be calculated as

$$d_{SFLS}(y, \overline{y_i^c y_j^c}) = \| y_i^c - y_j^c \| \tag{3}$$

In the worst case, there is no tagged feature line for a query sample y, and then SFLS uses the rule of NN to make the classification decision for the query y.

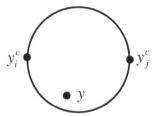

Fig. 3. The metric of SFLS

3   The Proposed Methods

NFL enhances the representational capacity of a sample set of limited size by using the lines passing through each pair of the samples belonging to the same class. However, the computational cost is large. In order to reduce the computational complexity, the nearest feature line with half face (NFLhalfface) classifier and extended nearest feature line with half face (ENFLhalfface) classifier are proposed in this section. The detailed information of NFLhalfface and ENFLhalfface are as follows.

Before introducing the new methods, several definitions are given. As shown in Fig. 4, the left half face and right half face are the two half faces of a full face.

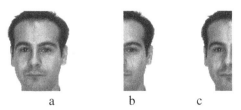

a   b   c

Fig. 4. Three definitions of a prototype face. (a) the full face of a prototype face. (b) the left half face of a prototype face. (c) the right half face of a prototype face.

3.1   The Idea of Nearest Feature Line with Half Face

NFLhalfface classifier does not calculate the distance between the query sample and the feature line composed by a pair of the samples belonging to the same class. Instead, NFLhalfface calculates the distances between the two half faces of query face and the feature line composed by two half face of a prototype face. The procedure of calculation is described in formula (5).

$$
\begin{aligned}
\mathrm{d}_{NFLhalfface}(y, y_i^c) &= \min(d_{NFL}(y_L, \overline{y_{i,L}^c y_{i,R}^c}), d_{NFL}(y_R, \overline{y_{i,L}^c y_{i,R}^c}) \\
&= \min(\parallel y_L - y_{L,p}^{i,c} \parallel, \parallel y_R - y_{R,p}^{i,c} \parallel)
\end{aligned}
\tag{4}
$$

where $y_{i,L}^c$ is the left half face of prototype y_i^c, $y_{i,R}^c$ is the right half face of prototype y_i^c, y_L is the left half face of the query face y, y_R is the right half face of the query face y, $y_{L,p}^{i,c}$ is the projection point of y_L on the feature line $\overline{y_{i,L}^c y_{i,R}^c}$, $y_{R,p}^{i,c}$ is the projection point of y_R on the feature line $\overline{y_{i,L}^c y_{i,R}^c}$. And the symbols are also used in formula (6).

The classification procedure of NFLhalfface is as follows. Firstly, the NFLhalfface distance between the query sample y and each prototype sample y_i^c is calculated by the formula (5), which generates a number of distances. Secondly, the distances are sorted in ascending order and each is associated with a class identifier and a prototype. The first rank gives the best matched class c* and the best matched prototypes i* of the c* class. The query sample y will be classified into the class c*.

3.2   The Idea of Extended Nearest Feature Line with Half Face

ENFLhalfface classifier calculates the product of the distances between query face's left half face and two half face of a prototype face. Then ENFLhalfface classifier calculates the product of the distances between query face's right half face and two half face of the same prototype face. At last the two results are divided by the distance between two half faces of the corresponding prototype face. The procedure of calculating is also described in formula (6).

$$
\begin{aligned}
\mathrm{d}_{ENFLhalfface}(y, y_i^c) &= \min(d_{ENFL}(y_L, \overline{y_{i,L}^c y_{i,R}^c}), d_{ENFL}(y_R, \overline{y_{i,L}^c y_{i,R}^c}) \\
&= \min(\frac{\parallel y_L - y_{i,L}^c \parallel \times \parallel y_L - y_{i,R}^c \parallel}{\parallel y_{i,L}^c - y_{i,R}^c \parallel}, \frac{\parallel y_R - y_{i,L}^c \parallel \times \parallel y_R - y_{i,R}^c \parallel}{\parallel y_{i,L}^c - y_{i,R}^c \parallel})
\end{aligned}
\tag{5}
$$

The classification procedure of ENFLhalfface is described as follows. Firstly, the ENFLhalfface distance between the query sample y and each prototype sample y_i^c is calculated by the formula (6), which generates a number of distances. Secondly, the distances are sorted in ascending order and each is associated with a class identifier and a prototype. The first rank gives the best matched class c* and the best matched prototypes i* of the c* class. The query sample y will be classified into the class c*.

4   Complexity Analysis

Suppose there are N_c prototype samples in the cth class, L is the dimensionality of each sample. With the NFL and ENFL classifiers, there are $N_c(N_c-1)/2$ feature lines in the cth class. So the complexities of NFL and ENFL are both $O(N_c^2 L)$ in the cth class. With the NFM classifiers, there is $N_c(N_c-1)/2$ feature mid-point in the cth class. So the complexity of NFM is $O(N_c^2 L)$ in the cth class. With the SFLS classifiers, there are $N_c(N_c-1)/2$ feature lines segment in the cth class. So the complexity of SFLS is $O(N_c^2 L)$ in the cth class. With the NFC classifiers, there are $N_c(N_c-1)/2$ feature centers in the cth class. So the complexity of NFC is $O(N_c^2 L)$ in the cth class. From the formula (2) and (3), we can know that there are $2N_c$ feature lines in the cth class. So the complexities of NFLhalfface and ENFLhalfface are both $O(N_c L)$ in the cth class. There are N_c samples in the cth class, so the complexity of NN is $O(N_c L)$.

5   Experimental Results

The classification performance of the proposed classifiers is compared with NN, NFL, ENFL, NFM and SFLS classification approach. The experiments are executed on Yale face database and AR face database.

Yale [22] face database contains 165 greyscale images in GIF format of 15 persons. There are 11 images per people, one per different facial expression or configuration: center-light, w/glasses, happy, left-light, w/no glasses, normal, right-light, sad, sleepy, surprised, and wink. All images are copped in 100×100.

The AR [23] database contains over 4000 face images of 126 subjects (70 men and 56 women). To reduce the computational complexity, the subset of AR database includes 1680 face images of 120 individuals with fourteen face images of different expressions and lighting conditions except wearing sun glasses and wearing scarf per subject, and all images in AR database were manually cropped into 50×40 pixels.

"Randomly-chose-N" scheme is taken for comparison: N images per person are randomly chosen from the Yale face database or AR face database as prototype set. The rest images of Yale face database or AR face database are used for testing. The whole system runs 20 times. To test the robustness of new algorithms, the average recognition rate (ARR) and the average running time are used to assess the performance of new algorithms.

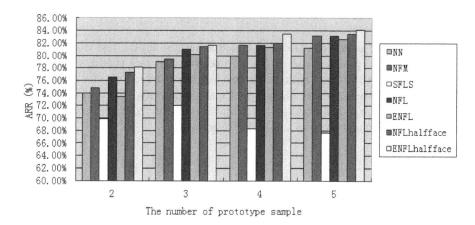

Fig. 5. The recognition rate of several classifiers using "randomly-choose-*N*" scheme on Yale face database

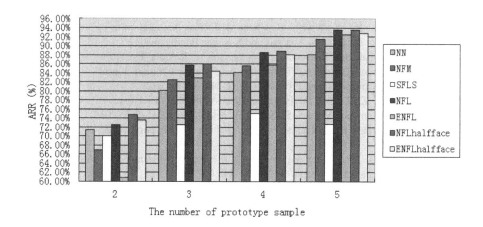

Fig. 6. The recognition rate of several classifiers using "randomly-choose-*N*" scheme on AR face database

In the first experiment, we adopt the "randomly-choose-*N*" scheme on Yale face database. The average recognition rates of NN, NFL, ENFL, NFM, SFLS, NFC, NFLhalfface and ENFLhalfface are shown in Fig. 5. In the second experiment, we adopt the "randomly-choose-*N*" scheme on AR face database. The average recognition rates of NN, NFL, ENFL, NFM, SFLS, NFLhalfface and ENFLhalfface are shown in Figure 6. From the Figure 5 and Figure 6, we can know that the recognition rate of the proposed is better than the recognition rate of NN, NFL, ENFL, NFM and SFLS on Yale face database and AR face database.

6   Conclusion

In this paper, two improved classifiers, called as nearest feature line with half face (NFLhalfface) and extended nearest feature line with half face (ENFLhalfface), are proposed. The two classifiers both use the two half faces of a full face for constituting the feature line, which reduces the computational complexity. And the experimental results on Yale face database and AR face database confirm the better performance of the two proposed classifiers than some other classifiers.

References

1. Turk, M., Pentland, A.: Eigenfaces for recognition. Journal of Cognitive Neuroscience 3(1), 71–86 (1991)
2. Belhumeur, P., Hespanha, J., Kriegman, D.: Eigenfaces vs. fisherfaces: recognition using class specific linear projection. IEEE Transactions on Pattern Analysis and Machine Intelligence 19(7), 711–720 (1997)
3. Bartlett, M.S., Movellan, J.R., Sejnowski, T.J: Face recognition by independent component analysis. IEEE Transactions on Neural Networks 13(6), 1450–1464 (2002)
4. He, X., Yan, S., Hu, Y., Niyogi, P., Zhang, H.J.: Face recognition using laplacianfaces. IEEE Transactions on Pattern Analysis and Machine Intelligence 27(3), 1–13 (2005)
5. Kekre, H.B., Shah, K.: Performance Comparison of Kekre's Transform with PCA and Other Conventional Orthogonal Transforms for Face Recognition. Journal of Information Hiding and Multimedia Signal Processing 3(3), 240—247 (2012)
6. Zhou, X., Nie, Z., Li, Y.: Statistical analysis of human facial expressions. Journal of Information Hiding and Multimedia Signal Processing 1(3), 241–260 (2010)
7. Cover, T.M., Hart, P.E.: Nearest neighbor pattern classification. IEEE Trans. Inform. Theory 13(1), 21–27 (1967)
8. Li, S.Z., Lu, J.: Face Recognition Using the Nearest Feature Line Method. IEEE Transactionson Neural Networks 10(2), 439–443 (1999)
9. Chien, J.T., Wu, C.C.: Discriminant waveletfaces and nearest feature classifiers for face recognition. IEEE Trans. Pattern Anal. Machine Intell. 24(12), 1644–1649 (2002)
10. Lu, J., Tan, Y.-P.: Uncorrelated discriminant nearest feature line analysis for face recognition. IEEE Signal Process. Lett. 17(2), 185–188 (2010)
11. Feng, Q., Pan, J.-S., Yan, L.: Restricted Nearest Feature Line with Ellipse for Face Recognition. Journal of Information Hiding and Multimedia Signal Processing 3(3), 297—305 (2012)
12. Feng, Q., Huang, C.-T., Yan, L.: Resprentation-based Nearest Feature Plane for Pattern Recognition. Journal of Information Hiding and Multimedia Signal Processing 4(3), 178–191 (2013)
13. Li, S.Z.: Content-based audio classification and retrieval using the nearest feature line method. IEEE Trans. Speech Audio Process. 8(5), 619–625 (2000)
14. Chen, K., Wu, T.Y., Zhang, H.J.: On the use of nearest feature line for speaker identification. Pattern Recognition Lett. 23(14), 1735 –1746 (2002)
15. Li, S.Z., Chan, K.L., Wang, C.L.: Performance evaluation of the nearest feature line method in image classification and retrieval. IEEE Trans. Pattern Anal. Machine Intell. 22(11), 1335–1339 (2000)

16. Chen, J.H., Chen, C.S.: Object recognition based on image sequences by using inter-feature-line consistencies. Pattern Recognition 37(9), 1913–1923 (2004)
17. Zheng, W., Zhao, L., Zou, C.: Locally nearest neighbor classifiers for pattern classification. Pattern Recognition 37(6), 1307–1309 (2004)
18. Zhou, Y., Zhang, C., Wang, J.: Extended nearest feature line classifier. In: Zhang, C., W. Guesgen, H., Yeap, W.-K. (eds.) PRICAI 2004. LNCS (LNAI), vol. 3157, pp. 183–190. Springer, Heidelberg (2004)
19. Zhou, Z., Kwoh, C.K.: The pattern classification based on the nearest feature midpoints. In: International Conference on Pattern Recognition, vol. 3, pp. 446–449 (August 2000)
20. Han, D.Q., Han, C.Z., Yang, Y.: A Novel Classifier based on Shortest Feature Line Segment. Pattern Recognition Letters 32(3), 485–493 (2011)
21. Feng, Q., Pan, J.S., Yan, L.: Nearest feature centre classifier for face recognition. Electronics Letters 48(18), 1120—1122 (2012)
22. The Yale faces database (2001),
    http://cvc.yale.edu/projects/yalefaces/yalefaces.html
23. Martinez, A.M., Benavente, R.: The AR Face Database. CVC Technical Report, vol. 24 (June 1998)

Studying Common Developmental Genomes in Hybrid and Symbiotic Formations

Konstantinos Antonakopoulos

Norwegian University of Science and Technology,
Department of Computer and Information Science,
Sem Sælandsvei 7-9, NO-7491, Trondheim, Norway
kostas@idi.ntnu.no

Abstract. One of the main challenges in developmental systems is the design of a method for building complex systems with a structural or computational goal. In previous work, we studied the common properties of several computational architectures consisting of connected computational elements. Their common property of sparsely connected networks, envisages how universal properties and processes can be included in a developmental mapping through an *EvoDevo* approach. The potentiality of using the same developmental mapping, to develop more than one class of computational architectures was also investigated through Common Developmental Genomes. In this work, the focus is towards development of intra-connected computational architectures, forming a common biological entity - a Hybrid architecture. Also, we explore how common developmental genomes operate under symbiosis and their effect on the evolutionary performance of the partners involved. The results are enlightening gaining a deeper understanding of the capabilities and the limitations of common developmental genomes in these original formations.

Keywords: Hybrid architectures, Symbiosis, Common developmental genomes, cellular automata, boolean network, L-systems.

1 Introduction

Artificial development has been widely used for designing complex structures [1],[2],[3] and as a means to increase the complexity of an artifact (aiming at a structural or computational goal) [4], [5]. Most evolutionary developmental systems, have an artificial setting having a specific genotype (genetic representation), a genotype-phenotype mapping process and they are usually targeting special phenotypic structures (i.e., structures comprising connected computational elements or computational architectures). Computational architectures can, for example, be cellular automata [6] or boolean networks [7].

In biology, organisms with different DNA, morphology or ecological niche, are considered of different species [8]. In the artificial metaphor, a species can be linked to a certain computational architecture (i.e., cellular automata). Computational architectures, such as, CA and BN can be considered as different species, since they have different structural and organizational properties.

J.-S. Pan et al. (eds.), *Genetic and Evolutionary Computing,*
Advances in Intelligent Systems and Computing 238,
DOI: 10.1007/978-3-319-01796-9_9, © Springer International Publishing Switzerland 2014

One central challenge in artificial development is to understand how a mapping process could work on a class of architectures (species) in a more general way. The hypothesis for going this way forward, is that it may be possible to exploit the most favorable properties from each architecture or be able to combine efficiently more than one architectures (i.e., a true multicellular approach), towards more scalable systems able of complex computation. Such a mapping process was investigated in [9], [10], [11], which gave rise to common developmental genomes. It was shown that common genomes successfully developed and evolved both CA, with a regular connectivity pattern [6] and BN, as NK networks [7].

In this work, we are changing the way we consider computational architectures: we explore them not as different species but as organs of a common biological entity. In this case, architectures need to be connected somehow (as biological organs do). In the quest of generating scalable systems able of complex computation, intra-connected architectures may have the required potential of a candidate method. Intra-connected architectures can be evolved in a number of ways: (a) exploit them as one computational entity, which from now one we will call by the name *Hybrid* architecture, or (b) explore close associations between architectures in symbiotic formations. Symbiosis is the phenomenon in which organisms of different species live together in close association, resulting in a raised level of fitness for one or more of the organisms [12]. A symbiotic environment assumes–among others–a host organism and a symbiont which is dependent on its host for its survival.

Using the hypothesis mentioned previously, this work has a twofold scope. First, to identify whether common developmental genomes are able to evolve hybrid architectures by exploiting the most favorable properties from each partner. Second, to understand how common developmental genomes function under symbiosis and their effect on the evolutionary performance of the co-evolving partners. Symbiosis here is of interest only from a computational point of view, meaning that no analysis on emerged morphologies or physiologies is given.

The rest of the article is laid out as follows. The genetic representation is briefly described in Section 2. The detailed description can be found in [9]. The hybrid architecture model and the symbiotic model are described in Sections 3 and 4, respectively. Experimental setup come in Section 5. In Section 6, we present the results and discussion with the conclusion in Section 7.

2   The Common Genetic Representation

A species is often used as the basic unit for biological classification and for taxonomic ranking [13]. As such, an organism with unifying properties and same characteristics can be of the same species. Figure 1, shows how the genome looks like. The genome is split into two parts (*chromosomes*). The first chromosome is responsible for creating the cells/nodes. The second chromosome is responsible for creating the connectivity (i.e., for the BNs). Each chromosome is built out of rules. Each rule has sufficient information for cell/node creation and connectivity. Also, the rules are of certain length. Those destined for cell/node creation

Fig. 1. This is how the genome looks like with the genome split into two *chromosomes*. The first chromosome is responsible for generating the cells/nodes whereas the second chromosome is responsible for generating the connectivity of the network.

are different from the ones for connectivity. Consequently, chromosomes contain different rules.

2.1 An L-system for the Genetic Representation

L-systems [14], [15] were chosen due to the ease of defining a specific rule set, that can target the rewriting of specific features of a structure, e.g., connections or node functions that enable a way of splitting genetic information into separate information carrying units (i.e., chromosomes). The first L-system processes the rules of the first chromosome, while the second L-system deals with the connectivity rules of the second chromosome.

2.2 The L-system for the First Chromosome

The L-system used here is context-sensitive. As such, development is using the strict predecessor/ancestor to determine the applicable production rule. The rules are able to incorporate all the cell processes of a computational architecture. Table 1(a), shows the type of symbols used by the L-system of the first chromosome. The length of each rule is 4 symbols (i.e., `4x8bits=32bits`). For node/cell generation the L-system runs for 100 timesteps and then it stops. As such, the intermediate phenotypes generated by development are of variable size.

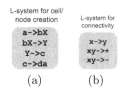

Fig. 2. (a) Example of L-system rules for the first chromosome, (b) Example of L-system rules for the second chromosome

Figure 2(a), gives an example of a L-system for the first chromosome. A detailed example of step-by-step development of both CA and BN architectures is described in [9].

2.3   The L-system for the Second Chromosome

The rules are able to generate the connections necessary for the wiring of the nodes. The length of the rules here is also 4 symbols / rule. In addition, we make sure that the chromosome has sufficient information for the developmental processes (i.e., growth, differentiation and apoptosis). The L-system uses a D0L (i.e., with zero-sided interactions). An example L-system for the second chromosome is shown in Figure 2(b), and the symbols used are explained at Table 1(b).

Table 1. (a) Symbol table for Node generation, (b) Symbol table for Connectivity generation

(a)		(b)	
Symbol	Description	Symbol	Description
a (AXIOM)	Add (growth)	x	Node (different from y)
b	Add (growth)	y	Node (different from x)
c	Add (growth)	+	Connect forward
d	Delete (apoptosis)	−	Connect backwards
X	Substitute (differentiation)	→	Production
Y	Substitute (differentiation)		
→	Production		

The *axiom* rule for the second chromosome is x→y. It means that development initially searches if the axiom exists. If so, development continues and looks for rules of type xy→+value, or xy→-value. In short, these two rules imply that if two different (i.e., distinct) nodes are found (x≠y), then it creates a connection forward (if the rule includes a '+'), or backwards (if the rule includes a '-'). The field value is encoded in the genotype and denotes the node number for the generated connection. For example, rule xy→+3 denotes that a connection will be created from the current node (node 0), to the one being three nodes forward. If value=0, a self-connection is created to the current node. The modularity of the genome, gives the possibility to develop itself to enable or disable parts of it (chromosomes), when this is required and driven by the goal set. For example, if the target architecture is a 2D-CA, the second chromosome (i.e., connectivity) is disabled, since connectivity is predetermined. Similarly for BN development, both chromosomes are enabled (i.e., nodes and connectivity).

3   The Hybrid Architecture Model

The investigation of how CA and BN respond as different species, in different environments, was studied in [16]. Evolving artificial species to explore the evolution of species [17] or the ecology [18], is of great importance as it provides a platform to address many important questions about natural systems [19].

Bull [20], studied the dynamics of multiple, mutually coupled traditional random boolean networks with varying degree of inter-network connectivity and differing intra-network connectivity.

Here, we study how *different* species can be evolved in close-association, forming one integrated architecture – a *Hybrid* architecture. The development and evolution of hybrid architectures with a common genome, will enable the exploitation of properties that are most favorable for both architectures, towards a common goal. Also, the emerging topology of hybrid architectures may help address many real-life multiple networks, each with their own internal structure to the external world (economic markets, immune networks, ecologies, etc). Figure 3, shows the merging of a CA (left) and a BN (right), with intra-connectivity $K = 2$.

Intra-connectivity is constant and does not change throughout evolution. The connections departing from BN, always connect to specific cells at the 2-D CA lattice. Destination cells of the CA comprise the first and the last cells. Similarly, connections departing from the CA, will always connect to specific nodes at the BN. Destination nodes at BN, since there is no specific geometry, depend on the dynamic size of the first chromosome (nodes chromosome). So, the nodes' location cannot be predefined but CA will always connect to the first and last active nodes of the BN.

At Figure 3, we see that the BN is connected to the first and last cells of our exemplified 3x3 2D CA (cell numbers 0 and 8). On the other hand, CA is connected to a 9-node BN through the connections located at nodes 0 and 8. The numbering of cells and nodes, is only an example. The architecture sizes of CA and BN, are given at the experimental setup (Section 5).

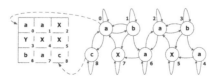

Fig. 3. The Hybrid architecture

Increasing the degree of intra-network connectivity will potentially have an impact on the topology and performance of the hybrid model. Also, the intra-network connectivity acts as a constraint to the external environment set by the system. Mutation rate is .005 and the crossover .001. The overall evaluation fitness is measured by averaging CA's and BN's partial fitnesses. Further details about the number of rules in the genome, developmental steps, species sizes, evolutionary mechanisms, fitness function etc., are common to the symbiotic model and are described in Section 5.

4   The Symbiosis Model

The word *symbiosis* describes the phenomenon in which organisms of different species live together in close association. Even though symbiosis is often used to describe mutually beneficial interactions between different species, it also describes close and long-term associations [21]. Bull [22], used a modified version of Kauffman and Johnsen's abstract *NKCS* model [7], to study symbiogenesis in two aspects: by exploring the endosymbiotic relationship between two genetically separated species with differing reproduction rates and by studying the evolutionary performance of endosymbionts that transfer increasing fractions of their genome to their partner (a.k.a. horizontal gene transfer).

In this work, we consider CA and BN as different species coevolving in an endosymbiotic environment. We study endosymbiosis and compare the evolutionary progress of the partner organisms. Partner organisms have a constant number of intra-connections meaning, they are loosely integrated.

In our model, endosymbionts exist as cooperating symbionts, evolving with their respective separate genome and populations, that is, as heterogeneous species. Close and long-term association is obtained by the continuous exchange of their outputs – the output of a species acts as an input to the other species, for each evolutionary step. Mutation and crossover operators act upon the respective individuals of each population. The fitness goal is though common for both species.

Here, we study a specific symbiotic association, where the 'host' has a purely external association with the 'symbiont' organism (ectosymbiosis). As such, we first examine the symbiotic case where CA is the 'host' and the BN, the 'symbiont'. At the second symbiotic case, we switch roles; the BN becomes the 'host' and the CA, the 'symbiont'. The genomes of the two coevolving organisms are quite independent from each other (i.e., epistatically independent) and all participating species are benefited from the relationship (i.e., mutualism).

Because of the inherent constraint in the way genomes are generated in our model, one would expect a facultative-obligative association with the CA as the 'host', but an obligative-obligative association for the case where the BN is the 'host'. These kinds of associations will potentially become evident from the evolutionary behavior of the coevolving organisms for each symbiotic case.

Species are evaluated and evolved at the same evolutionary time within populations of many individuals. This allows any association between species to become immediately evident. Each population in endosymbiosis receives its own fitness evaluation. Mutation rate is .0005 and crossover .0001. Further details about the genome's structure, the genetic algorithm, fitness function, etc., are provided in Section 5.

5   Experimental Setup

We run 20 experiments in total, resulting in 20 different organisms. The developmental process was apportioned of 1000 state steps. A total number of 70 rules were used for node and connectivity generation (total size `70x32=2248bits`). In this setup, each rule can be used more than once during development of the genome. The evaluation of CA and BN phenotypes is based on Table 2.

Table 2. Cell types with their functionality

Cell Type	Function name
a	NAND
b	OR
c	AND
d	IDENTITY CELL
X	XOR
Y	NOT

A 2D-CA of size **6x6** and a BN of size of $N = 36$ nodes are used. The number of outgoing connections per node is $K = 5$. The fitness function evaluates cycle attractors of size 2 - 800; best score 100 is assigned to individuals obtaining cycle attractor of 400. The fitness evaluation has a bell-curve "distribution" with the worst score (4) assigned at limit values. *Generational mixing* protocol was used as the GA's global selection mechanism and *Fitness proportionate* for parental selection. The population size is 20. The GA run for 10000 generations. Note that, the computational problem here is of minor importance.

6   Results and Discussion

In [16], it was studied the ability of our genetic representation to deal with the dynamic behavior of CA and BN. Their ability was assessed by looking at cycle attractors of certain size, both when evolved using separate genomes but also with a common genome (as different species). Figure 4, shows the average fitness evaluation of the common genome used to develop and evolve both CA and BN.

Here, we address the same problem and keep the same state space, as in [16], for comparison purposes.

Figure 5, shows the results of the hybrid architecture model. Comparing the fitness evaluations of Figure 4 and 5(a), we see a better overall performance of the hybrid model. The big evolutionary 'jump' in fitness around generation 4000, is potentially caused by the continuous information exchange between CA and BN on each evolutionary step, which has a direct impact in their dynamic behavior.

Figure 5(b), shows the average cycle attractors obtained per generation exposing also the worst individuals for each generation. We note a nearly-best cycle attractor (380) after generation 6000, but evolution was not able to maintain this solution. In addition, increasingly better fitness evaluations are slowly obtained after generation 7000.

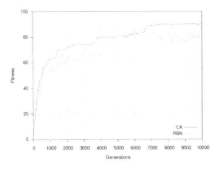

Fig. 4. Average fitness of CA and BN in random environment using common genome

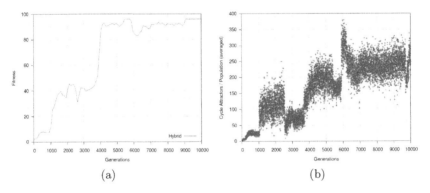

| (a) | (b) |

Fig. 5. Hybrid architecture model results: (a) Fitness evaluation, (b) Cycle attractors averaged across generations

Regarding the symbiotic model, we initially run the model with the same reproduction rate (rounds of mutation and selection) for the CA and BN. Both species obtained surprisingly good fitness evaluations (results not shown). To experience a phase-transition-like dynamic phenomena, it is more interesting to evolve the symbiotic model having different reproduction rates for the partners. As such, we run the model with the host having a relative rate of reproduction 4 times larger than the symbiont.

Figure 6(a), shows one of the best fitness plots (out of 20 runs), with the 'host' BN and the 'symbiont' CA. The host was not able to evolve at all and this had an impact on the evolvability of the symbiont. The promising solution found at generation 5100 by the host, was not able to maintain it due to the average performance of the symbiont. This had a negative impact resulting in fitness drop for the host organism (generation 6500). A random fitness drop for the host organism, had a major fitness impact (sudden drop) for the symbiont (generation 9700). Figure 6(b), shows the cycle attractors for both the host and symbiont organisms per generation. Here, the worst individuals per generation are also shown.

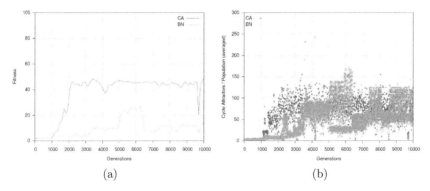

(a)   (b)

Fig. 6. Symbiotic model results with the 'host' boolean network and the 'symbiont' cellular automata: (a) Fitness evaluation, (b) Cycle attractors averaged across generations

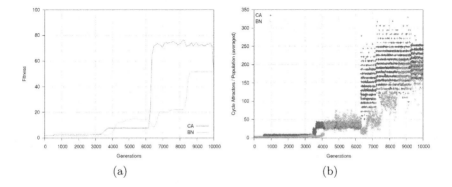

(a)   (b)

Fig. 7. Symbiotic model results with the 'host' cellular automata and the 'symbiont' boolean network: (a) Fitness evaluation, (b) Cycle attractor averaged across generations

Figure 7(a), shows one of the best fitness runs (out of 20) for the second symbiotic setup, with the CA being the 'host' and the BN, the 'symbiont'. Here, again, the host has a greater relative reproduction rate of 4. This is clearly mapped during fitness evaluation. Also, the impact of the host to the symbiont is considerable. The BN obtained only fair performance. From generation 4200 to 6200, it benefited greatly by the host resulting in higher fitness evaluation. The cycle attractors are shown in Figure 7(b). The plot shows the averaged individual fitnesses from each generation. As such, and due to the GA's selection mechanism (fitness proportionate), the plot has a scatter-like look after generation 6300, with relatively higher deviation. Though, the best individuals dominate the population after generation 9200 (with less dispersion from the population mean).

From the symbiotic model experiment, it was evident that who is the 'host' organism has a direct impact on the evolvability of the partners. The first chromosome responsible for node generation, inevitably acquires a facultative role over the second chromosome, whose association is rather obligative (i.e., a faculative-obligative symbiotic association).

Extending this evidence to the phenotypic level, the 'host' BN did not succeed in leveraging the evolvability of the 'symbiont' CA. Random destructive behavior had a direct impact to the obligative organism. On the other hand, the 'host' CA managed to evolve in a step-wise mode with radical evolutionary changes while feeding the 'symbiont' BN with potentially better genetic information allowing for better adaption.

7   Conclusion

The overall goal of this study is to target more adaptive scalable systems able of complex computation. We studied how different species can be evolved in close-association forming a hybrid architecture. The different species were evolved in cooperation having the same genetic information. The hybrid architecture model showed in average a better performance than in [16].

We also considered a symbiotic model with two types of associations. In the first case, the BN was not able to evolve as expected. This was mainly because the genetic information for node generation, is based on the development of the first chromosome (i.e., is CA dependent). This was due to the way the genetic representation is constructed by the system. In addition, the 'symbiont' CA had $\frac{1}{4}$ the reproduction rate of the host, which greatly constrained the evolutionary ability of the 'host'. In other words, the slower species were unable to track down the own optima within their own landscape. In the next case of endosymbiosis with the 'host' CA, both species showed fairly good performance. Here, the influence of the host organism had an impact at the evolutionary performance of the symbiont.

For future work, we will study how varying degrees of intra-connectivity (greater level of integration between partners) affects endosymbiosis. Taking advantage of the genome's modular structure, we could introduce horizontal gene transfer between coevolving species, when, any of the species show detriment behavior. Gene transfer can be applied in a selective way: transferring genes from a specific chromosome only or, in a holistic way, copying the entire genome from one species to its partner. Transfered genes under a new context (computational goal, constraints, etc.), may become better exploited by the receiving species or even utilize additional genetic information not previously exploited. In this way, it may be possible to better control and regulate the dynamic properties of any developing organism (phenotypic shaping).

References

1. Kitano, H.: Designing Neural Networks Using Genetic Algorithms with Graph Generation Systems. Complex Systems 4(3), 461–476 (1990)
2. Miller, J.F., Thomson, P.: A Developmental Method for Growing Graphs and Circuits. In: Tyrrell, A.M., Haddow, P.C., Torresen, J. (eds.) ICES 2003. LNCS, vol. 2606, pp. 93–104. Springer, Heidelberg (2003)

3. Gordon, T.G.W., Bentley, P.J.: Bias and Scalability in Evolutionary Development. In: Genetic and Evolutionary Computation Conference (GECCO), pp. 83–90 (2005)
4. Harding, S.L., Miller, J.F., Banzhaf, W.: Self-modifying cartesian genetic programming. In: Proceedings of the 9th Annual Conference on Genetic and Evolutionary Computation (GECCO), pp. 1021–1028. ACM Press (2007)
5. Steiner, T., Trommler, J., Brenn, M., Jin, Y., Sendhoff, B.: Global Shape with Morphogen Gradients and Motile Polarized Cells. In: Congress on Evolutionary Computation, pp. 2225–2232. IEEE Press (2009)
6. Von Neumann, J.: The Theory of Self-reproducing Automata. University of Illinois Press (1966)
7. Kauffman, S.A., Johnsen, S.: Co-evolution to the edge of chaos: Coupled fitness landscapes, poised states and co-evolutionary avalanches. Artificial Life II, pp. 325–370. Addison-Wesley (1992)
8. Robert, J.S.: Embryology, Epigenesis and Evolution: Taking Development Seriously. Cambridge Studies in Philosophy and Biology. Cambridge University Press (2004)
9. Antonakopoulos, K., Tufte, G.: A Common Genetic Representation Capable of Developing Distinct Computational Architectures. In: IEEE Congress on Evolutionary Computation (CEC 2011), pp. 1264–1271 (2011)
10. Antonakopoulos, K., Tufte, G.: On the Evolvability of Different Computational Architectures using a Common Developmental Genome. In: Rosa, A., Dourado, A., Madani, K., Filipe, J., Kacprzyk, J. (eds.) IJCCI 2012, vol. 2012, pp. 122–129. SciTePress Publishing (2012)
11. Antonakopoulos, K., Tufte, G.: Is Common Developmental Genome a Panacea Towards More Complex Problems? In: 13th IEEE International Symposium on Computational Intelligence and Informatics (CINTI 2012), pp. 55–61 (2012)
12. Bull, L., Fogarty, L.C.: Artificial Symbiogenesis. Artificial Life 2(3), 269–292 (1995)
13. Wilkins, J.: What is a species? Essences and generation. Theory in Biosciences 129(2), 141–148 (2010)
14. Lindenmayer, A., Prusinkiewicz, P.: Developmental Models of Multicellular Organisms: A Computer Graphics Perspective. In: Langton, C.G. (ed.) Proceedings of ALife, pp. 221–249. Addison-Wesley Publishing (1989)
15. Stauffer, A., Sipper, M.: Modeling Cellular Development Using L-Systems. In: Sipper, M., Mange, D., Pérez-Uribe, A. (eds.) ICES 1998. LNCS, vol. 1478, pp. 196–205. Springer, Heidelberg (1998)
16. Antonakopoulos, K., Tufte, G.: Investigation of Developmental Mechanisms in Common Developmental Genomes. In: 7th International ICST Conference on Bio-Inspired Models of Network, Information and Computing Systems (BIONETICS), pp. xxx–xxx. Springer (2012)
17. Thompson, J.N., Medel, R.: The coevolving web of life. Evolution: Education and Outreach 3(1), 6 (2009)
18. Momeni, B., Chen, C.C., Hillesland, K.L., Waite, A., Shou, W.: Using artificial systems to explore the ecology and evolution of symbioses. Cellular and Molecular Life Sciences 68(8), 1353–1368 (2011)
19. Nowak, M., Tarnita, C., Antal, T.: Evolutionary dynamics in structured populations. Philos. Trans. of the Royal Society of London - B Biological Sciences 365(1537), 19–30 (2010)

20. Bull, L., Alonso-Sanz, R.: On Coupling Random Boolean Networks. Automata 2008: Theory and Applications of Cellular Automata, pp. 292–301. Luniver Press (2008)
21. Wilkinson, D.M.: At cross purposes. Nature, 412–485 (2001)
22. Bull, L.: Artificial Symbiogenesis and Differing Reproduction Rates. Artificial Life 16(1), 65–72 (2010)

Routing and Wavelength Assignment in Optical Networks from Maximum Edge-Disjoint Paths

Chia-Chun Hsu[1, 2], Hsun-Jung Cho[1], and Shu-Cherng Fang[2]

[1] Department of Transportation Technology and Management, National Chiao Tung University, Hsinchu, 300, Taiwan, ROC
[2] Department of Industril and Systems Engineering, North Carolina State University, Raleigh, NC, USA
{chsu4,fang}@ncsu.edu, hjcho@cc.nctu.edu.tw

Abstract. The routing and wavelength assignment (RWA) problem is NP-hard and also an important issue in wavelength-division multiplexing (WDM) optical networks. We propose an algorithm which is based on the maximum number of edge-disjoint paths (MEDP) to solve the RWA problem. The performance of the proposed method has been verified by experiments on several realistic network topologies.

Keywords: edge-disjoint paths, MEDP, routing and wavelength assignment problem, RWA.

1 Introduction

In optical networks, wavelength-division multiplexing (WDM) is a technology which multiplexes a number of optical carrier signals onto a single optical fiber by using different wavelengths of laser light, thus further provides users more efficient use of the huge bandwidths. In such networks, sending data from one node to another requires an establishment of connection, which can be realized by determining a path in the network between the two nodes and allocating a wavelength on all of the edges traversed by the path. The all-optical path is commonly called a lightpath and the pair of nodes asking a lightpath for information transmission is a connection request.

Given an undirected network $G = (V, E)$ and a set of connection requests $T = \{(s_i, t_i) \in V \times V | i = 1, \dots, I \text{ and } s_i \neq t_i\}$, the routing and wavelength assignment (RWA) problem attempts to route each request and to assign wavelengths to these routes subject to two constraints: (i) in the absence of wavelength converters, a lightpath must use the same wavelength on all fiber links which it traverses, this is known as the wavelength continuity constraint; (ii) lightpaths that share a common physical link cannot use the same wavelength, which is known as the wavelength clash constraint. A feasible solution of the RWA problem is composed of a set of lightpaths $\pi = \{\pi_i | i = 1, \dots, I\}$ in G, where each path π_i connects the request $(s_i, t_i) \in T$; and a set of wavelengths $w = \{w_i | i = 1, \dots, I\}$ for these paths. Path π_i and π_j, $i \neq j$, cannot use the same wavelength if they share a common edge. The objective is to minimize the number of wavelengths used to satisfy all requests in T.

J.-S. Pan et al. (eds.), *Genetic and Evolutionary Computing*,
Advances in Intelligent Systems and Computing 238,
DOI: 10.1007/978-3-319-01796-9_10, © Springer International Publishing Switzerland 2014

Observe that the lightpaths can utilize the same wavelength as long as they do not traverse the same physical link (i.e., these paths are edge-disjoint). Maximizing the number of requests which can be simultaneously realized by edge-disjoint paths is useful for minimizing the number of required wavelengths. The idea of the proposed method is to tackle the RWA problem by solving the maximum edge-disjoint paths (MEDP) problem iteratively. Given an undirected graph G and a set of connection requests T, the MEDP problem maximizes the number of requests that are simultaneously realizable as edge-disjoint paths (EDPs). A feasible solution of MEDP is given by a subset $R \subseteq T$, such that each request in R is assigned a path and these paths are pairwise edge-disjoint. The path set is denoted by $S = \{p_j | j \in J\}$, where $J = \{k | (s_k, t_k) \in R\}$ is the set of indices of the accepted requests. The goal of the MEDP problem is to maximize the cardinality of R.

In this paper, we developed an algorithm for solving the RWA problem. The experiments results confirm the effectiveness of the proposed method. The rest of the paper is organized as follows. In Section 2, we review related works on RWA and provide some details of the bin-packing based algorithms. The proposed method is presented in Section 3. In section 4, the benchmark instances are briefly introduced and the computational results are provided in section 5. Finally, we make a conclusion and point out possible directions for future research.

2   Related Works

The RWA problem was proven to be NP-complete [5] in 1992. In the literature, the approaches for solving the RWA problem can be divided into two main categories. One decomposes the problem into two subproblems, the routing problem and wavelength assignment problem [3, 10] to be solved separately. The other one solves the two subproblems simultaneously [7, 8, 11]. Reference [4] covers different approaches and variants developed in the 1990s for RWA. A functional classification of RWA heuristics can be found in [6]. The bin-packing-based algorithm [7] is the state-of-art heuristic for RWA. The author adapted some ideas from bin-packing heuristics to the RWA problem by considering each connection request as an item and copies of the original graph as bins. Four bin-packing based heuristics were proposed in [7]: (i) first fit heuristic (FF-RWA), (ii) best fit heuristic (BF-RWA), (iii) first fit decreasing heuristic (FFD-RWA) and (iv) best fit decreasing heuristic (BFD-RWA). Computational results showed that FFD-RWA and BFD-RWA both outperform Greed-EDP-RWA [8]. Several local search and heuristic methods [1, 9, 11] which are based on the solution constructed by the bin-packing-based methods were proposed. In [11], a memetic algorithm which includes the iterated local search, mutation, and recombination operators was proposed. In addition, a multilevel algorithm was applied to address large size instances. The results showed that this method can be considered as the most sophisticated heuristic algorithm known in the literature.

2.1   Bin-Packing Based Methods for RWA

The bin packing problem is a classical combinatorial optimization problem that has been widely studied in the literature. Given is a list of I items of various sizes, and identical bins with a limited capacity. To solve the problem, it is necessary to pack these items into the minimum number of bins, without violating the capacity constraints. Four classical algorithms for the bin packing problem are the First Fit (FF), Best Fit (BF), First Fit Decreasing (FFD) and Best Fit Decreasing (BFD) algorithms. The FF algorithm packs each item into the bin with the lowest index. On the other hand, the BF algorithm packs each item into the bin which leaves the least room left over after packing the item. The FFD and BFD algorithm first place larger items into bins and then fill up remaining space with smaller items.

To apply bin-packing methods to solve the RWA problem, Skorin-Kapov of [7] considered using lightpath requests to represent items and using duplicates of graph G to represent bins. Each copy of G, i.e., G_j, $j = 1,2,...$, corresponds to one wavelength. Let the size of each lightpath π_i be represented by the length of its shortest path SP_i in graph G_j. To solve the RWA problem, packing maximum units of items (lightpaths) into a minimum number of bins (copies of G) may minimize the number of used wavelengths.

The FF algorithm runs as follows. Fisrt, a copy of G, bin G_1, is created. Higher indexed bins are created as needed. Lightpath requests are selected and routed on the lowest indexed copy of G if the length of the shortest path on such graph is less than a given threshold. If a lightpath is routed in bin G_j, the lightpath is assigned wavelength j and the edges along such path are removed from G_j. A new bin is created if no existing bin can accommodate the request.

On the other hand, the Best Fit bin-packing algorithm routes requests in the bin which they fit "best". The best bin is considered to be the one in which the request can be routed on the "shortest" shortest path. The FFD and BFD sorts the requests in a nonincreasing order in terms of the lengths of their shortest paths in G. The motivation is that, the connection request with the longest shortest path is usually harder to route. Thus considering these requests first then filling up the remaining space with the requests having the shortest routes may lead to fewer wavelengths used.

3   Solving RWA Using Edge-Disjoint Paths

A set of EDPs can be assigned the same wavelength, maximizing the number of EDPs may lead to fewer wavelengths required to satisfy all requests. Thereby a good method for solving the maximum edge-disjoint paths problem is useful for tackling the RWA. In [3], we have developed a GA-based method to solve the MEDP problem. The subroutine is called GA_MEDP, which has two inputs (G, T) and two outputs (R, S), where R denotes the set of accepted requests and S the corresponding edge-disjoint paths set.

Similar to the preprocessing on the order of T in bin-packing algorithm (FFD and BFD), the shortest path of each request in G is precomputed, then T is sorted in a non-increasing order of the shortest path distances. Although these paths are unlikely to be the final routes, they still provide good information about the minimum units of resources (edges) they occupy in G. Thus a better solution could be secured if we first consider the request with longer shortest path distance and then fill up the remaining space with the request with shorter shortest path length.

After the adjustment of T is made, the first B requests in T are selected. The current wavelength is denoted by the variable l initialized to be 1. Then *GA_MEDP* is executed to find the maximum number of edge-disjoint paths among the selected requests in G. The current wavelength l is assigned to the accepted requests and each of the obtained edge-disjoint paths is assigned to the corresponding lightpaths. The residual graph, where all edges used by the paths are removed, is stored in the variable G_l. The rejected requests at this stage remain in T and will be included in the next batch.

Before starting *GA_MEDP* with the next batch, the algorithm scans all the remaining requests in T in a backward manner. Starting from the last request, which has the shortest distance of shortest path in G, the algorithm tries to find a shortest path to route the request in G_l. If such path exists, the request is assigned wavelength l and removed from T. After the backward-scanning process is done, l is increased by 1. Another batch of requests is selected and *GA_MEDP* is executed again. The algorithm halts when T is empty. We call this proposed method the *GA_MEDP_RWA* algorithm, whose pseudocode is shown as follows.

Input: $G = (V, E)$, $T = \{(s_i, t_i)|i = 1, \dots, I\}$ and batch size B
Begin:
1. $w = \emptyset, \pi = \emptyset, \sigma = \{\sigma_i = i|i = 1, \dots, I\}$;
2. $l = 1$;
3.   Sort T in non-increasing order of their shortest paths distance in G
4.   **While** $T \neq \emptyset$
5. $T' = Get\_Req(T, B)$;
6. $(R, S, G_l) = GA\_MEDP(G, T')$;
7. $w_j = l, \forall j \in \{k|(s_k, t_k) \in R\}$;
8. $\pi_j = p_j, \forall j \in \{k|(s_k, t_k) \in R\}$;
9. $T = T \backslash (s_j, t_j), \forall j \in \{k|(s_k, t_k) \in R\}$;
10. $\sigma = \sigma \backslash j, \forall j \in \{k|(s_k, t_k) \in R\}$;
11.   **for** $m = |T|$ to 1
12.   Find shortest path SP_{σ_m} for $(s_{\sigma_m}, t_{\sigma_m})$ on G_l;
13.   **If** $|SP_{\sigma_m}| < \infty$ then
14. $w_{\sigma_m} = l$;
15. $\pi_{\sigma_m} = SP_{\sigma_m}$;
16. $T = T \backslash (s_{\sigma_m}, t_{\sigma_m})$;
17. $\sigma = \sigma \backslash \sigma_m$;
18.   **end if**
19.   **end for**
20. $l = l + 1$;
21.   **end while**
End
Output: π and w

4   Benchmark Instances

In order to evaluate the performance of the proposed method, 7 benchmark networks provided in [11] are collected. They assume the patterns and sizes of some real-life telecommunication networks. For each network, different numbers of connection requests are randomly generated according to a given probability, i.e., the probability that there is a request between a pair of nodes is p. There are 31 testing instances in total. The last two numbers of the instance's name indicate the value of p which is used to generate the instance. For example, instance EON_02 is generated on network EON with $p = 0.2$ and EON_10 is generated with $p = 1.0$.

Table 1. Main quantitative characteristics of the instances

Graph	$|V|$	$|E|$	Min.	Avg.	Max.	Diameter
NSFNET	16	25	2	3.1	4	4
EON	20	39	2	3.9	7	5
Norway	27	51	2	3.8	6	7
janos-us-ca	39	61	2	3.1	5	10
Germany50	50	88	2	3.5	5	9
zib54	54	80	1	3.0	10	8
ta2	65	108	1	3.3	10	8

(Min., Avg. and Max. denote the minimum, average and maximum degree, respectively)

5   Computational Experiments

The computational results and comparisons with the bin-packing based heuristic are reported in this section. The three algorithms were implemented in MATLAB and the experiments were conducted on a PC with Intel® Core i7 CPU @1.6GHz and 4 Gb of memory running the Windows 7 operating system. The experiments were conducted by applying PSO and the proposed method on each instance for 30 runs to obtain the best, worst, mean and standard deviation of the objective values. The average computational time are also recorded. Regarding the four bin-packing methods, each method only needs to be executed once, and the results and computational time are recorded.

The comparison of the proposed method and the four bin-packing based methods is shown in Table 2. The first column gives the name of the tested instance. For the FF, FFD, BF and BFD methods, the objective value, which is denoted by # wl, and the computation time t_{FF}, t_{FFD}, t_{BF} and t_{BFD} were recorded, respectively. The first three columns under *GA_MEDP_RWA* show the maximum, average, and minimum objective values that were obtained in 30 runs. The fourth column is the mean computation time (in seconds). The objective value is underlined and in boldface when it is the best among the five.

Table 2. Results of *GA_MEDP_RWA* and bin-packing based methods (time unit: sec)

instance	GA_MEDP_RWA				FF		FFD		BF		BFD	
	max	avg	min	avg time	# wl	time	# wl	time	# wl	time	# wl	time
NSF_02	5	5.0	**5**	2.41	6	0.34	**5**	0.17	6	0.19	6	0.18
NSF_04	8	7.3	**7**	5.80	9	0.37	9	0.35	8	0.52	8	0.50
NSF_06	9	8.8	**8**	7.97	10	0.46	10	0.49	9	0.59	9	0.64
NSF_08	14	13.2	**13**	15.11	17	1.06	15	1.10	14	1.50	14	1.47
NSF_10	17	16.6	**16**	22.89	19	1.70	17	1.45	17	1.96	17	2.08
EON_02	4	3.1	**3**	3.19	**3**	0.22	4	0.17	4	0.34	4	0.36
EON_04	9	8.3	**8**	14.74	10	1.10	9	1.14	**8**	1.53	**8**	1.54
EON_06	11	11.0	**11**	17.52	13	1.21	**11**	1.18	**11**	1.94	**11**	1.90
EON_08	14	13.7	**13**	26.63	16	2.20	**13**	2.99	**13**	3.61	**13**	3.72
EON_10	19	18.1	**18**	38.17	22	3.64	**18**	3.58	**18**	5.77	**18**	5.73
Norway_02	9	8.6	**8**	9.26	10	1.67	**8**	1.78	9	1.90	**8**	2.17
Norway_04	15	14.6	**14**	18.47	15	3.75	15	4.12	15	4.56	15	5.43
Norway_06	22	21.4	**21**	32.84	22	7.87	22	8.20	22	9.73	22	12.32
Norway_08	30	29.5	**29**	46.41	31	15.05	30	16.59	31	19.79	30	24.59
Norway_10	37	36.6	**36**	63.77	37	21.99	**36**	23.75	38	28.58	**36**	36.13
janos-us-ca_02	26	26.0	**26**	54.75	27	16.61	**26**	16.53	27	27.40	27	27.58
janos-us-ca_04	40	39.6	**39**	97.42	42	49.52	**39**	59.49	43	70.75	40	88.45
janos-us-ca_06	68	67.8	**67**	210.36	71	110.99	69	124.10	72	157.50	68	200.32
janos-us-ca_08	89	88.2	**88**	326.39	92	199.09	92	212.16	93	257.39	91	345.67
Germany50_02	21	20.7	**20**	109.28	21	35.86	21	36.10	21	46.10	22	58.58
Germany50_04	45	44.2	**43**	350.24	45	152.90	44	163.00	48	217.60	48	276.50
Germany50_06	61	60.5	**60**	584.92	63	347.70	61	410.61	63	430.89	65	544.74
Germany50_08	76	74.7	**73**	921.59	77	552.50	75	725.90	79	843.70	76	1098.40
zib54_02	32	31.1	**30**	149.44	33	52.28	31	55.01	35	77.70	33	91.67
zib54_04	67	65.7	**65**	406.90	67	209.80	**65**	220.50	73	304.64	71	379.55
zib54_06	92	90.0	**88**	762.56	91	442.95	89	522.85	99	696.78	98	889.98
zib54_08	122	119.2	**117**	1446.41	**117**	994.60	**117**	939.60	130	1229.60	127	1457.22
ta2_02	35	34.6	**34**	411.59	35	148.93	**34**	163.00	35	188.75	35	250.99
ta2_04	72	70.1	**69**	945.79	72	507.30	**69**	542.50	73	635.57	70	813.04
ta2_06	99	97.4	**96**	1865.20	98	1311.70	98	1484.00	100	1881.30	97	2047.40
ta2_08	131	129.7	**128**	2835.57	130	1935.30	129	2577.90	135	2644.43	**128**	3713.46

The proposed method can find the best solution among the five methods on all instances. However, bin-packing based methods have clearly advantages in terms of the computation time, especially on those relatively small instances. Nevertheless, the difference of computation time becomes smaller as the problem size grows. We take the instances on the three networks: Norway, Germany50, and ta2, whose sized are in an ascending order, to demonstrate how the computation time changes along with the problem size. Define the relative differences to be $(\bar{t} - t_{bin})/\bar{t}$, where $\bar{t}$ is the mean computation time of *GA_MEDP_RWA* and t_{bin} is the computation time of the bin-packing based methods (*bin* can be FF, FFD, BF or BFD, etc). The relative differences for all the instances on the three networks are shown in Fig. 1-3, respectively.

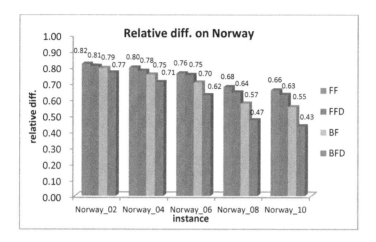

Fig. 1. Relative difference of computational time on graph "Norway"

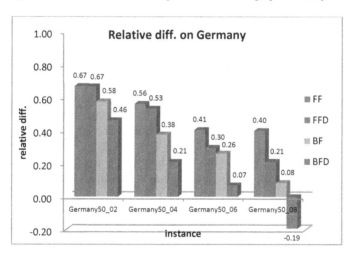

Fig. 2. Relative difference of computational time on graph "Germany"

In Fig. 1, the relative difference tends to go down as the number of connection requests increases. The trend also exists in Fig. 2 and Fig. 3 and most of other instances. In addition, the relative difference becomes smaller as the network size grows. In Fig. 2, the proposed method spent less time on Germany50_08 than BFD did. In Fig. 3, the same situation happened on ta2_06 and ta2_08. Since the three networks of Norway, Germany50 and ta2 are in an ascending order of the network size, by observing Figure 1-3, a conclusion can be drawn that the relative differences drop as the network size grows.

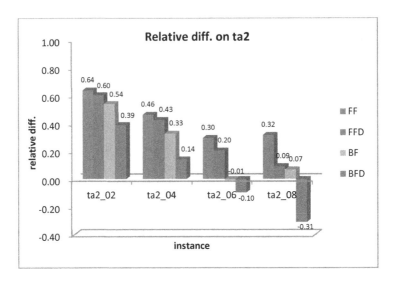

Fig. 3. Relative difference of computational time on graph "ta2"

6   Conclusions and Future Work

In this paper, an algorithm which combines the ideas of the bin-packing method and edge-disjoint paths is proposed for solving the routing and wavelength assignment (RWA) problem. The performance of the proposed GA_MEDP_RWA has been verified by several experiments on realistic network topologies. Compared with the bin-packing based methods, GA_MEDP_RWA can find the best solution among all testing instances. Although the proposed method takes longer computational time, the relative difference of computational time decreases as the problem size grows. In the future, we will test the proposed method on more instances and compare the results with other heuristics, e.g., PSO, ABC, etc. Moreover, a divide-and-conquer approach called the multilevel algorithm is worth studying for solving large-size problems.

References

1. Martins, A.X., Duhamel, C., Mahey, P., Saldanha, R.R., de Souza, M.C.: Variable neighborhood descent with iterated local search for routing and wavelength assignment. Computers and Operations Research 39(9), 2133–2141 (2012)
2. Hsu, C.-C., Cho, H.-J.: A Genetic Algorithm for the Maximum Edge-disjoint Paths Problem. Accepted by Neurocomputing
3. Li, G., Shmha, R.: The partition coloring problem and its application to wavelength routing and assignment. In: Proceedings of the First Workshop on Optical Networks (2000)
4. Zang, H., Jue, J.P., Mukherjee, B.: A review of routing and wavelength assignment approaches for wavelength-routed optical WDM networks. Optical Networks Magazine 1, 47–60 (2000)

5. Chlamtac, I., Ganz, A., Karmi, G.: Lightpath communications: an approach to high bandwidth optical WAN's. IEEE Transactions on Communications 40(7), 1171–1182 (1992)
6. Choi, J.S., Golmie, N., Lapeyrere, F., Mouveaux, F., Su, D.: A functional classification of routing and wavelength assignment schemes in DWDM networks: Static Case. In: Proceedings of the 7th International Conference on Optical Communications and Networks, pp. 1109–1115 (2000)
7. Skorin-Kapov, N.: Routing and wavelength assignment in optical networks using bin packing based algorithms. European Journal of Operational Research 177, 1167–1179 (2007)
8. Manohar, P., Manjunath, D., Shevgaonkar, R.K.: Routing and wavelength assignment in optical networks from edge disjoint path algorithms. IEEE Communications Letters 6(5), 211–213 (2002)
9. Noronha, T.F., Resende, M.G.C., Ribeiro, C.C.: A biased random-key genetic algorithm for routing and wavelength assignment. Journal of Global Optimization 50(3), 503–518 (2011)
10. Noraha, T.F., Ribeiro, C.C.: Routing and wavelength assignment by partition colouring. European Journal of Operational Research 171(3), 797–810 (2006)
11. Fischer, T., Bauer, K., Merz, P., Bauer, K.: Solving the routing and wavelength assignment problem with a multilevel distributed memetic algorithm. Memetic Computing 1(2), 101–123 (2009)

Robust Self-organized Wireless Sensor Network: A Gene Regulatory Network Bio-Inspired Approach

Nour El-Mawass[1], Nada Chendeb[1], and Nazim Agoulmine[2]

[1] LaSTRe Laboratory, Lebanese University, Tripoli, Lebanon
nour.elmawass@ieee.org, nchendeb@ul.edu.lb
[2] Networks and Multimedia Systems Group, University of Evry Val d'Essonne, Evry, France
nazim.agoulmine@iup.univ-evry.fr

Abstract. Minimal energy consumption and maximal event detection rate are among the main objectives in Wireless Sensor Networks (WSN). Sensor nodes are constrained units that have limited energy and low processing capabilities. Some challenging applications aim to spread a large number of nodes randomly in a geographical location to monitor it. Since it is difficult to access frequently and physically these sensors, an independent, failures resistant and distributed control, that is non-assisted by humans is mandatory. However, any intelligent strategy in WSN should have minimal requirements and low overhead. In this paper, we exploit the cell/node analogy to introduce a bio-inspired controller based on the principles of Gene Regulatory Network (GRN). This controller is adapted by the Genetic Algorithm. By implementing this controller in each node, the emergent network is characterized by an auto-organized, robust and adaptive behavior similar to a biological system. We compare the approach to a classical approach that uses redundancy as a failure resistance strategy, and found a significant increase in lifetime and event detection rates of the entire network.

Keywords: Wireless Sensor Network (WSN), Gene Regulatory Network, Self-Organization, Biological Inspiration, Genetic Algorithm, Robustness.

1   Introduction

Wireless Sensor Networks (WSNs) are a kind of communication networks based on sensor nodes that are characterized by a limited, generally non rechargeable battery and a processor with limited capacities. [1].

A large part of research in the WSN area has been motivated by two constraints: (1) modest intelligence of its processing unit (that simply senses and transmits events) and (2) limited battery capacity that should be saved in order to extend the lifetime of the network (in many cases, the sensor is considered "dead" once its energy is vanished). Many solutions drawn from the classic optimization world have been proposed. They include mainly solution based on spatial and temporal correlation, routing to minimize energy consumption [1] [2], etc...

Bio-inspired techniques have proved to be a successful source of solutions for many computer-related problems. They are characterized by attractive properties such

J.-S. Pan et al. (eds.), *Genetic and Evolutionary Computing*,
Advances in Intelligent Systems and Computing 238,
DOI: 10.1007/978-3-319-01796-9_11, © Springer International Publishing Switzerland 2014

as: adaptation to environmental changes, resilience, collaborative yet simple behavior, self-organization, distribution, evolution and survival [3].

This research aims to introduce a particular biological inspiration, the Gene Regulatory Network (GRN), to address the challenge of autonomous and distributed control in Wireless Sensor Networks (WSN). Like biological cell networks, a WSN is formed of identical simple components (cells/nodes) with limited capabilities that are able to communicate together in order to drive the network toward the required state. In biological systems, cells adapt to external environment perturbations through the optimized and evolvable structure of its Gene Regulatory Network (GRN). In this research, the objective is to build a WSN that inspires from the GRN to automatically find the optimized configuration that allows for its survival.

Each sensor is assimilated to a living cell and implements a GRN controller inspired from the principles of genetic regulation, with the aim to emerge a global optimal behavior at the network level. With this, the network will work in a distributed and dynamic manner, without a central failure point, and without the need to be manually configured at the initialization or during its lifetime.

The rest of the paper is organized as follows: Section 2 presents an overview of related works inspired from GRNs. Section 3 provides the details of the proposed Wireless Genomic Sensor controller while section 4 gives an overview of the main methods used to implement and adapt the proposed system. Section 5 presents results of the simulation, and shows the advantages of the system. We conclude with section 6 where we summarize the research issues and present some future work.

2   Related Work

Bio-inspired systems and applications have been extensively used and discussed in the literature [3] [4]. GRN inspirations, however, have received little attention. In the following, we discuss few GRN based works.

Das et al. (2008) [5] use a GRN model based on differential equations to solve the problem of coverage in wireless sensor networks. They show that the GRN approach is characterized by a performance similar to a Genetic Algorithm (NSGA II) applied in a distributed system. Their approach, however, is applied at the initialization only and does not consider the future evolution of the system. Besides, the approach is a central one, and although the distribution is cited as one possible aspect of the system, it has not been studied or discussed further.

A similar approach to our work is presented by Markham and Trigoni (2010) [6] [7]. The authors use a discrete model of GRN (called dGRN) in order to offer to every sensor the ability to regulate its sampling rate (and therefore energy conumption) based on its data and its neighbors' data. One of the main ideas in the work is that the structure of the dGRN controller embedded in every sensor is determined by the genetic algorithm that adapts it to an application where sensors have to track a mobile target. Authors argue that by using this technique, they can avoid the need for an expert, and limit the repetitive cycles of design, test, and adaptation.

Ghosh et al. propose the use of the attractor theory in GRN to achieve a fault-tolerant WSN routing. Preliminary results can be found in [8].

In a different orientation, Quick et al. (2003) propose GRN based real time controllers coupled to the environment. The approach is used in artificial organisms

[9] but it doesn't involve more than one agent. In Taylor work: "Biosys" [10], submarines robots are configured using a GRN controller. This work shows how a GRN controller allows a group of distributed entities to communicate and achieve a distributed task. Knabe et al. use GRN based artificial controllers to build biological clocks able to respond to periodic environmental stimuli [11].

3   Controller Structure

Genes do rarely encode information in a direct manner. Instead, it is the genes interaction at the cell or the multicellular level that implies the behavior of a living organism. The activation of a gene at a given point can regulate the expressions levels of genes in other points or even regulate the expression of the gene itself. In a multicellular organism, intracellular interactions add more complexity to the regulatory network.

In unicellular organisms, gene regulatory network optimizes the cell to make it survive its environment conditions at a given time. Gene regulatory network is composed of DNA segments. These segments will interact through their products: RNA and proteins and with other substances in the cell to determine the rates of transcription of the genes [12]. As in the previous cell example, the aim of our approach is to offer every sensor a phenotypic plasticity, i.e. the ability to choose the most adapted behavior to its environment [13].

Random Boolean Network (RBN) is one of the first models used to represent GRN. The model is based on an oriented graph where genes and their products are represented by nodes and the regulation is modeled by edges [14]. Each node has a binary state updated synchronously and periodically using regulation functions at discrete time steps [15]. Although we use a random network to form our controller, we don't use the Boolean state to represent the gene. Instead, we directly use the level of protein concentration as a proxy to the gene expression. A regulatory network has two types of influence: internal and external depending on whether the influence is performed by a protein concentration inside the cell or is a result of environment conditions or concentration of migrant proteins, coming from other cells.

A sensor has two states: ON (meaning normal sampling and communication) and OFF (in sleep mode with negligible energy consumption). A sensor state is decided dynamically based on its environment and the contribution it can add to this environment. A node turned off is judged non vital at the current instant, and doesn't make part of the functional network. It has periodical distant wake ups, where its controller can update its information and decide whether to switch ON or not.

3.1   Genome

We consider simplified controller genome representation as a sequence of Ng genes. Each gene has one activation site (formed of two proteins) and one output protein (its concentration reflects the gene expression level). An input protein is either an inhibitor or an amplifier (fig.1). Each gene is therefore representing a simple law, having proteins concentrations as inputs and outputs. Table 1 resumes the proteins used in the controller.

Table 1. Controller Proteins and their Description

Protein	Type	Description
P1	Sensory	Intersection between the sensor and its neighbors
P2	Sensory	Sum of the remaining energy in neighbors sensors
P3	Diffusion	Remaining energy in the sensor
P4	Actioner	State of the sensor
P5-P8	Intern	

A/I	Input Protein P1	A/I	Input Protein P2	Output Protein P3	Output function form

Fig. 1. Schema of the gene structure. A genome contains Ng genes.

3.2   Proteins

Each gene, through its output protein, can potentially act as inhibitor or amplifier of the expression of another gene. Moreover, some proteins act as sensory inputs or as actuators. Table 1 presents the number and the description of proteins used in our design. Three sensory proteins are used. These proteins represent environmental inputs including (1) node overlap with its on neighbors (2) the sum of neighbors' energies and (3) the energy left in the sensor at a given time. P4 is considered an output protein or actioner, its value determines the state of the node (on or off).

Overlap. Protein P1 is associated with the total overlap zone of a node. The intersection of two identical sensors coverage zones (fig.2) is given by the following equation (coverage radius is R and centers are Zi and Zj):

$$A = 2R^2 \left(\mathrm{acos}(di) - di\sqrt{1 - di^2} \right) \tag{1}$$

Where $di = D/2R$ and $D = \|Zi - Zj\|$.

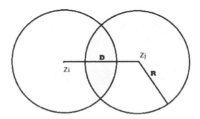

Fig. 2. Intersection of two coverage zones

Energy. Proteins P2 and P3 are associated with energy. P2 value of a node i is given by equation (2) where we sum the energies left in the neighbor sensors of i and divide the sum by the energy left in i. P3 is the ratio of residual energy in i over its initial energy (eq.3).

$$P_{2i} = \frac{1}{E_i} \sum_{k=1}^{V} E_K , \tag{2}$$

$$P_{3i} = \frac{E_i}{E_{i0}} \tag{3}$$

Activation. Protein P4 concentration is the parameter that determines the activation or deactivation of a sensor. If the value of the concentration exceeds a given activation threshold *Sa* (0.5 for example), the sensor is considered ON, otherwise, it is in OFF state.

Other Proteins. In order to have indirect and non-intuitive sensory protein-actuator relations, we use intern proteins P5 - P8 that add more complexity to the system.

3.3   Output Function

Each gene is associated with an output function. All Ng output functions have the same sigmoid form. They differ, however, in their parameters. The sigmoid form is issued from genes activation and proteins expression dynamics. The output is function of the activation level, defined as the sum of proteins concentrations at the input of the gene as shown in the following equation:

$$La = \sum_{i=1}^{M} \pm P_i$$

Where La is the activation level at the input of the gene, M is the number of proteins at the input (in this case 2), and Pi is an input protein to this gene. The $\pm$ sign corresponds to the protein effect: « + » for an amplifier and « − » for an inhibitor.

The resulting output is added to the output protein concentration in the cell. This concentration will decrease with a fixed decay rate (decay=0.1). Pi is therefore updated using the following equation:

$$Pi(t + 1) = (Pi(t) + \sum_{j=1}^{G} \Delta Pi)(1 - decay)$$

Output Function Types. A gene is « On » by default if its output function is positive for La = 0. The protein is therefore produced at a negative or null activation level. In such a case, the gene doesn't directly become inactive in response to an inhibitor input. The inhibitor effect should exceed the amplifier effect in order for the activation level to reach the negative input that corresponds to a null output. A gene is « Off » by default if its output function is null for La = 0.

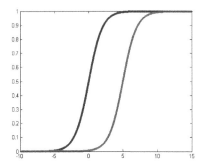 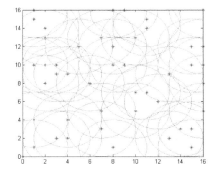

Fig. 3. Gene Output function. The blue sigmoid (left) shows a gene that is «on» by default, while the red one (right) shows a gene that is «off» by default.

Fig. 4. Geometric representation of the random coverage of 30 sensors of radius 3 units. over a 16 units. x 16 units square space.

4   System Simulation

4.1   Scenario

To test the proposed system, we introduce the following scenario of usage: N wireless sensors are randomly and uniformly placed in a rectangular area of dimensions (L x L), where events and failures occur randomly. Each sensor is characterized by two radiuses: (1) coverage radius R (indicating the zone where the sensor is able to detect events) and (2) Transmission radius R_T. These sensors should be able to (1) detect the events occurring in the square space and (2) transmit gathered information to a gateway node constantly fed by a power source.

At every time step, an event occurs with a probability of 0.1. It has a random position over the (L x L) space and its duration follows a normal law (mean= 7s and standard deviation = 4s.). The choice of these values should be adapted to the chosen application. The number of failures during the network lifetime is imposed by *PFailure*, the probability that a node does not work properly. For N=30 and *PFailure* = 0.1, three sensors will fail. The instant at which the sensor fails is randomly chosen between 0 and *Tmax*, the maximum lifetime allowed for the entire system.

4.2   Energetic Considerations

We continuously update the energy value in each sensor depending on its consumption which has mainly two parts: (1) the constant consumption due to sampling and processing and (2) the communication (transmitting/receiving) consumption that has a periodical component and a routing component. Our calculations are based on wireless temperature sensor, Monnit WIT™. The lifetime of a sensor having 270 joules of initial energy is estimated to be around 3600 s. We choose the Ad-Hoc DSR (Dynamic Source Routing) protocol to simulate routing, the cost being the energy consumption over the chosen route. Sensors energy levels are updated according to the selected routes.

4.3   Genetic Algorithm

We use the genetic algorithm to reach an optimized form of the structure proposed in the previous section. The genetic algorithm, an evolutionary algorithm inspired from natural selection concepts, is extensively used as an optimization technique in NP complete problems. Its implementation includes successive iterations where newer populations take the place of older ones. The individuals forming this population are obtained from the most adapted "parents" in the last generation using the process of cross-over [16]. In our implementation, the individual genome size is 6xNg, and the one point cross-over is performed with a 0.9 rate. To assess the adaptability of an individual we use a two-objective fitness function. We implement the tested individual genome/controller in each node of the network and we run a simulation over the whole network where random events and failures occur. The fitness function is then deduced from the network performance in terms of coverage and lifetime.

5   Results

Here, we provide the results of the simulation implemented in Matlab for the following parameters: L=16 units, N=50, R=3 units and RT=6 units (fig. 4). Fig. 7 shows the convergence of protein P4 values during the initialization stage. Initially, all nodes are turned on and every node is able to decide autonomously to remain ON or to sleep. We noted two "life cycles" based on the winner GRN controller.

Figure 5 (green line) shows an average initial coverage rate of ~ 93%, associated with an average of 25 sensors turned on (50% of the nodes) during the first life cycle of the network that has duration of 3600 s.

The coverage rate drops around t=3600s (fig.5, right), this is associated with a collective drain of energy among sensors that were turned on initially. This can also be seen in fig.5 (left) where the number of sensors turned on falls at the same time. Figure 6 (left) equally shows the drop in P4 concentration in these sensors. The network becomes quasi-inactive during the time needed to recover from this collective failure. The recovery can be clearly noted in fig.6 (right) where the P4 concentration of several nodes increase marking the "awakening" process of 14 nodes that were initially turned off. The coverage rate during the second life cycle is around 73%. The second drop appears near t = 7600 s and the same recovery behavior described above reappears (fig.5). Since the sensors left are not enough to have an acceptable coverage rate, we can no longer consider the network functional for a third life cycle.

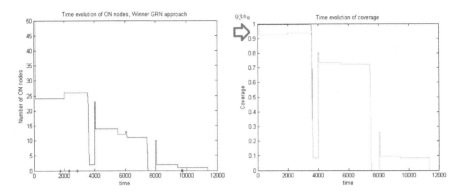

Fig. 5. Evolution of coverage rate, and number of ON nodes in the network over time (s.)

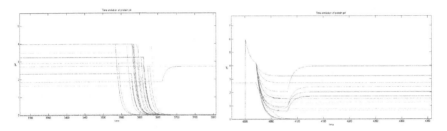

Fig. 6. Sudden drop of P4 concentration in "ON" sensors during the 1st life cycle (left figure) and collective increase of P4 concentration in the "ON"sensors at the beginning of the 2nd life cycle (right figure). Each line corresponds to a different senor.

A particular node failure occurs around t= 1730 s. As shown in fig.7, two lines representing the P4 concentrations in two sensors will take an increasing shape before stabilizing at a positive value higher than the activation point, which indicates that the sensors are now turned on to adjust the coverage rate.

We compare our approach to a redundancy based network where redundant sensors are used to ensure that the coverage rate is not influenced by nodes failures (red line in fig.5). The enhancement of the proposed network can be seen in two different ways involving enhancement of coverage rate and life time.

Formulated together: while keeping the coverage rate high enough (average of 83% compared to 99%), the enhanced network has its lifetime increased by 100%.

Concerning the computational overhead of our solution, we state that the computation added by the controller is resumed in Ng simple laws: one law per gene. When a law is executed, concentration values of two input proteins are summed and the output function value associated with this input is located in a matching table. These laws do not involve multiplication or any other complex or processor consuming functions.

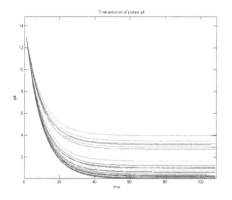

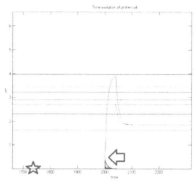

Fig. 7. P4 concentration convergence at the initialization

Fig. 8. Increase in P4 concentrations in two sensors

The only information to be exchanged during a communication period is the residual energy value at the communicating sensor. The state of a neighbor node can be deduced from the presence or absence of its information during a communication period. Additionally, the information can be contained in one byte of data and encapsulated in a data packet.

6   Conclusion and Future Work

Using GRN principles, we offer the sensor nodes in a WSN the ability to perform an autonomous and distributed behavior based on a set of simple laws. Despite their limited intelligence and constrained processor, sensors equipped with the GRN controller are able to self-organize and apply a strategy that increases the network efficiency and lifetime.

Our GRN based controller is proposed and optimized by the genetic algorithm using a fitness function that reflects our objectives (optimum coverage and lifetime).

In addition to the automatic recovery after individual and collective failures, the improvements of our enhanced network compared to a normal network can be seen in two ways: (1) improving the coverage rate by 66%, if the extended lifetime was considered and (2) increasing the lifetime of the network by 100%.

As a future work, we plan to study the GRN approach in the frame of mobile sensor network. We think that it will be equally beneficial to study the contributions of GRN to the routing domain in WSNs. Nowadays; biological networks are believed to have a scale free structure, and thus it may be better to explore an alternative controller based on this model. For additional justification of the approach that we proposed, we consider comparing it to a similar proposition that has the same optimization objectives.

References

1. Akyildiz, I., Su, W., Sankarasubramaniam, Y., et al.: Wireless sensor networks: a survey. Elsevier Computer Networks 38, 393–422 (2002)
2. Akkaya, K., Younis, M.: A survey on routing protocols for wireless sensor networks. Ad Hoc Networks 3(3), 325–349 (2005)
3. Dressler, F., Akan, O.B.: A survey on bio-inspired networking. Computer Networks 54(6), 881–900 (2010)
4. Dressler, F., Akan, O.B.: Bio-Inspired Networking: From Theory to Practice. IEEE Communications Magazine 48(11), 176–183 (2010)
5. Das, S., Koduru, P., Cai, X., Welch, S., et al.: The gene regulatory network: an application to optimal coverage in sensor networks. In: GECCO 2008 Proceedings of the 10th Annual Conference on Genetic and Evolutionary Computation (2008)
6. Markham, A., Trigoni, N.: Discrete Gene Regulatory Networks (dGRNs): A novel approach to configuring sensor networks. In: IEEE INFOCOM (2010)
7. Markham, A., Trigoni, N.: The Automatic Evolution of Distributed Controllers to Configure Sensor Network. Oxford Computer J. 54(3), 421–438 (2011)
8. Ghosh, P., Mayo, M., Chaitankar, V., et al.: Principles of Genomic Robustness Inspire Fault-Tolerant WSN Topologies: a Network Science Based Case Study. In: Seventh IEEE International Workshop on Sensor Networks and Systems for Prevasive Computing (2012)
9. Quick, T., Nehaniv, C.L., Dautenhahn, K., Roberts, G.: Evolving Embodied Genetic Regulatory Network-Driven Control Systems. In: Banzhaf, W., Ziegler, J., Christaller, T., Dittrich, P., Kim, J.T. (eds.) ECAL 2003. LNCS (LNAI), vol. 2801, pp. 266–277. Springer, Heidelberg (2003)
10. Taylor, T.: A Genetic Regulatory Network-Inspired Real-Time Controller for a Group of Underwater Robots. In: Proceedings of the Eighth Conference on Intelligent Autonomous Systems (2004)
11. Knabe, J.F., Nehaniv, C.L., Schilstra, M.J., et al.: Evolving Biological Clocks using Genetic Regulatory Networks. In: Artificial Life X Conference (Alife 10) (2006)
12. Albert, R.: Boolean modeling of genetic regulatory networks. Complex Networks, 459–479 (2004)
13. Bradshaw, A.: Evolutionary significance of phenotypic plasticity in plants. Advances in Genetics 13, 115–155 (1965)
14. Knabe, J.F.: Evolvability of Computational Genetic Regulatory Networks, PhD diss., University of Hertfordshire (2009)
15. Kauffman, S.: Metabolic stability and epigenesis in randomly constructed genetic nets. J. of Theoretical Biolog. 22(3), 437–467 (1969)
16. Haupt, R.L., Haupt, S.E.: Practical Genetic Algorithms, 2nd edn. John Wiley & Sons, Inc. (2004)

Automated Test Data Generation for Coupling Based Integration Testing of Object Oriented Programs Using Particle Swarm Optimization (PSO)

Shaukat Ali Khan and Aamer Nadeem

Center for Software Dependability
Mohammad Ali Jinnah University (MAJU),
Islamabad, Pakistan
shaukatali74@gmail.com, anadeem@jinnah.edu.pk

Abstract. Automated test data generation is a challenging problem for researchers in the area of software testing. Up until now, most of the work on test data generation is at unit level. Until level test data generation involves the execution of test path at unit level where interaction with other components is minimum. Test data generation for unit testing involves a single path and there is no usage of formal and actual parameters. The problem of automated test data generation becomes very challenging when we move to other levels of testing including integration testing and system level testing. At integration level, the variables are passed as arguments to other components and variables change their names; also multiple paths are executed from different components to ensure proper functionality. Recently evolutionary approaches have been proven a powerful tool for test data generation. In this paper, we have proposed a novel approach for test data generation for coupling based integration testing using particle swarm optimization. Up until now, there is no research for test data generation for coupling based integration testing using particle swarm optimization. Our approach takes the coupling path as input, containing different sub paths, and generates the test data using particle swarm optimization. We have also proposed architecture of tool for automation of our approach. In future, we will implement our proposed approach and will perform different experiments to prove its significance.

Keywords: Coupling Path; Consequent Method; Antecedent Method; Coupling Variable; Coupling Type; Particle Swarm Optimization (PSO).

1   Introduction

Software testing is one of the most important phases in software development cycle. Testing can be performed at different levels including unit, integration, or system level. Unit level testing validates the functionality of individual units. Integration level testing, tests the interaction of different components after integration with other components. System level testing treats the system as black box and checks the functionality of the system as a whole. Integration testing is an important level of testing which verifies the different components interactions and message passing through interfaces. Unit level testing is a base for integration testing, if the units work correctly then different units are integrated together using different interfaces exposed by different components. Integration level testing verifies that the interfaces are correctly integrated and message passing through interfaces is correct. Integration testing is concerned with the interactions among components. Does a component call other components correctly? Are the right parameters with right types and ranges are passed? Does the called method return the proper type and the value is in the correct range? These questions are focus of the integration testing. Unfortunately, very little research has been done in the area test data generation for integration testing using evolutionary approaches. Coupling based integration testing is based upon coupling relationships that exist among variables across call sites in procedures. In the same way as unit level testing is a base for integration testing, integration testing is a base for system level testing. System level testing is difficult to achieve before integration testing [20, 21].

Test data are an important part of test case, without test data test case execution is not possible. Research has explored several techniques for test data generation using evolutionary approaches. A number of test data generation approaches have been developed and automated. Random test data generation, generates test data based on selective random inputs form some distribution. Path- oriented and structural approaches use the program's control flow graph for test data generation; they select a path, and use a technique such as symbolic execution for generation of test data. Goal-oriented test-data generation approaches select inputs to execute the selected goal, such as statement, condition coverage, decision coverage, irrespective of the path taken. Evolutionary test approaches use evolutionary algorithms i.e. genetic algorithm, for selection and generation of test data by applying evolutionary operators, i.e., crossover and mutation. Most of the work on test data generation has been done at unit level. Unit level test data generation involves the test data that executes the test case for unit level testing. [20, 21]

This paper presents a novel approach for automated test data generation for coupling based integration testing of object oriented programs using particle swarm optimization. Our approach takes the coupling path as input, containing different sub paths, and generates the test data using particle swarm optimization.

The rest of the paper is organized as follows: Section II elaborates the background knowledge of evolutionary approaches and coupling based integration testing. Section III describes the proposed approach of test data generation for coupling based integration testing of object oriented programs using particle swarm optimization and section IV represents the architecture of tool for test data generation of coupling based integration testing of object oriented programs. Section V represents the related work. Section VI concludes the paper and presents the future work.

2   Background

PSO was first invented by Kennedy and Eberhart in 1995 [44]. PSO shows few similarities with genetic algorithms, it does not use evolution operators such as crossover and mutation. Each member in the swarm, called a particle, adjusts its position by learning from its own experience and from other members of the swarm.

During the iteration process, each particle maintains its own current position, its present velocity and its personal best position. In general, the personal best position of particle is denoted by pbest and the global best position of the entire population is called gbest.

Jin and Offutt [41] proposed an approach for integration testing of procedural languages that is based upon coupling relationships among variables across different call sites in different procedures. They defined three types of coupling relationship that must be tested: parameter coupling, shared data coupling and external device coupling. When one procedure passes parameters to other procedure, then parameter couplings occur. When two procedures references the same global variables, then shared data couplings occur. External device couplings occur when two procedures accesses the same external storage medium. Their approach requires the execution of programs from each definition of a variable in a caller to a call site and then to the uses using formal arguments in the called procedures.

Up until now, most of the work on test data generation is for unit testing of object oriented programs. Integration testing is an important part of testing phase of software development life cycle. Coupling based testing is an integration testing approach that is based upon coupling relationships that exist among different variables across different call sites in functions. Different types of coupling exist between variables across different call sites. Up until now, test data generation approaches cater only unit level testing. There is no work for test data generation for coupling based integration testing.

Alexander and Offutt [40] extend the approach of Jin and Offutt [41] of coupling based testing for object oriented programs. They identified four types of coupling relationships among object oriented programs. They also identified four structural types of coupling sequences for coupling based testing of object oriented programs. We used the classification of coupling types of object oriented programs identified by Alexander and Offutt and generates the test data for each coupling type using particle swarm optimization. Our proposed approach takes the coupling path as input and generates the test data for that path [39, 40, 41].

Test data generation for unit testing involves the single path, execution and monitoring of single path is required in unit testing. In integration testing, as different components interact with each other so execution and monitoring of multiple paths are required. Test data generation for unit testing involves a single path and there is no usage of formal parameters. In integration testing, different methods interact with each other via passing message passing through actual and formal parameters so there is a requirement of mapping between actual and formal parameters. We have to maintain a mapping table for that contains mapping between actual and formal parameters as variable change names in formal parameters. Def use analysis is very difficult to achieve as variables change name in formal parameters so a mapping must be maintain for actual to formal parameters.

3   Proposed Approach for Test Data Generation

We have proposed a novel approach for test data generation for coupling based integration testing using particle swarm optimization. Working of particle swarm optimization for test data generation for coupling based integration testing has been depicted in figure 2. Our approach starts with random data, generator by random test data generator, based upon the coupling variables and other variables involved in the execution of coupling path. In our proposed technique, each particle presents the input variable required for executing a specific coupling path. We have encoded each particle, each input variable required for executing a specific coupling path, in binary format.

Random population of n numbers is produced to initialize n particles. Each number is converted into binary format before initializing particles. After initialization phase, now each input variable required for executing a coupling path is presented by a particle containing an array of 0's and 1's. We have initialized each dimension of the position vector between 1 and maximum length required for presenting any input variable as have been shown in the particle below.

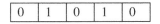

Fig. 1. Particle encoding in PSO

To assess the fitness f(x) of each particle x in the population. We have calculated fitness of particle based upon the coverage analysis. By coverage analysis, we mean that how many nodes in the coupling path are covered by current particle. For every particle in the population we see whether its current fitness is better than its previous personal best. If this is the case, we set its personal best to its current value, otherwise it has same personal best.

Once we have updated all the personal best positions, we will determine the global best by using local best of all particles. If this global best is having better fitness than previous global best, we update the global best otherwise we do not change the global best. After determining the personal best and the global best, we have changed the current velocity of each particle using PSO standard equations [44] and cost function proposed Tracey et al. [15]. After having updated velocities for each particle, we now calculate the new position of each particle. If end condition is fulfilled, stop and return the global best of the population. Otherwise go to step of fitness evaluation. There are three different types of completion criteria, Maximum number of generations configured for each particle, if there is no improvement in the global fitness of the swarm for certain number of generations and time can also be used as terminating criteria.

To illustrate the working of our proposed approach for test data generation for coupling based integration testing, consider the following example. Coupling path contains three sub paths. One path is coupling sub path where context variable, which is used to define and use coupling variable, is defined. Second path is antecedent path where coupling variable is defined by using context variable defined in coupling sub path. Third path is consequent path where coupling variable is used by using context variable defined in coupling sub path. This example is to show the working of PSO evolutionary algorithm for coupling based integration test data generation. We have chosen a simple path for illustration of our proposed approach.

Coupling path= Coupling sub path + Antecedent path+ Consequent path

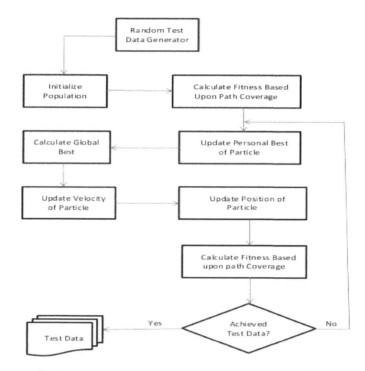

Fig. 2. Proposed approach for test data generation using PSO

Table 1. Showing Test Path Encoding in PSO

P1	P11	1, 3, 4, 5 ,6 ,7	Coupling sub path
	P12	1, 3, 5, 6,7, 9, 10	Antecedent path
	P13	1, 3, 4 ,5 ,7, 8	Consequent Path

Our proposed approach encoded particle in PSO in the form of following structure. Let suppose a particle 'P1', we encoded the particle after breaking into its sub paths, the first is coupling sub path, second is antecedent path and third is consequent path.

We have taken one particle 'P1' and encoded it in PSO after splitting into its sub paths. Each path has its own dimensions and fitness. The dimension of each path is its statement id in path e.g. '1' is id for some statement in program used in the path and fitness of each particle is calculated on the basis of number of statements covered by each path in execution. In table1, the particle P1 is encoded after splitting into three sub particles i.e. P11, P12, and P13. Each particle has its own dimensions and fitness. Local best of each particle is calculated in the following way:

$$lb\_P11 = 1, 3, 4, 5, 6, 7 \qquad 0/5$$

$$lb\_P12 = 1, 3, 5, 6, 7, 9, 10 \qquad 0/7$$

lb_P13= 1, 3, 4, 5, 7, 8   0/7

Every sub particle is showing its local best at the start and global best is calculated, by using the above local best values of every sub particle, for particle 'P'

$$Gb\_P= (lb\_P11+lb\_P12+ lb\_P13)/3$$

4   The Proposed Tool

We have proposed a high level architecture of the tool for our proposed approach and is depicted in Figure 3.

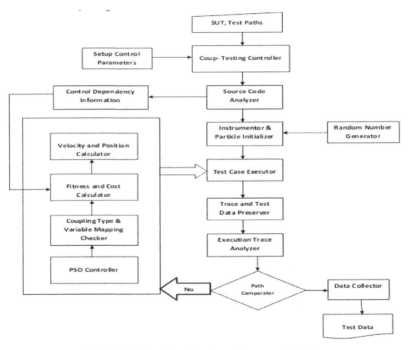

Fig. 3. High Level architecture of proposed Tool

Coup Testing Controller is the main component of our tool. This component receives all the inputs from the external sources including source code of system under test, test paths and the control parameters of the tool including number of iterations for test data generation, stopping condition etc.

PSO controller checks the coupling type and mapping information for the variables because variables change names when passed from one method to other methods. Instrumentor extract the information about actual and formal parameters from the source code and stored it in mapping table for further usage. The fitness calculator calculates the fitness of each data value using the cost function proposed Tracey et al. [15] and then these values are passed to velocity and position calculator for calculation of new position based on velocity of the particle.

5   Related Work

Evolutionary approaches are mostly used in the area of automated test data generation [8, 11, 13, 14, 15, 17] for unit testing. Cheon et al. [3, 4] proposed a specification based fitness function for evolutionary testing of object oriented program. They also proposed automation of Java program testing at unit level using evolutionary approaches. Dharsana et al. [5] generated test cases for Java based programs and also performed optimization of test cases using genetic algorithm. Jones et al. [6] performed automatic structural testing using genetic algorithm in their approach. Bilal and Nadeem [2] proposed a state based fitness function for object oriented programs using genetic algorithm. Smith and Robson observe that OO classes cannot be tested directly and classes must be tested indirectly by sampling among their instances [23]. Fiedler approach uses a combination of black and white-box testing techniques for class testing [22]. Edwards created test cases based on specifications of the system [26].Perry and Kaiser [27] concluded that single inheritance, method overriding, and multiple inheritance do not reduce the amount of testing effort, and in many cases, increase the required effort to achieve test adequacy. Jorgensen and Erickson describe an approach to integration testing that is similar to many black box testing techniques [14]. They define paths through a collection of classes that are traversed through method calls.

Most research in OO testing has been at the intra-class level. This includes work by Hong et al. [28], Parrish et al. [29], Turner and Robson [30], Doong and Franklin [31], and Chen et al. [32]. Intra-class testing strategies focus on one class at a time; hence, it does not find problems that exist in the interfaces between classes, or in inheritance, dynamic binding, and polymorphism among classes. Chen and Kao [25] describe an approach to testing OO programs called Object Flow Testing, in which testing is guided by data definitions and uses in pairs of methods that are called by the same caller, and testing should cover all possible type bindings in the presence of polymorphism. Though their work is similar, there are significant differences. First, their criteria are coarse-grained. The criteria presented by Alexander and Offutt [39, 40] are a superset of those of Chen and Kao. Kung et al. observe that one of the key difficulties in testing OO software understands the relationships that exist among the components [25].

6   Conclusion and Future Work

In this paper, we have proposed a novel approach for automatic test data generation for coupling based integration testing using particle swarm optimization. Our proposed approach takes coupling path as input and then generates test data for coupling path using particle swarm optimization. We have also proposed tool's architecture for automation of our proposed approach. In future, we will implement our proposed tool and will perform different experiments to prove the significance of our approach.

References

1. Baresel, A., Sthamer, H., Schmidt, M.: Fitness Function Design to improve Evolutionary Structural Testing. In: Proceedings of the Genetic and Evolutionary Computation Conference (GECCO 2002), New York, USA (July 2002)
2. Bashir, M.B., Nadeem, A.: A State Based Fitness Function for Evolutionary Testing of Object-Oriented Programs. In: Lee, R., Ishii, N. (eds.) Software Engineering Research, Management and Applications 2009. SCI, vol. 253, pp. 83–94. Springer, Heidelberg (2009)
3. Cheon, Y., Kim, M.Y., Perumandla, A.: A Complete Automation of Unit Testing for Java Programs. In: The 2005 International Conference on Software Engineering Research and Practice (SERP), Las Vegas, Nevada, USA (June 2005)
4. Cheon, Y., Kim, M.: A specification-based fitness function for evolutionary testing of object-oriented programs. In: Proceedings of the 8th Annual Conference on Genetic and Evolutionary Computation, Washington, USA (July 2006)
5. Dharsana, C.S.S., Askarunisha, A.: Java based Test case Generation and Optimization Using Evolutionary Testing. In: International Conference on Computational Intelligence and Multimedia Applications, Sivakasi, India (December 2007)
6. Jones, B., Sthamer, H., Eyres, D.: Automatic structural testing using genetic algorithms. Software Engineering Journal 11(5), 299–306 (1996)
7. Liaskos, K., Roper, M., Wood, M.: Investigating data-flow coverage of classes using evolutionary algorithms. In: Proceedings of the 9th Annual Conference on Genetic and Evolutionary Computation, London, England (July 2007)
8. McGraw, G., Michael, C., Schatz, M.: Generating software test data by evolution. IEEE Transactions on Software Engineering 27(12), 1085–1110 (2001)
9. McMinn, P., Holcombe, M.: The state problem for evolutionary testing. In: Cantú-Paz, E., et al. (eds.) GECCO 2003. LNCS, vol. 2724, pp. 2488–2498. Springer, Heidelberg (2003)
10. McMinn, P.: Search-based Software Test Data Generation: a Survey. Journal of Software Testing, Verifications, and Reliability 14(2), 105–156 (2004)
11. Pargas, R., Harrold, M., Peck, R.: Test-data generation using genetic algorithms. Software Testing. Verification and Reliability 9(4), 263–282 (1999)
12. Roper, M.: Computer aided software testing using genetic algorithms. In: 10th International Software Quality Week, San Francisco, USA (May 1997)
13. Sthamer, H.: The automatic generation of software test data using genetic algorithms. PhD Thesis, University of Ghamorgan, Pontyprid, Wales, Great Britain (1996)
14. Seesing, A., Gross, H.: A Genetic Programming Approach to Automated Test Generation for Object-Oriented Software. International Transactions on Systems Science and Applications 1(2), 127–134 (2006)
15. Tracey, N., Clark, J., Mander, K., McDermid, J.: Automated test-data generation for exception conditions. Software—Practice and Experience, 61–79 (January 2000)
16. Tonella, P.: Evolutionary Testing of Classes. In: Proceedings of the ACM SIGSOFT International Symposium of Software Testing and Analysis, Boston, MA, pp. 119–128 (July 2004)
17. Watkins, A.: The automatic generation of test data using genetic algorithms. In: Proceedings of the Fourth Software Quality Conference, pp. 300–309. ACM (1995)
18. Wegener, J., Baresel, A., Sthamer, H.: Evolutionary test environment for automatic structural testing. Information and Software Technology Special Issue on Software Engineering using Metaheuristic Innovative Algorithms 43, 841–854 (2001)

19. Wegener, J., Buhr, K., Pohlheim, H.: Automatic test data generation for structural testing of embedded software systems by evolutionary testing. In: Proceedings of the Genetic and Evolutionary Computation Conference (GECCO 2002), New York, USA, pp. 1233–1240. Morgan Kaufmann (July 2002)
20. Copeland, L.: A Practitioner's Guide to Software Test Design. STQE Publishing (2004)
21. Beizer, B.: Software Testing Techniques. International Thomson Computer Press (1990)
22. Fiedler, S.P.: Object-oriented unit testing. Hewlett-Packard Journal 40(2), 69–75 (1989)
23. Smith, M.D., Robson, D.J.: Object-oriented programming: The problems of validation. In: Sixth International Conference onSoftware Maintenance, pp. 272–282. IEEE Computer Society Press, Los Alamitos (1990); Overbeck, J.: Integration testing for object-oriented software. PhD Dissertation, Vienna University of Technology (1994)
24. Pande, H.D., Landi, W.A., Ryder, B.G.: Interprocedural def–use associations for C systems with single level pointers. IEEE Transactions on Software Engineering 20(5), 385–403 (1994)
25. Chen, M.-H., Kao, M.-H.: Testing object-oriented programs—An integrated approach. In: Proceedings of the 10th International Symposium on Software Reliability Engineering, pp. 73–83. IEEE Computer Society Press, Boca Raton (1999); Kung, D., Gao, J., Hsia, P., Toyoshima, Y., Chen, C.: A test strategy for object-oriented systems. In: Nineteenth Annual International Computer Software and Applications Conference, pp. 239–244. IEEE Computer Society Press, Los Alamitos (1995)
26. Edwards, S.H.: Black-box testing using flowgraphs: An experimental assessment of effectiveness and automation potential. Software Testing, Verification and Reliability 10(4), 249–262 (2000)
27. Perry, D.E., Kaiser, G.E.: Adequate testing and object-oriented programming. Journal of Object-oriented Programming 2(5), 13–19 (1990)
28. Hong, H.S., Kwon, Y.R., Cha, S.D.: Testing of object-oriented programs based on finite state machines. In: The 1995 Asia Pacific Software Engineering Conference, pp. 234–241. IEEE Computer Society Press, Los Alamitos (1995)
29. Parrish, A.S., Borie, R.B., Cordes, D.W.: Automated flow graph-based testing of object-oriented software modules. Journal of Systems and Software 23(2), 95–109 (1993)
30. Turner, C.D., Robson, D.J.: The state-based testing of object-oriented programs. In: Conference on Software Maintenance, pp. 302–310. IEEE Computer Society Press, Los Alamitos (1993)
31. Doong, R.-K., Frankl, P.: Case studies on testing object-oriented programs. In: Fourth Symposium on Software Testing, Analysis and Verification, pp. 165–177. ACM Press, New York (1991)
32. Chen, H.Y., Tse, T.H., Chan, F.T., Chen, T.Y.: In black and white: An integrated approach to class-level testing of object-oriented programs. ACM Transactions on Software Engineering and Methodology 7(3), 250–295 (1998)
33. Meyer, B.: Object-Oriented Software Construction, 2nd edn. Prentice-Hall, Englewood Cliffs (1997)
34. Harrold, M.J., Rothermel, G.: Performing data flow testing on classes. In: Second ACM SIGSOFT Symposium on Foundations of Software Engineering, pp. 154–163. ACM Press, New York (1994)
35. Alexander, R.T.: Testing the polymorphic relationships of object-oriented components. Technical Report ISE-TR-99-02, Department of Information and Software Engineering, George Mason University (February 1999)

36. Alexander, R.T., Offutt, J.: Analysis techniques for testing polymorphic relationships. In: Thirtieth International Conference on Technology of Object-oriented Languages and Systems (TOOLS30), Santa Barbara, CA, pp. 104–114 (1999)
37. Frankl, P.G., Weyuker, E.J.: An applicable family of data flow testing criteria. IEEE Transactions on Software Engineering 14(10), 1483–1498 (1988)
38. Rapps, S., Weyuker, W.J.: Selecting software test data using data flow information. IEEE Transactions on Software Engineering 11(4), 367–375 (1985)
39. Alexander, R.T.: Testing the polymorphic relationships of object-oriented programs. Dissertation, George Mason University (2001)
40. Alexander, R.T., Offutt, J.: Criteria for testing polymorphic relationships. In: Proceedings of the International Symposium on Software Reliability and Engineering (ISSRE 2000). IEEE Computer Society, SanJose (2000)
41. Jin, Z., Jefferson Offutt, A.: Coupling-based Criteria for Integration Testing. The Journal of Software Testing, Verification, and Reliability 8(3), 133–154 (1998)
42. Liu, X., Zhang, M., Bai, Z., Wang, L., Du, W., Wang, Y.: Function Call Flow based Fitness Function Design in Evolutionary Testing. In: APSEC 2007 Proceedings of the 14th Asia-Pacific Software Engineering Conference, Nagoya, Japan, December 5-7, pp. 57–64 (2007)
43. Baresel, A., Sthamer, H., Schmidt, M.: Fitness function design to improve evolutionary structural testing. In: Proceedings of the Genetic and Evolutionary Computation Conference (GECCO 2002), New York, USA, pp. 1329–1336 (July 2002)
44. Kennedy, J., Eberhart, R.C.: Particle Swam Optimization. In: Proc. of IEEE International Conference on Neural Networks (ICNN 1995), Path, Australia, pp. 1942–(1948)

Boosting Scheme for Detecting Region Duplication Forgery in Digital Images

Deng-Yuan Huang[1], Ta-Wei Lin[1], Wu-Chih Hu[2], and Chih-Hung Chou[1]

[1] Department of Electrical Engineering, Dayeh University, 515 Changhua, Taiwan
`kevin@mail.dyu.edu.tw, daweimailbox@gmail.com,`
`Manp5030@hotmail.com`
[2] Department of Computer Science and Information Engineering, National Penghu University of Science and Technology, 800 Penghu, Taiwan
`wchu@npu.edu.tw`

Abstract. The detection of copy-move forgery image is important in the field of blind image forensics because it is of pure image processing technique without any support of embedded security information. The proposed method consists of a boosting scheme, feature extraction and similarity matching for the detection of duplicated regions. The boosting scheme comprises an estimation of dark channel, histogram equalization and grayscale layering, by which the number of image blocks on each subimage layer can be dramatically reduced so that the time efficiency of subsequent lexicographical sorting and similarity matching can be greatly improved. Experimental results show that the proposed boosting scheme can significantly enhance the computation efficiency and have a good detection rate. Moreover, the propose method is robust to any angles rotation attack.

Keywords: Copy-move forgery detection, Similarity matching, Feature extraction, Dark channel.

1 Introduction

In recent years, the popularity of digital cameras, smart phones and tablet computers has made the acquisition of digital images become easier. In addition, modern photo-editing software package such as Photoshop and PhotoImpact makes it relatively simple to create digital image forgeries, on which people almost cannot perceive the difference between the original image and its tampered version. The most common approach used to create a digital image forgery is the so-called copy-move method, which copies a specific block of image and then pastes it into another region in the same image to achieve information hiding. Because the copied block comes from the same image, some features such as noise distribution, surface texture and lighting condition are very similar to the rest of the image and thus it leads to a great challenge in detecting and locating the tampered parts.

In general, the detection of copy-move forgery can be roughly categorized into keypoint-based methods and block-based methods [1]. The widely used features are

J.-S. Pan et al. (eds.), *Genetic and Evolutionary Computing*,
Advances in Intelligent Systems and Computing 238,
DOI: 10.1007/978-3-319-01796-9_13, © Springer International Publishing Switzerland 2014

SIFT [2] and SURF [3] for the keypoint-based methods and DCT [4], DWT [5], SVD [5, 6], PCA [7] and Zernike moments [8] for the block-based methods, respectively. The typical workflow for copy-move forgery detection (CMFD) is shown in Fig.1, but in this study, we only focus our attention on block-based methods for CMFD. First, the image is split into fixed-size overlapping blocks and feature extraction such as DCT, DWT, SVD, PCA or Zernike moments is performed to each block for representing their features. Then, the feature vectors are sorted by a lexicographical order method and similar blocks would be consecutive. Finally, the duplicated blocks are filtered out by the similarity measure of Euclidean distance or correlation coefficient.

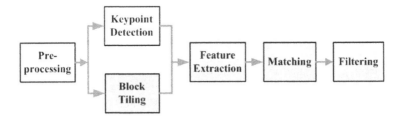

Fig. 1. Typical workflow of copy-move forgery detection methods [1]

In block-based CMFD methods, most of the algorithm frameworks are similar to that proposed by Popescu et al. [7], indicating that much computational effort are required to the task of dimension reduction, matching control and similarity filtering for large numbers of image blocks. In general, as test images grow larger, the computational complexity of CMFD becomes much higher. To reduce the computational cost, the task of CMFD is normally conducted on a reduced image size; information loss is, however, inevitable. In this paper, we propose a boosting scheme for CMFD not only to improve the computational efficiency but also to retain the detection accuracy especially when test images become considerably large.

This paper is organized as follows. In Section 2, we introduce our new algorithm for CMFD with a boosting scheme. Experiment results are presented in Section 3. Finally, we will give a brief summary in Section 4.

2   The Proposed Algorithm

The proposed method for CMFD with a boosting scheme is shown in Fig. 2. The boosting scheme consists of the procedures of dark channel evaluation, histogram equalization and grayscale layering. The remaining of the workflow for CMFD is similar the common pipeline shown in Fig. 1. In this paper, we use SVD to extract the features of image blocks and transform the diagonal matrix into a column vector for representing the feature of each image block. Lexicographical order using quick-sort algorithm is then conducted so that similar blocks would be consecutively sorted. Finally, the similarity measure of Euclidean distance is utilized and a visual map is then built with white intensity value for duplicated regions.

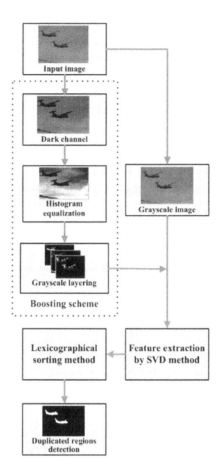

Fig. 2. Proposed method for copy-move forgery detection with a boosting scheme

Block-based CMFD is, however, an exhaustive search approach that detects the duplicated blocks by searching all pixels in the tampered image; therefore, this method would become computationally prohibitive when test images grow considerably large. Suppose a grayscale image is of size $m{\times}n$ pixels, it can be partitioned into enormous small overlapping blocks of size $B{\times}B$ pixels. Total image blocks of $N{=}(m{-}B{+}1){\times}(n{-}B{+}1)$ are then generated by sliding the window of B×B pixels over the whole image by one pixel each time from upper left to bottom right corner. For instance, suppose the sizes of tampered image and sliding window are $1024{\times}1024$ and $8{\times}8$ pixels, respectively, the number of image blocks will be high up to $(1024{-}8{+}1)2{=}1{,}034{,}289$. The large number of image blocks inevitably leads to extremely high computational burden for subsequent feature extraction and similarity matching.

The computation complexities for lexicographical order by quick sort algorithm and similarity matching are O(Nlog2N) and O(N), respectively, where N is the number of total image blocks. Therefore, the key point to increase computation efficiency is to greatly reduce the number of image blocks to be estimated. Based on the idea above, we propose a boosting scheme to reduce the number of estimated blocks each time and it turns out to increase time efficiency dramatically. The proposed method consists of the boosting scheme, feature extraction and similarity matching, which are described as follows.

2.1   Proposed Boosting Scheme

The boosting scheme consists of an estimation of dark channel, histogram equalization and grayscale layering. He et al. [9] used dark channel prior for single image haze removal, which is based on an observation that haze-free images except for sky regions have a very low and nearly close to zero intensity in at least one color channel. As haze becomes heavier, the dark channel is no longer dark. Therefore, the intensity of the dark channel can fully respond to the scattering properties of object surface due to atmospheric light. Since duplicated regions in a forgery image generally have the same lighting condition and surface texture, they should have the same values of the dark channel.

For an arbitrary image I(x, y), the dark channel can be estimated as.

$$I_{dark}(x, y) = \min_{(x,y)\in\Omega}\left[\min_{c\in(R,G,B)} I_c(x, y)\right] \tag{1}$$

where Ω is a local patch centered at (x,y), c represents one of the three color channels of R, G or B, and $I_c(x,y)$ denotes the color channel c of I(x,y), and $I_{dark}(x,y)$, as shown in Fig. 3(c), is the result of the estimated dark channel.

(a)   (b)   (c)

Fig. 3. Result of the estimated dark channel; (a) original image, (b) forgery image of (a), and (c) result of estimated dark channel.

In [9], He et al. found that the gray levels for 86% of the pixels in the dark channel are in the range [0, 16] by estimating 5,000 haze-free images, indicating that the components of the histogram are totally biased toward the low side of the grayscale. In general, the low contrast can be enhanced by the method of histogram equalization (see Eq. (2)), through which the components of the histogram tend to cover the entire range of the grayscale to achieve uniformly distributed pixels. As formulated in Eq. (2), for a given gray level of r_p, it can be transformed into r_q after the mapping of histogram equalization.

$$r_q = 255 \times \sum_{k=0}^{p} p(r_k) \tag{2}$$

where r_k is the kth gray level and $p(r_k)$ is the probability of occurrence of gray level r_k, for $k=0, 1,\ldots, p$.

As indicated previously, the key point to increase computation efficiency is to reduce the number of image blocks considered each time. This idea motivates us to partition the whole gray levels of an image into subgroups with a specific range of grayscale, which is similar to the concept of bin grouping used in image segmentation proposed by Huang et al. [10]. For instance, if an image is divided into 32 subgroups, we can assign the grayscale $\{f(x,y)|0 \le f \le 7\}$ to subgroup_0, grayscale $\{f(x,y)|8 \le f \le 15\}$ to subgroup_1,..., and $\{f(x,y)|248 \le f \le 255\}$ to subgroup_31. By this way, 32 binary subimage layers can be obtained if we use one new image to display the gray levels corresponding to one specific subgroup in the query image and mark them in white intensity value. In this manner, we call it as "grayscale layering." Therefore, the mapping of any gray level to a specific subimage layer can be given as.

$$I^k(x,y) = \begin{cases} 255 & if \ k = \lfloor f(x,y)/8 \rfloor \\ 0 & otherwise \end{cases} \tag{3}$$

where the symbol $\lfloor \cdot \rfloor$ denotes to take the floor (i.e., round towards minus infinity) of the value $f(x,y)/8$ to be an integer k, where $k=0, 1,\ldots, 31$, and $I^k(x,y)$ represents the kth subimage layer corresponding to grayscale subgroup_k.

2.2   Feature Extraction and Similarity Matching

In this work, SVD is used to extract the feature vectors of image blocks. The features of singular values are inherently rotation-invariant and mostly used in the applications of image forgery detection, data compression and noise reduction so on. The fundamental theory of SVD is described as follows.

Let A be an $m \times n$ image matrix, the decomposition of SVD can be expressed by the following form.

$$A = U \Sigma V^T = \sum_{i=0}^{Rank(A)} \sigma_i \mathbf{u_i} \mathbf{v_i^T} \tag{4}$$

where U is an $m \times m$ matrix whose columns are the vectors $\mathbf{u_i} \in \Re^m$, for $i=0,1,..,m-1$, $\Sigma = diag(\sigma_0, \sigma_1,\ldots,\sigma_{r-1},0,\ldots,0)$ is an $m \times n$ diagonal matrix whose entries σ_i, for $i=0, 1,\ldots, r-1$, are positive and ordered so that $\sigma_0 \ge \sigma_1 \ge \ldots \ge \sigma_{r-1} > 0$, each of which is called singular value, and V is an $n \times n$ matrix whose columns are $\mathbf{v_i} \in \Re^n$, for $i=0, 1,.., n-1$. The superscript T on the matrix V denotes the matrix transpose of V. The value r is the rank of the matrix A. Both U and V are orthonormal matrices. For the diagonal matrix Σ, it represents how much the vectors in V are stretched to give the vectors in U.

The similarity measure of Euclidean distance is exploited for blocks matching to find the duplicated regions of a forgery image. After extracting the features of image blocks by the SVD method, the entries of the diagonal matrix can be represented as an r-dimension feature vector. Let $\mathbf{s}^{(k)} = [\sigma_0^{(k)}, \sigma_1^{(k)}, ..., \sigma_{r-1}^{(k)}]^T$ and $\mathbf{s}^{(l)} = [\sigma_0^{(l)}, \sigma_1^{(l)}, ..., \sigma_{r-1}^{(l)}]^T$ be the feature vectors of the kth and lth image blocks, respectively, the Euclidean distance $d(\mathbf{s}^{(k)}, \mathbf{s}^{(l)})$ can be calculated as.

$$d(\mathbf{s}^{(k)}, \mathbf{s}^{(l)}) = \left\{ \sum_{i=0}^{r-1} \left(\sigma_i^{(k)} - \sigma_i^{(l)} \right)^2 \right\}^{1/2} \tag{5}$$

By this way, the similarity matching of image blocks is carried out to identify the duplicated regions.

2.3   The Proposed Algorithm

In this section, we propose a novel algorithm with a boosting scheme for the detection of copy-move forgery image. The boosting scheme comprises an estimation of dark channel, histogram equalization and grayscale layering. The details of the procedures for detecting tampered regions are described as follows.

1. Input a forgery image of size $m \times n$ pixels and use an image block of size $B \times B$ pixels.
2. Estimate the dark channel of the original image, and perform histogram equalization for the dark channel map to achieve a more uniform distribution of pixels.
3. Split the whole gray levels of the histogram equalization map into 32 subgroups, each of which has a specific range of gray levels that are marked as white intensity value in the subimage layer, where the white pixel has the same location as that of the histogram equalization map. By the similar manner, 32 separate subimage layers can be obtained.
4. For each subimage layer, apply SVD to every image block in the grayscale image with an upper left coordinate that corresponds to the location of white pixel in the subimage layer for feature extraction, and then converting the diagonal matrix into a column feature vector $\mathbf{s}^{(k)} = \left[\sigma_0^{(k)}, \sigma_1^{(k)}, \cdots, \sigma_{B-1}^{(k)} \right]^T$, for $k=0, 1,.., N_s-1$, where N_s is the total number of white pixels in the subimage layer. Therefore, we can construct a matrix $A = \left[\mathbf{s}^{(0)}, \mathbf{s}^{(1)}, \cdots, \mathbf{s}^{(Ns-1)} \right]$ using all the feature vectors of the subimage layer.
5. For each subimage layer, apply lexicographical order to the matrix A so that similar blocks are consecutively sorted.
6. For each subimage layer, copy-move blocks identification is performed. For two neighboring feature vectors $\mathbf{s}^{(k)} = [\sigma_0^{(k)}, \sigma_1^{(k)}, ..., \sigma_{B-1}^{(k)}]^T$ and $\mathbf{s}^{(k+1)} = [\sigma_0^{(k+1)}, \sigma_1^{(k+1)}, ..., \sigma_{B-1}^{(k+1)}]^T$, if $\left| \sigma_0^{(k+1)} - \sigma_0^{(k)} \right| < \varepsilon$, then the distance

$d = \sqrt{\sum_{i=1}^{\lfloor B/2 \rfloor} \left(\sigma_i^{k+1} - \sigma_i^k \right)^2}$ is calculated; otherwise, skip to the next record. Moreover,

if the distance $d < \alpha$, the two feature vectors $\mathbf{s}^{(k)}$ and $\mathbf{s}^{(k+1)}$ are considered to be similar and their upper left coordinates $\left(x^{(k)}, y^{(k)} \right)$ and $\left(x^{(k+1)}, y^{(k+1)} \right)$ are recorded, respectively, where the Euclidean distance of them is calculated as $l = \sqrt{\left(x^{(k+1)} - x^{(k)} \right)^2 + \left(y^{(k+1)} - y^{(k)} \right)^2}$, and if $l > \beta$, then these two blocks are identified as the copy-move ones, which will be marked as white pixels in a prepared black image.

7. Repeat step 4 to step 6 until all the image blocks in the 32 subimage layers have been identified.
8. Output the resulting visual map for the duplicated regions.

3   Experimental Results

The proposed method was implemented on Borland C++ 6.0. All experiments were performed on a personal computer of 2.33 GHz processor with 4GB memory. In these experiments, the setting parameters are listed as follows: Ω= 8×8 (dark channel), B=8 (block size), ε=0.01, α=0.001 and β=15. The query images are all downloaded from internet websites.

Table 1. Comparisons of time efficiency with and without the boosting scheme for different sizes of images (unit: Sec)

	512×512 pixels			1024×1024 pixels		
	with boosting	*without boosting*	*Ratio*	*with boosting*	*without boosting*	*Ratio*
Total time	13.06	114.05	8.73	131.38	1660.8	12.64

Two image sizes of 512×512 and 1024×1024 pixels were used to verify the time efficiency of the proposed boosting scheme. As shown in Table 1, comparisons of time efficiency are performed under the situations of using the boosting scheme or not. The time efficiency is improved by a factor of 8.73 for 512×512 pixels and 12.64 for 1024×1024 pixels, respectively, indicating the feasibility of the proposed boosting scheme.

To verify the robustness against rotation attack, one duplicated region with rotation angles of 90 degrees and 180 degrees, as shown in Fig. 4 and Fig. 5, respectively, were performed. Clearly, the tampered region can be successfully detected by the proposed method, indicating the effectiveness of the proposed method.

(a)   (b)   (c)

Fig. 4. Result of detecting duplicated region with a rotation angle of 90 degree; (a) original image, (b) forgery image of (a), and (c) result of duplicated region detection

(a)   (b)   (c)

Fig. 5. Result of detecting duplicated region with a rotation angle of 180 degree; (a) original image, (b) forgery image of (a), and (c) result of duplicated region detection

4   Conclusions

In this paper, we have presented a novel method to detect copy-move image forgery based on a boosting scheme and the SVD method. The boosting scheme comprises an estimation of dark channel, histogram equalization and grayscale layering. The time efficiency can be greatly improved by a factor of 8.73 for the image of 512×512 pixels and 12.64 for the image of 1024×1024 pixels, respectively, when the boosting scheme is applied. Due to the inherently rotation-invariant feature of the SVD method, large angle rotation of tampered region can be successfully detected, indicating the feasibility of the proposed method.

Acknowledgments. This research was supported by a grant from National Science Council, Taiwan, under the contracts of NSC-101-2221-E-212-020 and NSC-101-2221-E-346-011.

References

1. Wang, Y.W., Lin, C.N.: A Line-Based Skid Mark Segmentation System Using Image-Processing Methods. Transportation Research Part C 16, 390–409 (2008)
2. Thali, M.J., Braun, M., Brüschweiler, W., Dirnhofer, R.: Matching Tire Tracks on the Head Using Forensic Photogrammetry. Forensic Science International 113, 281–287 (2000)

3. Buck, U., Albertini, N., Naether, S., Thali, M.J.: 3D Documentation of Footwear Impressions and Tyre Tracks in Snow with High Resolution Optical Surface Scanning. Forensic Science International 171, 157–164 (2007)
4. Colbry, D., Cherba, D., Luchini, J.: Pattern Recognition for Classification and Matching of Car Tires. Tire Science and Technology 33, 2–17 (2005)
5. Moreno, P., Bernardino, A., Victor, J.S.: Gabor Parameter Selection for Local Feature Detection. In: 2nd Iberian Conference on Pattern Recognition and Image Analysis (IBPRIA), Estoril, Portugal, pp. 11–19 (2005)
6. Jung, S.W., Bae, S.W., Park, G.T.: A Design Scheme for a Hierarchical Fuzzy Pattern Matching Classifier and Its Application to the Tire Tread. Fuzzy Sets and Systems 65, 311–322 (1994)
7. Vapnik, V.N.: Statistical learning theory. John Wiley and Sons, New York (1998)

Robust Watermarking Scheme for Colour Images Using Radius-Weighted Mean Based on Integer Wavelet Transform

Ching-Yu Yang

Dept. of Computer Science and Information Engineering,
National Penghu University of Science and Technology, Penghu, Taiwan
chingyu@npu.edu.tw

Abstract. Based on integer wavelet transform (IWT), the author uses the radius-weighted mean (RWM) and proposes a robust color image watermarking scheme. More specifically, by employing the Euclidean distance of the RWM and mean, a watermark (or copyright logo) is effectively embedded into the low-low subband of the IWT domain. Experiments confirm that the marked images generated by the proposed method are robust against manipulations such as JPEG, JPEG2000, color quantization, noise additions, cropping, (edge) sharpening, and so on. In addition, the resultant perceived quality is good and the payload is not bad while our peak signal-to-noise ratio (PSNR) is better than existing techniques.

Keywords: Robust digital watermarking, radius-weighted mean, IWT domain.

1 Introduction

Due to the ubiquitous broadband services and fast information retrieve services, it is convenient for people to share their resources and surf on the internet. However, data can be tampered with and eavesdropped during transmission. In addition to encryption/decryption systems, data hiding techniques play an alternative option to protect intellectual property right, content authentication, copy control, and copyright protection [1-3]. In general, color images are more attractive than gray-level ones for human being. Namely, color images are popular for people and commonly circulated around the world. Therefore, several researchers have presented data hiding techniques for color images [4-6]. Based on the particle swarm optimization and k-nearest neighbor algorithm, Findik et al. [5] suggest a novel digital watermarking for color images. Simulations indicated that the resultant images are robust several image process operations. Niu et al. [6] presented a color image watermarking scheme based on support vector regression and nonsubsampled contourlet transform. One major merits of the scheme is the robustness of against geometric distortions. However, the aforementioned papers provide a limited payload size. This shortcoming may prohibit their functionality. For example, it is not feasible for the above two schemes if a watermark of size is larger than 4,096 bits.

J.-S. Pan et al. (eds.), *Genetic and Evolutionary Computing,*
Advances in Intelligent Systems and Computing 238,
DOI: 10.1007/978-3-319-01796-9_14, © Springer International Publishing Switzerland 2014

In this paper, the author proposes a robust digital watermarking for color images by applying the Euclidean distance of RWM and mean to the low-low (LL) subband of IWT domain. The rest of the paper is organized as follows. First, the radius-weighted mean (RWM) is briefly specified in Section 2. The process of bit embedding and bit extraction for the proposed method is described in Section 3. Experimental results are demonstrated in Section 4. Finally, a conclusion is summarized in Section 5.

2   Review of RWM

The RWM is a special kind of point originally introduced to register shapes [7-8]. Thereafter, the RWM has been used to generate economic block truncation coding (EBTC) for real-time compression [9]. Subsequently, the RWM has been applied for colour images quantization and data hiding [10-11]. Let $S = \{(r_i, g_i, b_i) \mid i = 1, 2, ..., MN\}$ be a RGB colour system. The RWM $R = (r', g', b')$ and mean $O = (\bar{r}, \bar{g}, \bar{b})$ of are defined by:

$$r' = \left[\sum_{i=1}^{MN} \mu_i r_i \right] \bigg/ \left[\sum_{i=1}^{MN} \mu_i \right], \tag{1}$$

$$g' = \left[\sum_{i=1}^{MN} \mu_i g_i \right] \bigg/ \left[\sum_{i=1}^{MN} \mu_i \right], \tag{2}$$

$$b' = \left[\sum_{i=1}^{MN} \mu_i b_i \right] \bigg/ \left[\sum_{i=1}^{MN} \mu_i \right], \tag{3}$$

where

$$\mu_i = \sum_{i=1}^{MN} \sqrt{\left[(r_i - \bar{r})^2 + (g_i - \bar{g})^2 + (b_i - \bar{b})^2 \right]}, \tag{4}$$

$$\bar{r} = \left[\sum_{i=1}^{MN} r_i \right] \bigg/ MN, \tag{5}$$

$$\bar{g} = \left[\sum_{i=1}^{MN} g_i \right] \bigg/ MN, \tag{6}$$

and

$$\bar{b} = \left[\sum_{i=1}^{MN} b_i\right]\Big/ MN.$$ (7)

3   The Proposed Method

An input image is first decomposed to the IWT domain by the following two formulas:

$$d_{1,k} = s_{0,2k+1} - s_{0,2k}$$ (8)

and

$$s_{1,k} = s_{0,2k} + \left\lfloor \frac{d_{1,k}}{2} \right\rfloor,$$ (9)

where $d_{j,k}$ and $s_{j,k}$ are the kth high-frequency and low-frequency wavelet coefficients at the jth level, respectively [12]. Then, data bits are embedded into the host blocks which derived from the LL-subband of the IWT domain. The $\lfloor x \rfloor$ is a floor function.

Two rules, namely, primary-rule and secondary-rule, are used to determine whether a host block can hide a data bit. Specifically, if a host block satisfies either of the two rules, then a data bit is embedded into the block, otherwise, the block would be skipped. Note that the skipped blocks contain no data bit. The specifications of the primary-/secondary-rule are described in the following subsections.

3.1   The Primary-Rule

Without loss of generality, let $P_j = \left\{(r_{kj}, g_{kj}, b_{kj})\right\}_{k=0}^{n^2-1}$ be the jth block of size $n \times n$ taken from the LL-subband of IWT domain. Also let $\Omega_1 = \left\{ p_{lj} \mid\mid\mid p_{lj} O \mid\mid < \tau \mid\mid OR \mid\mid, p_{lj} \in P_j \right\}$ and $\Omega_2 = \left\{ p_{mj} \mid\mid\mid p_{mj} O \mid\mid \geq \tau \mid\mid OR \mid\mid, p_{mj} \in P_j \right\}$ be the two subsets of P_j with $P_j = \Omega_1 \cup \Omega_2$, where τ is a control parameter. The $\mid\mid P_j O \mid\mid$ represents the Euclidean distance of P_j and O, and $\lVert OR \rVert$ represents the Euclidean distance of O and R. If an input bit is 1 and $\mid \Omega_1 \mid > \mid \Omega_2 \mid$, then do nothing, which meaning the block "carries" data bit 1; otherwise, we repeatedly decrease p_{mj} by the λ value each time until either $\mid \Omega_1 \mid > \mid \Omega_2 \mid$ or times η is encountered. Both η and λ are two integers. Conversely, if an input bit is 0 and $\mid \Omega_1 \mid \leq \mid \Omega_2 \mid$, then do nothing, which meaning the block "carries" bit 0; otherwise, we repeatedly increase p_{lj} by the λ

value each time until either $|\Omega_1| \leq |\Omega_2|$ or η times is encountered. Notice that if a block fails to satisfy the primary-rule, then the secondary-rule is subsequently applied to this block.

3.2   The Secondary-Rule

Let $P_j = \hat{C} \cup \tilde{C}$ be the host block with $\hat{C} = \{(r_{ij}, g_{ij}, b_{ij}) \mid i = 0,2,4,6,8\}$ and $\tilde{C} = \{(r_{tj}, g_{tj}, b_{tj}) \mid t = 1,3,5,7\}$, as shown in Fig. 1 when $n=3$. Also let $\| OR \|_x$ and $\| OR \|_d$ be the Euclidean distances of O and R that computed from two subsets $\hat{C}$ and $\tilde{C}$, respectively. If an input bit is 1 and $\| OR \|_x > \| OR \|_d$, then do nothing, which meaning the block "carries" data bit 1; otherwise, we repeatedly increase $\hat{C}$ by the λ value each time until either $\| OR \|_x > \| OR \|_d$ or times η is encountered. Conversely, if an input bit is 0 and $\| OR \|_x < \| OR \|_d$, then do nothing, which meaning the block "carries" bit 0; otherwise, we repeatedly increase $\tilde{C}$ by the λ value each time until either $\| OR \|_x < \| OR \|_d$ or η times is encountered. Finally, if a block fails to satisfy the secondary-rule, then the block is marked as a skipped block.

$(r_{0j}g_{0j}b_{0j})$	$(r_{1j}g_{1j}b_{1j})$	$(r_{2j}g_{2j}b_{2j})$
$(r_{3j}g_{3j}b_{3j})$	$(r_{4j}g_{4j}b_{4j})$	$(r_{5j}g_{5j}b_{5j})$
$(r_{6j}g_{6j}b_{6j})$	$(r_{7j}g_{7j}b_{7j})$	$(r_{8j}g_{8j}b_{8j})$

Fig. 1. A 3×3 block taken from the LL-subband of IWT

3.3   Bit Embedding

The major steps of bit embedding are specified as follows:

Step 1. Input a host block H from the LL-subband of IWT domain. If the end of input is encountered, then proceed to Step 5.

Step 2. If H satisfies the primary-rule, then Set 0 to the corresponding position of the block map B, and repeat from Step 1.

Step 3. If H satisfies the secondary-rule, then Set 1 to the corresponding position of the block map B, and repeat from Step 1.

Step 4. Set mark 2 to the corresponding position of the block map B, and return to Step 1.

Step 5. Stop.

To further promote security and help the decoder to extract the hidden watermark, the block map B, the RWM, the mean, λ, and η are sent to the receiver by out-of-bound transmission.

3.4   Bit Extraction

The process of bit extraction is much simper than that of bit embedding. Let $Q_j = \left\{ (r_{kj}, g_{kj}, b_{kj}) \right\}_{k=0}^{n^2-1}$ be the jth hidden block taken from the LL-subband of IWT domain,   which   derived   from   a   marked   image.   Also   let $\Omega_1' = \left\{ q_{lj} \parallel\parallel q_{lj}O \parallel< \tau \parallel OR \parallel, q_{lj} \in Q_j \right\}$ and $\Omega_2' = \left\{ q_{mj} \parallel\parallel q_{mj}O \parallel\geq \tau \parallel OR \parallel, q_{mj} \in Q_j \right\}$ be the two subsets of Q_j with $Q_j = \Omega_1' \cup \Omega_2'$. In addition, let $Q_j = \hat{C}_j \cup \tilde{C}_j$ be the hidden block with $\hat{C}_j = \{ (r_{ij}, g_{ij}, b_{ij}) \,|\, i = 0,2,4,6,8 \}$ and $\tilde{C}_j = \{ (r_{tj}, g_{tj}, b_{tj}) \,|\, t = 1,3,5,7 \}$, and $\parallel OR \parallel_x$ and $\parallel OR \parallel_d$ be the Euclidean distances that computed from two subsets $\hat{C}_j$ and $\tilde{C}_j$, respectively. The major steps of bit extraction are specified as follows:

Step   0.   Read   in   RWM,   mean, λ, η,   and   the   block   map
    $B = \{ b_{ij} \,|\, i < M / 2n, j < N / 2n \}$..

Step 1. Input a hidden block, say the jth hidden block, from the LL-subband of IWT domain. If the end of input is encountered, then proceed to Step 5.

Step 2. If both conditions of $|\Omega_1'| > |\Omega_2'|$ and the corresponding mark $b_{ij}=0$ are satisfied, then data bit 1 is extracted; otherwise, data bit 0 is extracted, and return to Step 1.

Step 3. If both conditions of $\parallel OR \parallel_x > \parallel OR \parallel_d$ and $b_{ij}=1$ are satisfied, then data bit 1 is extracted; otherwise, data bit 0 is extracted, and return to Step 1.

Step 4. If $b_{ij}=2$, then return to Step 1.

Step 5. Stop.

4   Experimental Results

Several 512×512 color images were used as host images. Each RGB pixel of the host images is represented by 24 bits, 8 bits per component. A binary watermark of size 70×70 was used as test data. The marked images generated by the proposed method are depicted in Fig. 2. From the figure we can see that the perceived quality is good. Apparently, no false colors exist in the figure. Their PSNR and payload are 49.89 dB/4,787 for Lean, 46.03/4,900 for Jet, 41.72/4,900 for Baboon, 48.01/4,513 for Peppers, 38.76/4,900 for Splash, and 47.25/4,263 for Couple, respectively. Figure 3 shows the PSNR and payload generated by the proposed method using various τ in host images. It can be seen that the PSNR and payload for images: Peppers and Couple are slightly inferior to those for other four images.

To demonstrate the robustness of the proposed method, the marked images are tested by a variety of attacks. The extracted watermarks and their bit correct ratio (BCR) are given in Table 1. The tested marked image is generated by the proposed method using $\tau=5$, $\eta=11$, and $\lambda=1$, respectively, on image Lena. The PSNR is defined by

$$PSNR = 10\times\log_{10}\frac{255^2}{MSE} \tag{10}$$

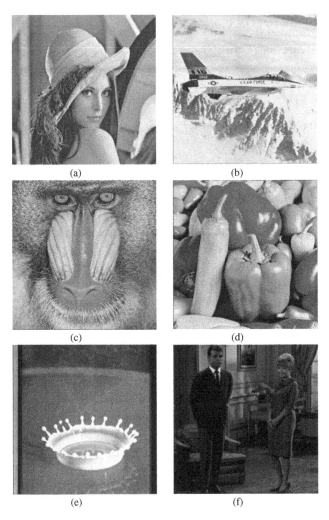

(a)   (b)

(c)   (d)

(e)   (f)

Fig. 2. The marked images generated by the proposed method using various τ in host images. (a) Lena ($\tau=5$), (b) Jet ($\tau=4$), (c) Baboon ($\tau=25$), (d) Peppers ($\tau=12$), (e) Splash ($\tau=9$), and (f) Couple ($\tau=2$).

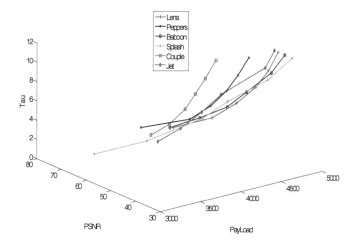

Fig. 3. The PSNR and payload generated by the proposed method using various τ in host images

with

$$MSE = \frac{1}{3MN} \sum_{i=1}^{MN} \left[(r_i - \hat{r}_i)^2 + (g_i - \hat{g}_i)^2 + (b_i - \hat{b}_i)^2 \right]$$ Here (r_i, g_i, b_i) and $(\hat{r}_i, \hat{g}_i, \hat{b}_i)$ denote

the RGB pixel values of the host image and the marked image. In addition, the BCR is defined by

$$BCR = \left(\frac{\sum_{i=0}^{ab-1} w_i \oplus \tilde{w}_i}{a \times b} \right) \times 100\%,$$ (11)

where w_i and $\tilde{w}_i$ represent the values of the original watermark and the extracted watermark respectively, as well as the size of a watermark is $a \times b$. The BCR for an extracted watermark is 100% if a marked image is not manipulated. Table 1 shows that most extracted watermarks are recognized. Notice that the extracted watermark is still recognizable when the marked image was compressed by JPEG2000 with a compression ratio (CR) of 110. Similar performance can be found in the case of JPEG compression with quality factor (QF) of 10, which is an approximated CR of 50. Figure 4 illustrates the BCR values of the extracted watermarks under attacks: JPEG2000 and JPEG with various CR, as well as colour quantized operations. From the figure we can find that the larger the CR imposed to the marked images, the less the BCRs are obtained. Conversely, the marked images quantized by less number of colours, the less the BCRs are acquired by the proposed method. Further, Table 1 confirms that BCR with value above 80%, the extracted watermarks are recognized.

Table 1. The survived watermarks extracted from the marked images which had been undergone various manipulations.

Attacks	Survived watermarks	Attacks	Survived watermarks
Null Attack BCR = 100%	DH&NL CSIE NPUST	Gaussian noise (4%) BCR =85.57%	DH&NL CSIE NPUST
JPEG2000 (CR=110) BCR=81.55%	DH&NL CSIE	Winding BCR =81.49 %	DH&NL CSIE
JPEG (QF=10) BCR=83.00%	DH&NL CSIE	Blurring BCR=90.14%	DH&NL CSIE NPUST
Color quantization (8-color), BCR=82.16%	DH&NL CSIE NPUST	Sharpening BCR=91.31%	DH&NL CSIE NPUST
Diagonal-cutting (50%) BCR=63.82%	DH&NL CSIE NPUST	Edge Sharpening BCR=97.92%	DH&NL CSIE NPUST
Anti-diagonal-cutting (50%), BCR =65.39%	DH&NL CSIE NPUST	Equalized BCR=72.06%	DH&NL CSIE NPUST
Uniform noise (4%) BCR =87.12%	DH&NL CSIE NPUST		

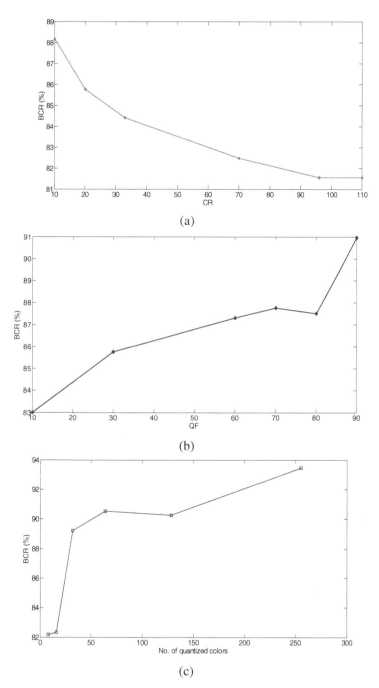

Fig. 4. The BCRs extracted from the marked images under various attacks. (a) JPEG2000, (b) JPEG, and (c) colour quantization.

Performance comparison between the proposed method and existing schemes: Findik *et al.*'s scheme [5] and Niu *et al.*'s technique [6] is given in Table 2. It can be seen that our method provides the largest payload (in bit) among these compared methods while the resultant PSNR (dB) is the best. Further, the robustness of resisting from JPEG and JPEG2000 attacks for the proposed method is superior to that for other two schemes.

Table 2. PSNR and payload comparison between our method and existing schems

Images	Methods		
	Findik et al. [5]	Niu et al. [6]	Our method
Lena	41.83/4,096	40.57/1,024	44.48/4,900
Jet	39.63/4,096	-	46.03/4,900
Baboon	42.76/4,096	41.67/1,024	44.44/4,815

5   Conclusions

In this paper, the author proposes a robust watermarking scheme for color images based on integer wavelet transform. By using the idea of the Euclidean distance of the mean and RWM, the secret bits are effectively embedded into the LL-subband of IWT domain. Experimental results indicate that the marked images generated by the proposed method are tolerant of manipulations such as JPEG, JPEG2000, color quantization, noise additions, cropping, (edge) sharpening, winding, burring, and equalized. In addition, the resultant perceived quality is good while both the PSNR and payload size are better than existing schemes. To further promote hiding capability, we are trying to employ different approach to embed data bits into the LH-/HL-subband of IWT domain. The work will be our future study.

References

1. Cox, I.J., Miller, M.L., Bloom, J.A., Fridrich, J., Kalker, T.: Digital Watermarking and Steganography, 2nd edn. Morgan Kaufmann, MA (2008)
2. Wang, S., Yang, B., Niu, X.: A secure steganography method based on genetic algorithm. Journal of Info. Hiding and Multim. Sig. Proces. 1(1), 28–35 (2010)
3. Noriega, R.M., Nakano, M., Kurkoski, B., Yamaguchi, K.: High payload audio watermarking: toward channel characterization of MP3 compression. Journal of Info. Hiding and Multim. Sig. Proces. 2(2), 91–107 (2011)
4. Li, L., Luo, B.: A color image steganography method by multiple embedding strategy based on sobel operator. In: 2009 Int. Conf. on Multimedia Information Network and Security, Wuhan, China, November 18-20, pp. 118–121 (2009)

5. Findik, O., Babaoglu, I., Ulker, E.: A color image watermarking scheme based on hybrid classification method: particle swarm optimization and k-nearest neighbor algorithm. Optics Commun. 283, 4916–4922 (2010)
6. Niu, P.P., Wang, X.Y., Yang, Y., Lu, M.Y.: A novel color image watermarking scheme in nonsampled contourlet-domain. Expert System with Applications 38, 2081–2098 (2011)
7. Mitiche, A., Aggarwal, J.K.: Contour registration by shape-specific point for shape matching. Comp. Vision, Graphics Image Process. 22, 396–408 (1983)
8. Lin, J.C., Chou, S.L., Tsai, W.H.: Detection of rotationally symmetric shape orientations by fold-invariant shape-specific points. Pattern Recog. 25, 473–482 (1992)
9. Yang, C.Y., Lin, J.C.: EBTC: An economical method for searching the threshold of BTC compression. Electron. Lett. 32, 1870–1871 (1996)
10. Yang, C.Y., Lin, J.C.: RWM-cut for color image quantization. Computer & Graphics 20, 577–588 (1996)
11. Yang, C.Y.: Use of radius weighted mean to hide data in colour images. In: The 5th IET Int. Conf. on Ubi-Media Computing, Xining, China, August 16-18 (2012)
12. Calderbank, A.R., Daubechies, I., Sweldens, W., Yeo, B.L.: Wavelet transforms that map integers to integers. Applied & Computational Harmonics Analysis 5(3), 332–369 (1998)

A Novel Data Hiding Method Using Sphere Encoding

Ching-Min Hu[1], Ran-Zan Wang[1,*], Shang-Kuan Chen[2], Wen-Pinn Fang[2],
Yu-Jie Chang[3], and Yeuan-Kuen Lee[4]

[1] Department of Computer Science & Engineering, Yuan Ze University, Taiwan
[2] Department of Computer Science & Information Engineering, Yuan Pei University, Taiwan
[3] Department of Computer & Communication Engineering,
National Kaohsiung First University of Science and Technology, Taiwan
[4] Department of Computer Science & Information Science, Ming Chuan University, Taiwan
rzwang@saturn.yzu.edu.tw

Abstract. In this paper a novel data hiding scheme using least square mapping is proposed. The method conceals a k-ary secret digit in n-pixel block in which the embedding function with minimum mean square error (MSE) between the cover block and the stego-block is searched. The constructions for the proposed method in three-pixel block (sphere encoding) under different k are demonstrated. Experiment results show that the proposed scheme achieves higher embedding efficiency than several reported steganographic schemes.

Keywords: Data Hiding, Steganography, Information Hiding, Sphere Encoding.

1 Introduction

Image steganography [1, 2] is the technique about concealing secret data in images. In a typical image steganographic scheme, the embedding function is designed for hiding secret data in a cover image to form the stego-image. The quality of the stego-image must be sufficiently high so as to make it visually indistinguishable from the cover image, thereby maintaining the secrecy to the embedded secret data. Given a high-quality stego-image, an unintended observer will not be aware of the very existence of the hidden secret data; however, the authorized recipient can extract the hidden data from the stego-image applying the corresponding data extraction function.

In the past decade many steganographic methods were proposed to hide moderate-size of secret data under acceptable degree of distortion to the cover image. The least significant bit (LSB) substitution methods [3, 4] replace the LSB planes of an image with the secret data, and conduct a pixel adjustment procedure to obtain higher quality of stego-images. The method is simple and acceptable quality of the generated stego-images can be obtained. To obtain higher embedding efficiency, many methods taking multiple pixels as an embedding unit to hide the secret data were explored. Mielikainen [5] proposed an improved LSB matching steganographic method. The

---

* Corresponding author.

method applies a pair of pixels as a unit to carry two secret bits, one in the LSB of the first pixel and the other by a function of the two pixel values. The mean square error (MSE) in this scheme is 0.375 when the payload is 1 bpp. It is better than LSB replacement method in which the MSE is 0.5 under 1.0 bpp payload. In 2006 Zhang and Wang [6] proposed a data embedding method by exploring modification direction (EMD). The method tries to reduce the amount of alterations to the cover image introduced in the embedding process. It embeds a $(2n+1)$-ary number in n cover pixels in which at most one pixel is increased or decreased by 1. Compare with LSB replacement hiding technique, the above two methods [5, 6] can obtain better quality of stego-images; however, the maximum payloads in the two methods are restrained. To embed more data in an image, Chao et al. [7] proposed a hiding method using diamond encoding (DE), it embeds a $(2k^2+2k+1)$-ary digit in a pair of cover pixels, which makes the application of hiding a large amount of data in an moderately-size image possible. In 2012 Hong and Chen [8] improved the DE hiding method and proposed the adaptive pixel pair matching (APPM) data hiding scheme, in which a k-ary $(k \geq 1)$ number is embedded in each pair of pixels of the cover image. APPM conceptualizes a pixel pair as a reference coordinate in 2D space, and searches for a stego-coordinate in the neighborhood set connoting the given message digit. The method provides more flexibility in the choice about the base of the hidden message, and achieves higher quality of stego-images compare with the DE embedding method. Both DE technique and AAPM method search for the best embedding solution in 2D space, the constraint for the searching space limits the performance of these schemes.

   In this paper a high-payload image steganography technique is proposed, it processes n-pixel block as an n-D vector, and a k-ary secret digit is embedded in each block. The embedding function is defined as the inner product of the pixel vector with a coefficient vector modulo k, and a procedure is designed to find the best embedding function that minimizes the MSE between the cover image and the stego-image. The remainder of this paper is organized as follows: The details of the proposed scheme are presented in Section 2. Experiment results are shown in Section 3, and a brief conclusion is made finally in Section 4.

2   The Proposed Method

Consider the instance about embedding a secret digit in the k-ary notational system in an image block with n pixels, with each pixel is t-bit. The block is represented in a n-D vector $\mathbf{B} = [p_1, p_2, ..., p_n]$, where $1 \leq p_j \leq 2^t-1$ for $j = 1, 2, ..., n$. Given a coefficient vector $\mathbf{C} = [c_1, c_2, ..., c_n]$, $1 \leq c_j \leq k-1$ for $j = 1, 2, ..., n$, the embedding function for hiding a k-ary secret digit in n-pixel block with coefficient vector $\mathbf{C}$ is defined below:

$$f_{k,n}^{\mathbf{C}}(\mathbf{B}) = (\mathbf{C} \cdot \mathbf{B}) \bmod k = \left[\sum_{i=1}^{n}(c_i \times p_i)\right] \bmod k, \quad i = 1, 2, ..., n. \qquad (1)$$

The embedding function evaluates the characteristic value for block $\mathbf{B}$, which is an integer in the range from 0 to $k-1$. The characteristic value $f_{k,n}^{\mathbf{C}}(\mathbf{B})$ represents for the message carried in block **B**. According to the value of the secret digit m to be hidden and the characteristic value of the cover block **B**, one of the following two embedding rules is conducted to embed m in **B**:

(a) If $f_{k,n}^{\mathbf{C}}(\mathbf{B}) = m$, do nothing.

(b) If $f_{k,n}^{\mathbf{C}}(\mathbf{B}) \neq m$, an image block $\mathbf{D} = [q_1, q_2, \ldots, q_n]$ with characteristic value m, i.e. $f_{k,n}^{\mathbf{C}}(\mathbf{D}) = m$, is selected as the stego-block to replace **B**.

In Eq. (1) the embedding function is defined according to certain coefficient vector **C**, the definitions for valid embedding function and valid coefficient vector are given below:

Definition 1. Let $f_{k,n}^{\mathbf{C}}$ be the embedding function for hiding a k-ary digit in n-pixel block based on coefficient vector **C**. $f_{k,n}^{\mathbf{C}}$ is defined as a valid embedding function if for each integer v in the range from 0 to $k-1$, there exists an image block $\mathbf{B}_v$ whose characteristic value is equal to v, i.e. $f_{k,n}^{\mathbf{C}}(\mathbf{B}_v) = v$.

Definition 2. If $f_{k,n}^{\mathbf{C}}$ is a valid embedding function, then **C** is a valid coefficient for embedding a k-ary digit in n-pixel block.

Definition 1 indicates that the coefficient vector **C** should carefully be determined to ensure that each message value from 0 to $k-1$ can be embedded in an image block, and promises the correct extraction to the embedded secrets.

Property 1. Given the embedding function in Eq. (1), there is at least one coefficient vector $\mathbf{C} = \{c_i\}$, $i = 1, 2, \ldots, n$, which forms a valid embedding function $f_{k,n}^{\mathbf{C}}$.

Proof: The property can be examined as follows. Let $\mathbf{1} = [1, 1, \ldots, 1]$ denote the n-D vector with all elements are set 1, it can be seen that the characteristic value for the embedding function $f_{k,n}^{1}$ is the sum of the n pixel values of the input block modulo k. Let $\mathbf{B}_0 = [0, 0, \ldots, 0]$ be the image block with all n pixels are 0, and $\mathbf{B}_i$ be the new block derived from $\mathbf{B}_{i-1}$ by adding 1 to one of the n pixels in $\mathbf{B}_{i-1}$. It can easily be verified that $f_{k,n}^{1}(\mathbf{B}_i) = i$, for $i = 0, 1, \ldots, k-1$, which proves the correctness to this property.

Property 1 shows that there is at least one way to establish the embedding function for embedding a k-ary secret digit in an n-pixel block in the proposed scheme. Given certain message base k, all of the valid coefficient vectors can be identified through an exhaustive search, and the best coefficient vector $\hat{\mathbf{C}}$ is selected to construct the best embedding function. Before introduce the procedure about finding the best coefficient vector, more characteristics about a valid embedding function are examined below.

Remind from the above discussion that to embed secret digit m in block $\mathbf{B} = [p_1, p_2, ..., p_n]$, if $f_{k,n}^{C}(\mathbf{B}) \neq m$ then $\mathbf{B}$ is replaced by a block $\mathbf{D} = [q_1, q_2, ..., q_n]$ with $f_{k,n}^{C}(\mathbf{D}) = m$. The MSE between the two blocks is evaluated by

$$MSE(\mathbf{B}, \mathbf{D}) = \sum_{i=1}^{n}(q_i - p_i)^2. \tag{2}$$

To obtain better quality of stego-image, if there are multiple blocks $\{\mathbf{D}_1, \mathbf{D}_2, ..., \mathbf{D}_w\}$ having characteristic value m, the block $\hat{\mathbf{D}}$ with minimum MSE between it and $\mathbf{B}$ is selected to be the stego-block:

$$\hat{\mathbf{D}} = \arg\min_{\mathbf{D}_i}\{MSE(\mathbf{B}, \mathbf{D}_i)\}. \tag{3}$$

It can be verified that he embedding error for embedding secret message m in block $\mathbf{B}$ applying embedding function $f_{k,n}^{C}(\mathbf{B})$ is determined by $(m - f_{k,n}^{C}(\mathbf{B}))$ mod k. That is, if $(m_1 - f_{k,n}^{C}(\mathbf{B}_1))$ mod k is equal to $(m_2 - f_{k,n}^{C}(\mathbf{B}_2))$ mod k, then the embedding error for embedding m_1 in $\mathbf{B}_1$ is equal to the embedding error for embedding m_2 in $\mathbf{B}_2$. Let $\hat{\mathbf{D}}(i)$ denote the embedding error for embedding m in $\mathbf{B}$ when $(m - f_{k,n}^{C}(\mathbf{B})) \bmod k = i$, the error for the proposed embedding function is defined below:

Definition 3. The embedding error for the embedding function $f_{k,n}^{C}$ is evaluated by the following equation:

$$E(f_{k,n}^{C}) = \sum_{i=0}^{k-1}\hat{\mathbf{D}}(i). \tag{4}$$

The objective function of the proposed scheme is to find the best coefficient vector $\hat{\mathbf{C}}$ to minimize the embedding error:

$$\hat{\mathbf{C}} = \arg\min_{\mathbf{C}}\{E(f_{k,n}^{C})\}. \tag{5}$$

Table 1 lists the best coefficient vector $\hat{\mathbf{C}}$ for small values of k and n is set 3 obtained by exhaustive search, it is entitled as the sphere encoding because the searching space for the best solution is in 3D space.

Based on the above discussion, the process about embedding t-bit secret message M in cover image I with size $h \times w$ is summarized the following steps:

Step 1. Select the parameter n, i.e. the number of pixels in a block.
Step 2. Convert M in a list of k-ary digits $M = \{m_1, m_2, ...\}$, where k in the minimum integer satisfies the requirement:

$$Ceiling(h \times w / n) \geq k. \tag{6}$$

Step 3: Find the best coefficient vector $\hat{\mathbf{C}}$ to construct the embedding function $f_{k,n}^{\hat{\mathbf{C}}}$.

Step 4: Take a secret digit m_i from the message list, and take n not-processed-yet pixels from the cover image I in scan order to from the cover block $\mathbf{B}$. Embed m_i in $\mathbf{B}$ aforementioned with the embedding function $f_{k,n}^{\hat{\mathbf{C}}}$.

Step 5. Repeat Step 4 until all secret digits are embedded. The stego-image SI is obtained.

The secret can be revealed from the stego-image SI using the following steps:

Step 1. Take n not-processed-yet pixels from the SI in scan order to from the stego-block $\mathbf{D}_i$ and evaluate the $m_i = f_{k,n}^{\hat{\mathbf{C}}}(\mathbf{D}_i)$, m_i is exactly the i-th secret digit in the secret message.

Step 2. Repeat Step 1 until all message digits are extracted. The message M can be obtained by converting the extracted secret digits into a binary bit stream.

3   Experimental Results

The experiments for hiding secret message in three-pixel block (n=3) under various message bases k are conducted to evaluate the performance to the proposed scheme. The six test images used in our experiments are 8-bit grayscale images with 512×512 pixels as shown in Fig. 1, and the messages embedded are random sequences generated by a pseudo random number generator (PRNG). In experiment I, 400,000 bits of secret message (about 1.5 bpp payload) are hidden in the cover images, the embedding function applied in this test is $f_{16,3}^{[1,2,6]}$. Table 2 summarizes the PSNR between the stego-image and the cover image of the proposed scheme and four hiding schemes include (1) the simple LSB replacement method, (2) the optimal pixel adjustment process (OPAP) method, (3) the diamond encoding (DE) method, and (4) the adaptive pixel pair matching (APPM) method, where 2 LSB replacement are applied in LSB replacement and OPAP methods, the parameter k is set 2 in DE embedding method, and the parameters k=9 and C_9=3 are set in APPM method. It can be seen that the PSNRs obtained in the proposed scheme are higher than those in the four methods, demonstrating the feasibility of the proposed method. Experiment II hides 650,000 bits of secret message (about 2.5 bpp payload) in the cover images, the embedding function applied in the test is $f_{41,3}^{[1,5,13]}$. Experiment III hides 1,000,000 bits of secret message (about 4.0 bpp payload) in the cover images, the embedding function applied in the test is $f_{256,3}^{[1,7,46]}$. The comparisons of these two experiments with the four hiding schemes are summarized in Tables 3 and 4, in that the proposed scheme also exhibits better quality of stego-images than the four hiding schemes. Figure 1 show the stego-images of hiding message in cover image 'Lena'. The PSNRs between the stego-images and the cover image are 51.14 dB, 48.31 dB, and 43.12 dB when the payloads are 400,000bit, 650,000 bits, and 1,000,000 bits, respectively. Figure 2 shows a similar experiment in which the cover image is "Jet". It can also be seen visually that the stego-images obtained in these experiments are with high quality.

Fig. 1. Six test images

Fig. 2. Stego-images of 'Lena'. (a) Payload=400,000 bits, PSNR = 51.14 dB. (b) Payload=650,000 bits, PSNR = 48.31 dB. (c) Payload=1,000,000 bits, PSNR = 43.12 dB.

Fig. 3. Stego-images of 'Jet'. (a) Payload=400,000 bits, PSNR = 51.15 dB. (b) Payload=650,000 bits, PSNR = 48.31 dB. (c)Payload=1,000,000 bits, PSNR = 43.13 dB.

Table 1. List of the best coefficient vector $\hat{\mathbf{C}}$ for small k when n is set 3

k	1,2,3	4,5	6-11	12,13	14,15	16-19	20	21-25	26-29	30-31
$\hat{\mathbf{C}}$	[1,1,1]	[1,1,2]	[1,2,3]	[1,2,4]	[1,2,5]	[1,2,6]	[1,2,7]	[1,3,8]	[1,3,9]	[1,3,11]

k	32	33	34	35	36	37	38	39	40	41
$\hat{\mathbf{C}}$	[1,4,10]	[1,6,15]	[1,2,4]	[1,11,16]	[1,6,9]	[1,3,14]	[1,6,9]	[1,12,18]	[1,4,14]	[1,5,13]

Table 2. Comparison of PSNR between stego-image and cover image for 400,000 payload in five hiding schemes. (unit: dB)

Image	2-bit LSB	2-bit OPAP	DE(k=2)	APPM(C_9=3)	Proposed method k=16, C=[1,2,6]
Lena	45.32	47.55	48.02	49.89	51.14
Jet	45.36	47.56	48.02	49.89	51.14
Boat	45.21	47.56	48.01	49.91	51.14
Elaine	45.31	47.56	48.01	49.89	51.14
House	45.31	47.55	48.02	49.90	51.14
Sailboat	45.32	47.55	48.00	49.90	51.14
Mean	45.30	47.55	48.01	49.89	51.14

Table 3. Comparison of PSNR between stego-image and cover image for 650,000 payload in five hiding schemes. (unit: dB)

Image	3-bit LSB	3-bit OPAP	DE(k=4)	APPM(C_{32}=7)	Proposed method k=41, C=[1,5,13]
Lena	38.76	41.56	43.11	43.95	48.31
Jet	38.77	41.56	43.12	44.02	48.31
Boat	38.75	41.55	43.11	43.97	48.31
Elaine	38.72	41.55	43.14	43.96	48.32
House	38.65	41.55	43.12	44.05	48.30
Sailboat	38.73	41.57	43.12	43.96	48.31
Average	38.73	41.56	43.12	43.98	48.31

Table 4. Comparison of PSNR between stego-image and cover image for 1,000,000 payload in five hiding schemes. (unit: dB)

Image	4-bit LSB	4-bit OPAP	DE(k=10)	APPM(C_{199}=37)	Proposed method k=256, C=[1,7,46]
Lena	31.99	35.01	35.42	36.04	43.12
Jet	32.07	35.04	35.89	36.05	43.13
Boat	32.15	35.00	35.50	36.07	43.11
Elaine	32.04	35.01	35.55	36.06	43,12
House	32.00	35.05	35.70	36.10	43.13
Sailboat	32.02	35.00	35.60	36.07	43.13
Average	32.04	35.01	35.60	36.06	43.12

4   Conclusion

In this paper a high-payload image steganographic scheme is proposed. It processes n pixels as a vector in n-D space, and searches for the best embedding function that minimizes the error introduced in the embedding process. The proposed scheme with three-pixel block ($n=3$) under various message bases are constructed and illustrated. Experimental results show that the proposed scheme has higher embedding efficiency than several reported hiding schemes.

References

1. Bender, W., Gruhl, D., Morimoto, N., Lu, A.: Techniques for data hiding. IBM Systems Journal 35(3&4), 313–336 (1996)
2. Johnson, N.F., Jajodia, S.: Exploring stegnography: seeing the unseen. IEEE Computer 31(2), 26–34 (1998)
3. Thien, C.C., Lin, J.C.: A simple and high-hiding capacity method for hiding digit-by-digit data in images based on modulus function. Pattern Recognition 36(12), 2875–2881 (2003)
4. Chan, C.K., Cheng, L.M.: Hiding data in images by simple LSB substitution. Pattern Recognition 37(3), 469–474 (2004)
5. Mielikainen, J.: LSB matching revisited. IEEE Signal Processing Letters 13(5), 285–287 (2006)
6. Zhang, X., Wang, S.: Efficient steganographic embedding by exploiting modification direction. IEEE Communication Letters 10(11), 782–783 (2006)
7. Chao, R.M., Wu, H.C., Lee, C.C., Chu, Y.P.: A novel image data hiding scheme with diamond encoding. EURASIP Journal on Information Security, Article ID 658047, 1–9 (2009)
8. Hong, W., Chen, T.S.: A novel data embedding method using adaptive pixel pair matching. IEEE Transactions on Information Forensics and Security 7(1), 176–184 (2012)

Non-expanding Friendly Visual Cryptography

Wen-Pinn Fang[1,*], Ran-Zan Wang[2], and Shang-Kuan Chen[1]

[1] Department of Computer Science & Information Engineering, Yuanpei University, No.306,
Yuanpei St., HsinChu, 30015, Taiwan
[2] Department of Computer Science & Engineering, Yuan Ze University, Taoyuan 320,
Taiwan 300, Taiwan
wpfang@mail.ypu.edu.tw

Abstract. This paper proposed a non-expanding visual sharing scheme. Based on probability method, a friendly visual secret sharing result is approached. Different from traditional friendly visual cryptography method with block expansion characteristic, the proposed method preserve the size of shares. The result is also demonstrates in the paper.

Index Terms: visual secret sharing, transparency, friendly, non-expanding.

1 Introduction

Visual cryptography is proposed by Shamir[1]. In traditional visual cryptography there are two properties: noisy like shares and pixel expansion step. Noisy like shares causes difficult to manage shares. Pixel expansion step causes the size of transparencies bigger than original image. There is another visual secret sharing approach without expanding method, named random-grid method [6-8]. The method also does not need extra code book in generating shares. In 2009, Fang[11] proposed a non-expansion visual secret sharing method with reversible property is proposed. The properties of the proposed method include security, fast decoding and small share size. However, it is not easy to implement (n,r) threshold sharing, access structure sharing, and friendly visual secret sharing. This paper proposed a friendly non-expanded sharing scheme. Although there are only (2,2) examples, it is possible to extend to another type of visual secret sharing.

The rest of this paper is organized as follows. Section 2 reviews the visual cryptography. Section 3 describes halftone method. Section 4 describes the proposed method. The Experimental results are demonstrated in Section 5. Finally, brief conclusions are given in Section 6.

2 Visual Cryptography

Visual cryptography was first proposed by M. Naor and A. Shamir [1] when the easiest version of producing two noise-like transparencies with each pixel is black or

---

[*] Corresponding author.

J.-S. Pan et al. (eds.), *Genetic and Evolutionary Computing*,
Advances in Intelligent Systems and Computing 238,
DOI: 10.1007/978-3-319-01796-9_16, © Springer International Publishing Switzerland 2014

white 50% each. With a single transparency, the content cannot be recognized to show the original image, while the two transparencies Fig.1 (b) and (c) overlaid the original image can be shown as in Fig.1 (d). The theory is based on human eyes' relative cognition toward dark and light colors. The dark spots in one area are seen black by human eyes, if the original image is black, then the locations of dark spots in the area will be reversed and vice versa is white. First it is to create two basis matrices, as shown in Table 1, two corresponding points will be expanded into blocks, if the original image is black, the locations of the black points are reversed, on the contrary, the same. The overlaying produces expansion, however it is close to the result of the overlaid original image.

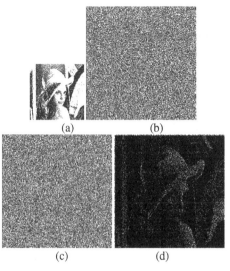

Fig. 1. Examples of traditional visual cryptography (a) Original image, (b) (c) for the stars kept, (d) the superposed (b) and (c) the result

Shamir also designed visual cryptography with fault-tolerance property, named (n, r) threshold scheme. The method is to create basis matrix, and then look up the table to generate transparencies. In the beginning, most of studies handle with single secret. Recently, there are a lot of studies handling multi secret images. For example, Ateniese, et al.[2] discuss access structure. Wu and Chang[3] proposed a method that someone can get two secret images with different stack angles. Fang and Lin[4]proposed shift style visual cryptography method in 2006 which has two secret images with different align location. Fang[5] proposed reversible visual cryptography scheme in 2007, which has two secret images; one secret image appears with just stacking two shares and the other secret image appears with stack two shares after reversing one of them. Because there are two secret images, it is more difficult to create a fake share.

Table 1. Visual cryptography Method Description

3   Halftone Method

Halftone is the reprographic technique that simulates continuous tone imagery through the use of dots, varying either in size, in shape or in spacing. Error diffusion[16] is a type of halftoning in which the quantization residual is distributed to neighboring pixels that have not yet been processed. The proposed method adopt adopts Floyd-Steinberg error-diffusion method to convert grey-value to black-and-white in order to realize binary VC. The weighting is shown as equation (1).

$$\frac{1}{16}\begin{bmatrix} - & - & 7 \\ 3 & 5 & 1 \end{bmatrix} \tag{1}$$

4   Proposed Method

There are two phases of the proposed method: encoding phase and decoding phase.Before encoding phase, the probability is predefined, based on the predefined values, eight basis matrices is generated. It is shown in the encoding phase algorithm.

Encoding phase
   Input : secret image I , host image H_1, H_2 which size are w×h, all of them are bi-level images
   Output: shares S_1, S_2

Let A(2,2,2,9)=
$\quad$ {{{{{1,0,1,1,0,0,1,1,0},{1,1,1,0,0,1,1,0,0}},
$\quad$ {$\quad$ {1,1,0,0,1,1,0,1,0},{1,1,1,1,1,1,0,1,0}}},
$\quad$ {{$\quad$ {1,1,1,1,0,1,0,1,1},{1,1,1,1,0,0,0,0,1}},
$\quad$ {$\quad$ {0,1,1,1,0,1,1,1,1},{0,1,1,1,0,1,1,1,1}}}},
$\quad$ {{{$\quad$ {0,1,1,0,1,1,1,0,0},{1,0,1,1,0,0,0,1,1}},
$\quad$ {$\quad$ {0,1,0,0,0,1,1,1,1},{1,1,1,1,1,1,1,0,1,0}}},
$\quad$ {{$\quad$ {0,1,1,1,1,1,1,1,0},{1,1,1,1,0,0,0,0,1}},
$\quad$ {$\quad$ {1,1,1,1,1,1,0,1,0},{0,1,1,1,0,1,1,1,1}}}}}}

For i=1 to w
$\quad$ For j=1 to h
$\qquad$ p=random select from 1 to 9;
$\qquad$ if I(i,j)=white then
$\quad$ c_1=1
else
$\quad$ c_1=0;
end if
if $H_1(i,j)$=white then
$\quad$ c_2=1
else
$\quad$ c_2=0;
end if
if $H_2(i,j)$=white then
$\quad$ c_3=1
else
$\quad$ c_3=0;
end if
if $A(c_1,c_2,c_3,1,p)$ =0 then
$\quad$ $S_1(i,j)$=white
Else
$\quad$ $S_1(i,j)$=black
End if

if $A(c_1,c_2,c_3,2,p)$ =0 then
$\quad$ $S_2(i,j)$=white
Else
$\quad$ $S_2(i,j)$=black
End if
$\qquad$ End for
End for
$\qquad\qquad$ ----- end of algorithm

Decoding phase

Users just put the shares overlapped with proper alignment. The secret image will be revealed.

5   Experimental Results

Two examples of experimental result is shown as in Figure.2 and Figure.3, . Fig.2(a) is the secret image, Fig.2 (b1) and (b2) are host images, Fig. 2 (c1) and (c2) are the shares, and Fig.2 (d) is the stacked result from Fig.2 (c1) and (c2). Fig 3. demonstrates gray-scale image sharing. Fig.3(a) is the secret image, Fig.3 (b1) and (b2) are host images, Fig. 3 (c1) and (c2) are the shares, and Fig. (d) is the stacked result from Fig.3 (c1) and (c2). Notice, The stego-images are gray-scale image.

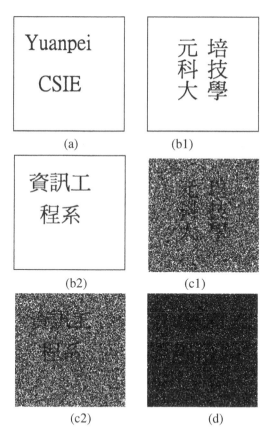

Fig. 2. Experimental result (a) is the secret image (b1) is host image1 (b2) is host image2 (c1) is the share1 (c2) is the share 2 (d) is the stacked image

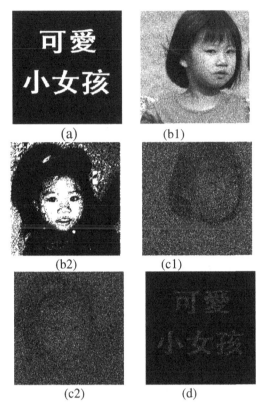

Fig. 3. Experimental result (a) is the secret image (b1) is host image1 (b2) is host image2 (c1) is the share1 (c2) is the share 2 (d) is the stacked image

6   Discussion and Conclusions

This paper proposed a non-expanded visual secret sharing scheme. The experiments include sharing binary image and gray level image (halftone image).

In the experiment, the probability is predefined: If the pixel value of secret image is white, the probability of black pixel is 7/9, the probability of white pixel is 2/9. The probability of black pixel in stacked image is 7/9 and the probability of black pixel in stacked image is 1. It is impossible to identify the pixel value of secret image by only one transparency. It is a safe sharing method. Using the idea of proposed method, it is easy to manage shares and easy to extend the result to (n, r) sharing.

References

1. Naor, M., Shamir, A.: Visual cryptography. In: De Santis, A. (ed.) EUROCRYPT 1994. LNCS, vol. 950, pp. 1–12. Springer, Heidelberg (1995)
2. Ateniese, G., Blundo, C., De Santis, A., Stinson, D.R.: Visual Cryptography for General Access Structure. Information and Computing 129, 86–106 (1996)

3. Wu, H.C., Chang, C.C.: Sharing visual multi-secrets using circle shares. Computer Standards & Interfaces 28, 123–135 (2005)
4. Fang, W.P., Lin, J.C.: Visual Cryptography with Extra Ability of Hiding Confidential Data. Journal of Electronic Imaging 15, 023020 (2006)
5. Fang, W.P.: Visual Cryptography in reversible style. In: IEEE Proceeding on the Third International Conference on Intelligent Information Hiding and Multimedia Signal Processing (IIHMSP 2007), Kaohsiung, Taiwan, R.O.C., November 26-28 (2007)
6. Kafri, O., Keren, E.: Encryption of pictures and shapes by random grids. Optics Letters 12(6), 377–379 (1987)
7. Shyu, S.J.: Image encryption by random grids. Pattern Recognition 40(3), 1014–1031 (2007)
8. Chen, T.H., Tsao, K.H.: Visual secret sharing by random grids revisited. Pattern Recognition (2008),
   `http://www.sciencedirect.com/science?_ob=MImg&_imagekey=`
   `B6V14-4V1TXMJ-1-1&_cdi=5664&_user=2414342&_orig=mlkt&_`
   `coverDate=11%2F30%2F2008&_sk=999999999&view=c&wchp=dGLzVtz-`
   `zSkzV&md5=0f9b092b81e841ed86e4a8c6eadd4a22&ie=/sdarticle.pdf`
9. Lukac, R., Plataniotis, K.N.: Bi-level based secret sharing for image encryption. Pattern Recognition 38, 767–772 (2005)
10. Fang, W.P., Lin, J.C.: Multi-channel Secret Image Transmission with Fast Decoding: by using Bit-level Sharing and Economic-size Shares. International Journal of Computer and Network Security 6, 228–234 (2006)
11. Fang, W.P.: Non-expansion Visual Secret Sharing in Reversible Style. International Journal of Computer and Network Security 9(2), 204–208 (2009)
12. Yang, C.N.: New visual secret sharing schemes using probabilistic method. Pattern Recognition Letters 25, 481–494 (2004)
13. Fang, W.-P.: A Survey for Visual Sharing Scheme with Geometry Property. Journal of Image Processing and Communication 2(1), 35–39 (2010)
14. Fang, W.P.: Maximizing the Secret Hiding Ratio in Visual Secret Sharing with Reversible Property. International Journal of Computer and Network Secure
15. Fang, W.P., Lin, J.C.: Multi-channel Secret Image Transmission with Fast Decoding: by using Bit-level Sharing and Economic-size Shares. International Journal of Computer and Network Security 6, 228–234 (2006)
16. `http://en.wikipedia.org/wiki/Error_diffusion`

An Embedded 3D Face Recognition System Using a Dual Prism and a Camera

Chuan-Yu Chang[1], Chuan-Wang Chang[2], and Min-Chien Chang[3]

[1,3] Department of Computer Science and Information Engineering,
National Yunlin University of Science and Technology, Yunlin, Taiwan
chuanyu@yuntech.edu.tw
[2] Department of Computer Science and Information Engineering,
Far East University, Tainan, Taiwan
chuan@cc.feu.edu.tw

Abstract. In this paper, a single camera and a dual prism are integrated to implement a three-dimensional face recognition system. The proposed system is implemented on an embedded development platform named UBIKIT6612. A dual prism placed in front of the camera is used to simulate human binocular vision. We then used the active appearance models (AAM) to find out the corresponding feature points and calculate the depth of the face by stereo vision. Accordingly, three-dimensional facial model of each member is constructed. Facial features extracted from the 3D facial models are used for identification. To promote the recognition accuracy, we first exclude most of non-members by support vector data description (SVDD), followed by conducting a multi-class support vector machines (SVM) for face recognition. Experimental results show that the proposed method of the exclusion of non-members works more efficiently than those of traditional methods.

Keywords: Three Dimensional Face Recognition, Active Appearance Model.

1   Introduction

The progress of biometric authentication has progressed from the traditional RFID card to the most popular face recognition nowadays [1]. Most of traditional face recognition methods only make a use of single camera, which means that it can only capture two-dimensional (2D) images, and is unable to distinguish between a real face and a photograph. However, face recognition requires a large amount of computation and still relies on PCs now. In contrast, embedded system has rapidly developed over these years. Also, the cost of embedded system is much lower than PC. Therefore, we hope that we can not only develop a real-time identification but also have a strong recognition rate for embedded face recognition system through the method of this paper.

Most of the three-dimensional facial recognition systems use dual cameras at different angles to simulate the left and right eyes [2][3]. Through the theory of stereo vision, we can obtain the depth of face and reconstruct the 3D face model

J.-S. Pan et al. (eds.), *Genetic and Evolutionary Computing,*
Advances in Intelligent Systems and Computing 238,
DOI: 10.1007/978-3-319-01796-9_17, © Springer International Publishing Switzerland 2014

for identification. However, those methods suffer from three major drawbacks: 1. Execution performance of embedded systems is limited, because the CPU usually lack for floating point unit. 2. The cost is high; it must bear the cost of two cameras. 3. Most of the non-members cannot be excluded. In order to exclude non-members' data accurately in the general identification system, it must be trained together with non-member information during the training phase to achieve the best classification results [4][5]. The members' data that we usually have in the identification system are limited, so the classifier cannot exclude most of non-member data efficiently, that's why we came up with a solution which can solve the three aforementioned shortcomings simultaneously. The experimental results show that the proposed method of the exclusion of non-members works more efficiently than that of the SVM classifier.

In this paper, the system overview is described in Section 2, while Section 3 details the methods and techniques. Section 4 presents the experimental results, and the conclusion is given in Section 5.

2   System Overview

2.1   Embedded Platform

In this paper, the embedded development platform named Ubikit6612 is used to implement the proposed method. There is an Omap3530 ARM Cortex with 600MHz on the development board. Fig. 1 shows the layout of the Ubikit6612.

Fig. 1. The layout of the Ubikit6612

2.2   Imaging Device

Fig. 2(a) is the schematic diagram of the dual prism module. A customized dual prism is placed in front of a single camera to simulate human binocular vision. According to the refraction of dual prism, two images (left and right) will project on the image plane. Fig. 2(b) shows the proposed dual prism module.

2.3   Camera Calibration

To obtain intrinsic and extrinsic parameters of the dual prism module, a calibration board is used for the device. The size of the board is $180mm \times 220mm$. The board consists of 9×11 grids, with each square being $20mm \times 20mm$. In this paper, the Matlab camera calibration toolbox is applied to obtain camera parameters [6]. Table 1 lists the obtained calibration parameters.

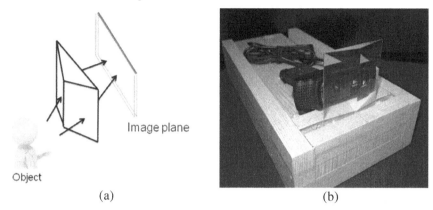

(a)   (b)

Fig. 2. (a) Schematic diagram of dual prism module (b) Dual prism module

Table 1. Calibration parameters

Intrinsic parameters	Parameters
fc_left	$(x , y) = (755.4 , 755.2)$
cc_left	$(x , y) = (188.5 , 99.8)$
fc_right	$(x , y) = (727.4 , 751.8)$
cc_right	$(x , y) = (137.5 , 90.1)$
Extrinsic parameters	**Parameters**
$Rotation$	$(x , y , z) = (0 , 0 , 0)$
$Translatio\,n$	$(x , y , z) = (6.6 , 0 , 0)$

where cc_left is the center of the image coordinates of left camera, fc_left is the focal length after the correction of left part of image. The cc_right is the center of the image coordinates of right camera, fc_right is the focal length after the correction of right part of image. The units are mm; the external parameters, $Rotation$ and $Translatio\,n$, are the rotational and translational position of the image that the left part of the image relative to the right half on dual prism.

3   Proposed Method

This paper proposes a 3D face recognition technique. The processes of the proposed system are depicted in Fig. 3. The details of each step are described in follows.

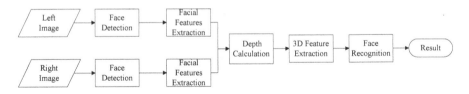

Fig. 3. Block diagram of the proposed system

3.1   Face Detection

In this paper, an Adaboost face detection algorithm proposed by Viola and Jones is adopted [7]. This method is one of the most widely used algorithms and can achieve a high detection rate with fast computation.

3.2   Facial Features Extraction

This paper uses the active appearance model (AAM) to detect facial feature points [8]. Numerous facial images with marked feature points are used to train AAM. To extract significant facial feature points and optimize the model, 68 feature points were marked manually, including six points on each eyebrow, five points for each eye, twelve points for the nose, nineteen points for the mouth, and fifteen points for the face contour. Fig. 4 shows the marked 68 feature points.

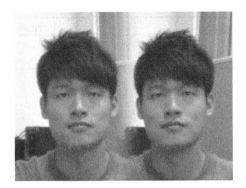

Fig. 4. Extracted facial feature points

3.3   Depth Calculation

Stereo vision is a manner similar to human binocular vision that extracts 3D information from two different views on a scene. By comparing these two images, the relative depth information can be obtained, in the form of disparities. According this disparity, we can reconstruct the 3D model. Figure 5 shows the schematic diagram of the stereo vision.

The 3D coordinate (x, y, z) projections correspond with coordinates (x'_l, y'_l) and (x'_r, y'_r) on the image plane. From Fig. 5, we can obtain the relation of the left image plane and the right image plane perspective via triangle geometric relations as Eq.(1) and Eq.(2).

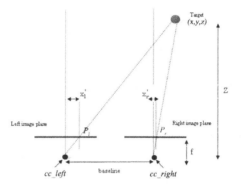

Fig. 5. The conceptual diagram for stereo vision

$$\frac{x}{z} = \frac{x'_l}{f} \tag{1}$$

$$\frac{x-b}{z} = \frac{x'_r}{f} \tag{2}$$

where b is the length of the baseline. x'_l and x'_r are the distances between P_l and cc_left , P_r and cc_right, respectively. Accordingly, the depth z is obtained by

$$z = \frac{bf}{x'_l - x'_r} = \frac{bf}{d} \tag{3}$$

where d is the disparity between x'_l and x'_r. This depth is used to reconstruct the three- dimensional models.

3.4   3D Feature Extraction

To diminish the influence of expressions, 12 features that include 4 distance features and 8 area features are extracted from the normalized 3D face model (see Fig. 7). The four distance features are obtained from four pairs of points (31,46), (36,40), (40,46), and (48,49). The distance between feature points a and b is calculated as below.

$$dis(a,b) = \sqrt{(a_x - b_x)^2 + (a_y - b_y)^2 + (a_z - b_z)^2} \qquad (4)$$

where a and b are feature points in a 3D coordinate (x, y, z).

The eight area features are obtained from point pairs (31,36,43), (40,43,46), (40,48,49), (40,41,48), (41,44,45), (41,42,45), (45,46,49), and (46,48,49). The area of the points pair (a, b, c) can be calculated by Heron's formula:

$$area(a,b,c) = \sqrt{P(P - A) \times (P - B) \times (P - C)} \qquad (5)$$

where $P = (A + B + C)/2$ is the semi-perimeter of the triangle's perimeter, $A = dis(a,b)$, $B = dis(b,c)$ and $C = dis(a,c)$.

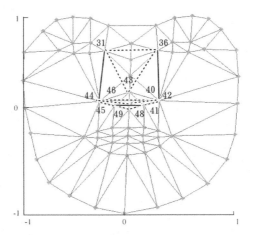

Fig. 6. Selected features of the 3D face model

3.5   Face Recognition

Due to the fickleness of the huge and unknown information of non-members in face recognition system, this paper makes the use of support vector machine and support vector data description [9] to be capable of excluding non-members of face database. The following sections describe these two methods

a.   Support Vector Machine

Support vector machine (SVM) is a popular and robust classifier in classification task. The purpose of SVM is to search for the best hyperplane to separate patterns into two classes. **Fig. 7** is a schematic diagram of hyperplane. In order to avoid patterns can't be linearly separated by the optimum hyperplane, a nonlinear kernel function, radial basis function (RBF), is adopted to map the original vectors to a higher dimensional space. The decision function of classification for a new pattern $\mathbf{x}$ is defined as follow:

$$f(\mathbf{x}) = \text{sgn}\left(\sum_{i=1}^{l} \alpha_i y_i \exp\left(\frac{-\|\mathbf{x} - \mathbf{x}_i\|^2}{2\sigma^2} \right) + b \right) \tag{6}$$

where l is the number of support vectors, α_i is the Lagrange multiplier. y_i is the corresponding target, $\mathbf{x}_i$ is the support vector, and σ is the band width of the RBF kernel function. In this paper, a multi-classes SVM is applied for emotional classification. $n \times (n-1)/2$ one-against-one (OAO) SVMs are implemented, in which, n is the number of member. In this paper, the multi-class SVM is implemented by the LIBSVM [10]. The grid search method and 5-fold cross validation is performed to find the best parameters for SVMs.

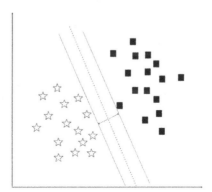

Fig. 7. Schematic diagram of hyperplane

b.   Support Vector Data Description

The SVDD was inspired from the support vector machines. It is able to find all of the covered training data and has a minimum volume (or minimum radius) of the best hypersphere. It can also calculate a decision boundary with a set of the surrounding training data in order to make the right description. Fig. 8 shows the schematic diagram of hypersphere. In this paper, the SVDD is applied for face recognition, and are implemented by LIBSVM [10].

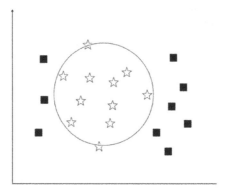

Fig. 8. Schematic diagram of hypersphere

4   Experimental Results

The proposed 3D face recognition system was developed by Ubikit6612. The stereo face database contains 10 members. Each member has 10 stereo images with different expressions and poses. There are 100 stereo images in total. The size of each image is 320×240 pixels. Fig. 9 shows the stereo images of a member.

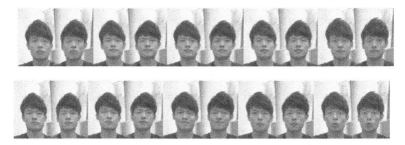

Fig. 9. Stereo images of a member

4.1   Measurement of the Execution Time

The floating-point accelerator (FPA) unit was implemented only in very few ARM cores. The floating-point operation instructions are emulated in kernel. It is very inefficient when it perform a context switch. In order to improve the efficiency of floating-point operation, ARM provided a new application binary interface (ABI) called embedded application binary interface (EABI). In this paper, the EABI was compiled into the BSP Linux kernel such that improves the overall operation efficiency. The execution time for the face detection algorithms in traditional OABI and the proposed EABI are measured and listed in Table 2.

Table 2. Performance measure in terms of execution time

Number of Image	OABI	EABI
1	3.5sec	0.9sec
10	37sec	9.8sec

Since EABI could raise the performance of floating point operation, the execution time on the EABI is much fast than those of on the OABI.

4.2   Face Recognition

We calculated the accuracy of the proposed method with 5-fold cross-validation. The recognition rates are listed in Table 3. We compared our proposed with Sun's method [11]. Sun's method makes a use of stereo vision to obtain three-dimensional information, applying principal component analysis (PCA) for recognition. From Table 3, the proposed method obtains better recognition results.

Table 3. Comparison of the recognition results

Number of Image	Our method	Sun
5	95%	92%
10	91%	88%

4.3   Classification Performance Evaluation

In order to improve the accuracy and security of our facial recognition system, excluding non-members data is crucial. To evaluate the capacity to exclude non-members, the false positive rate is calculated:

$$False\ positive\ rate = \frac{FP}{FP + TN} \tag{7}$$

where FP is the number of the non-members incorrectly identified as members, and TN is the number of non-members correctly identified as non-members. The false positive rate of our system was lower than those of the SVM, which is shown in Table 4.

Table 4. Comparison of the false positive rate

Number of Members	Our method	SVM
5	4%	10%
10	6%	16%

4.4   Testing of 2D Image

Traditional face recognition only uses a single camera to capture images; it is easily cheated through a photo. This paper demonstrates that the proposed 3D face recognition method can effectively avoid this problem. Fig. 10(a) and 10(b) shows a 3D face model constructed from a real face and a face photo, respectively. Obviously, the constructed 3D face model tends to a plane for a face photo. The total depth of the 3D model is close to zero. The results illustrate that we can avoid from photo fraud.

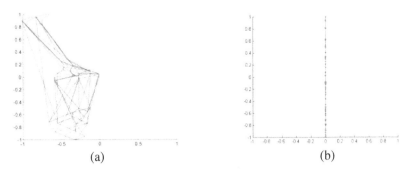

(a)   (b)

Fig. 10. 3D face model constructed from (a) a real face. (b) a face photo.

5   Conclusion

This paper applied the dual prism module to extract two images of simulated human eyes, and then used Active Appearance Models to find out the corresponding feature points and calculate the depth of the face by stereo vision. The part of identification makes use of the reconstructed three-dimensional facial models to extract the triangular area feature and distance feature. We built up the feature vectors and then classified them through SVDD and SVM. The experimental results show that the proposed method of the exclusion of non-members works more efficiently than the method of SVM classifier, and the recognition rate is higher than Sun's method.

Acknowledgment. This work was supported by the National Science Council, Taiwan, under the grants NSC 100-2218-E-224 -007 -MY3.

References

1. Vijaya Kumar, B.V.K., Savvides, M., Venkataramani, K., Xie, C.: Spatial frequency domain image processing for biometric recognition. In: Proceeding of IEEE ICIP, vol. 1, pp. 22–25 (September 2002)
2. Shimizu, M., Yoshizuka, T., Miyamoto, H.: A gesture recognition system using stereo vision and arm model fitting. International Congress Series, vol. 1301, pp. 89–92 (2007)

3. Chang, C.Y., Huang, C.S.: Application of active appearance model for dual-camera face recognition. In: Proceeding of International Conference on Information Security and Intelligence Control, pp. 333–336 (2012)
4. Tang, H., Yin, B., Sun, Y., Hu, Y.: 3D face recognition using local binary patterns. Signal Processing 93, 2190–2198 (2013)
5. Cortes, Vapnik, V.: Support-Vector Network. Machine Learning 20, 273–297 (1995)
6. Bouguet, J.Y.: Camera Calibration Toolbox for Matlab,
   `http://www.vision.caltech.edu/bouguetj/index.html`
7. Viola, P., Jones, M.J.: Robust Real-time Face Detection. International Journal of Computer Vision 57, 137–154 (2004)
8. Cootes, T.F., Edwards, G.J., Taylor, C.J.: Active Appearance Models. IEEE Transactions on Pattern Analysis and Machine Intelligence 23, 681–685 (2001)
9. Tax, D., Duin, R.: Support vector data description. Machine Learning 54, 45–66 (2004)
10. Chang, C.C., Lin, C.J.: LIBSVM: A library for support vector machines (2001),
    `http://www.csie.ntu.edu.tw/~cjlin/libsvm`
11. Sun, T.H., Chen, M., Lo, S., Tien, F.-C.: Face recognition using 2D and disparity eigenface. Expert Systems with Applications 33, 265–273 (2007)

Robust Watermarking for Multiple Images and Users Based on Visual Cryptography

Sheng-Shiang Chang[1], Chih-Hung Lin[2], Tzung-Her Chen[1,*], and Kai-Siang Lin[1]

[1] Department of Computer Science and Information Engineering,
National Chiayi University,
No. 300 University Rd., Chia-Yi City 60004, Taiwan
[2] Graduate Institute of Mathematics and Science Education,
National Chiayi University,
No. 85 Wenlong Vil., Minxiong Township, Chiayi County 62103, Taiwan
thchen@mail.ncyu.edu.tw

Abstract. In this paper, we proposed a robust watermarking worked for multiple images and multiple users. This scheme adopted polynomial-based image secret sharing, DWT transform, and chaos technique. Secret sharing scheme provides better efficient on protecting copyright with multiple users, specifically in some situation, we are not easy to recover watermark by all participants. In order to verify copyright easier, we use polynomial-based image secret sharing scheme to reach a supervised verification. Any two users whose owns capability can recover the watermark with the trust authority. The experimental result shows that this scheme is efficient and robust.

Keywords: watermarking, visual cryptography, wavelet transform.

1 Introduction

During the last decade, the rapid development of the Internet made life more and more convenient. When the Internet goes around us every day, the more digital data will be transmitted. However, the security, authentication, and copyright protection of digital data, have become an importation issue. About the copyright protection of image, the watermarking technique is the popular mechanism and widely used to protect image.

Visual cryptography (VC)-based watermarking schemes were proposed in many researches in recent years [2,3,6]. By VC-based algorithm, it can achieve large embedding capacity, and share a secret image between multi users with meaningless when only one image. However, Liu [4] points out the robustness of traditional VC-based algorithms are not stable enough, therefore, they adopt some techniques to enhance the robustness by transform domain technique, chaotic technique and noise reduction. The main contribution of the Liu's scheme is it can deal with multiple images and multiple users. For secret, it's an ideal assumption that a key is stored dispersedly in difference places and difference user. In real life, it may be really

---

* Corresponding author.

J.-S. Pan et al. (eds.), *Genetic and Evolutionary Computing*,
Advances in Intelligent Systems and Computing 238,
DOI: 10.1007/978-3-319-01796-9_18, © Springer International Publishing Switzerland 2014

difficult to gather all users. Specifically when all users are the members of the same company, but are distributed to different country, it's a problem for recovering the watermark in order to protect the copyright.

In this research, we proposed a secret sharing method for easier reconstruct the secret key image. By polynomial-based secret sharing scheme, we can solve the problem which is difficult to gather keys in dispersed locations from VC-based scheme. The algorithm can recover the watermark easily by an assumed threshold number of users.

This paper is organized as follows. In section 2 ,we discuss some technique in this paper will used likes polynomial-based secret sharing scheme, discrete wavelet transform (DWT), chaotic map. In section 3, we propose our watermarking scheme. Section 4 and section 5, some experimental results and discussion are shown, respectively.

2   Related Work

In this section, we discuss some related works in this paper will used likes polynomial-based secret sharing scheme, discrete wavelet transform (DWT), chaotic map, median filter.

2.1   Polynomial-Based Secret Share

Secret sharing refers to method for distributing a secret amongst a group of participants, each of whom is allocated a share of the secret. The secret can be reconstructed only when a sufficient number, of possibly different types, of shares are combined together; individual shares are of no use on their own.

Shamir's Secret Sharing is an algorithm in cryptography. It is a form of secret sharing, where a secret is divided into parts. The main idea of Shamir's threshold scheme is that 2 points are sufficient to define a line, 3 points are sufficient to define a parabola, 4 points to define a cubic curve and so forth. That is, it takes k points to define a polynomial of degree $k - 1$ [5].

Shamir's Secret Sharing also call polynomial-based secret sharing scheme. Counting on all participants to combine together the secret might be impractical, and therefore sometimes the threshold scheme is used.

In polynomial-based secret sharing scheme, user can decide the number of shares and how many participants is sufficient number. Figure 1 shows an example about (2,2) polynomial-based scheme.

 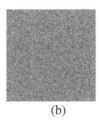  

         (a)   (b)   (c)   (d)

Fig. 1. Example of polynomial-based secret sharing scheme.(a) is original image, (b) and (c) are secret image, (d) is recovered image from (b) and (c).

2.2   Discrete Wavelet Transform

Discrete wavelet transform (DWT) is a kind of transform domain technique. Compared to spatial domain processing, it has more robustness. The transform domain technique can extract feature image from original image, therefore it has more ability to resist compress attack, like JPEG compress [4]. First, DWT decomposed the input image to four subbands LL_1, HL_1, LH_1 and HH_1. Subbands of HL_1, LH_1, HH_1 represent the finest scale wavelet coefficients, and those called detailed image. Subband of LL_1 is approximation image, because of LL_1 stands for the coarse level coefficients. In our scheme, we use LL_1 for Two-level wavelet decomposition. Finally, it obtains LL_2 subband for the watermarking system [4,8].

2.3   Chaotic Map

The chaos system is widely used in the study of image security like encryption algorithm and watermarking system. Deal with the multimedia data, chaos system can be treated as an information transmission system, which can output a data after scrambled. In proposed scheme, the position of pixel in watermark will be disturbed by the chaos system, therefore, the watermark will has more ability to defend modify illegal from large error pixels, like cropping attack.

In this paper, we use torus automorphism [7] to implement chaos system. The position of original pixel will moved to the new position by the following formula.

$$X_{new} = \left[X_{original} + X_{original} \times k\right] mod\ N$$

$$Y_{new} = \left[Y_{original} + Y_{original} \times (k + 1)\right] mod\ N$$

(x_{new}, y_{new}) is new position after the torus auotmorphism, $(x_{original}, y_{original})$ is the original position, N is the size of watermark. Actually, the torus automorphism must run a large number times in order to obtain enough entropy. The Figure 2 show difference running times of the torus automotphism.

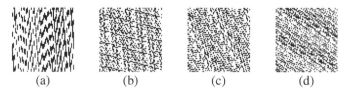

| (a) | (b) | (c) | (d) |

Fig. 2. Difference rounds in torus automorphism. (a) Image with one round in torus automorphism, (b) Image with two rounds in torus automorphism, (c) Image with three rounds in torus automorphism, (d) Image with four rounds in torus automorphism.

2.4   Median Filter

In image processing, the smooth filter is a noise reduction technique. Some attacks of the watermarking scheme are attacked by salt-and-pepper noise. By the processing of smooth filter, we can improve the visual quality when recovering watermark.

3   Proposed Method

The proposed method has three parts which include key generation, embedding and extraction.

First, the users want to protect the copyright of n images. Trust authority generates t key images by polynomial-based image secret sharing scheme. In embedding, trust authority calculates with watermark, n images and t key images by XOR operation. Finally gets a secret share S and store it in trust authority. Users will get t keys (if the numbers of user are t) from trust authority. After embedding, trust authority publishes all images on Internet. Illegally users on Internet will modify or attacks these images. In extracting, users get all attacked images and enough number of key images, and then can success recover the watermark from the trust authority.

The extracting algorithm of some VC-based watermarking schemes needs all the key images of users to recover the watermark, such as [4]. In the proposed scheme, the extracting can recover the watermark just at least two users. It can recover easily but still suit supervised working, in other words, only one user will not work it successfully. If the number of user is t, we can decide how many users can recover watermarking successfully. The range of t is 2 to m, which is the maximum of users.

When users extracting watermarking, the proposed scheme does not require original image, and can uses the attacked images to extract watermark. Therefore, it has the blindness property for watermarking criteria, and extracting also is not required other storage for the original images.

3.1   Key Generation

In this part, users have to decide the number of all candidate t, that is, who has the part of key. The second step is to decide the number of threshold k Extracting operation only worked when the number of users reached the threshold k The Figure 3 shows the key management by polynomial-based secret sharing.

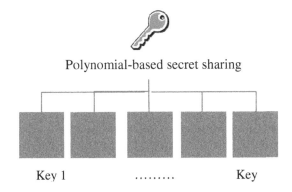

Polynomial-based secret sharing

Key 1   ....   Key

Fig. 3. Key management by polynomial-based secret sharing

3.2   Embedding

In this section, we will obtain the secret image by calculate with original image, watermark and the key of users. The key of users must include all users. We will describe the method by the following six steps.

All parameter as following: n images I_1 to I_n , t users, watermark image W.

1. In order to get feature image of original image, we use 2-levels DWT. Processing images by two steps DWT , and obtain LL_2 image of each original images.
2. Convert the obtained LL_2 images into binary images. Our scheme will use 128 as the default threshold.
3. Change watermark into chaotic image by applying torus automorphism with k (user decide).
4. Combine all keys to obtain a correct key image which can represent all candidates by polynomial-based secret sharing scheme.
5. Deal with LL_2, watermark, correct key, the XOR operation is adopted to obtain the secret image S.
6. Release all original images for internet.

The flow chart of the embedding part is show in Figure 4.

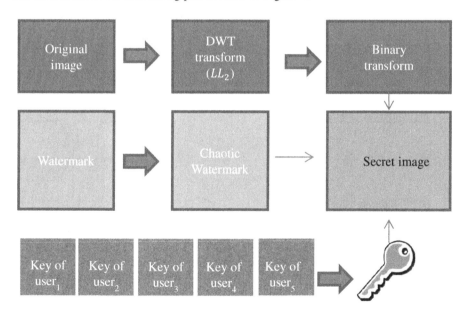

Fig. 4. Embedding procedure

3.3   Extracting

System will extracting watermark from attacked image, secret image and key of users. The numbers of key must conform the threshold which decided in embedding procedure. The following six steps describe the method.

All parameter as following: attacked images, the secret image S and the key of users.

1. Using 2-levels DWT. Processing images by two steps DWT, and obtain LL_2 image of each attacked images.
2. Convert the obtained LL_2 images into binary images. In our scheme, will use the 128 as the default threshold.
3. Obtain key images from users, and generate the correct key by polynomial-based secret sharing scheme.
4. Deal with LL_2, correct key and secret image S, do XOR operation to obtain the attacked watermark.
5. Inverse operation of torus automorphism is adopted to recover watermark.
6. Reducing the noise and improve visual quality on extracted watermark.

The flow chart of the extracting is show in Figure 5.

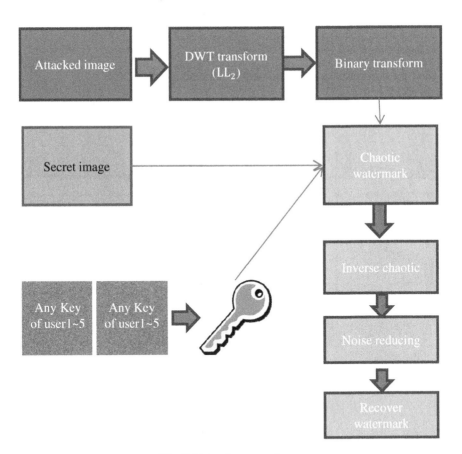

Fig. 5. Extracting procedure

4   Experimental Results

In our simulation, we using an example to show this algorithm can work well. Parameters are all described as below. We use Lena as the protected image, name of our university abbreviation *NCYU* by handwriting, number of key image(users) are two, the output secret image as *S*. And after extracting, we will recover watermark image successfully. The experimental result is listed in Figure 6.

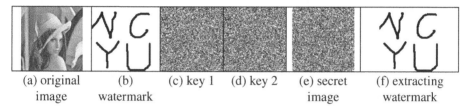

| (a) original image | (b) watermark | (c) key 1 | (d) key 2 | (e) secret image | (f) extracting watermark |

Fig. 6. Experimental results, (a) original image, (b) watermark, (c) and (d) key images calculate from Key generation, (e) secret image, (f) extracting watermark

About the robustness of our scheme, we use well-known attacks to simulate some situations. The attacks include blurring attack, sharpening attack, scaling attack, cropping attack, distortion attack. Attacked image and extracted watermark are listed in Figure 7. The results shows our scheme still recover successfully after attacks.

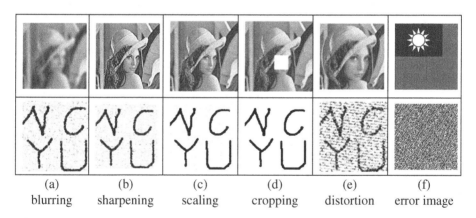

| (a) blurring | (b) sharpening | (c) scaling | (d) cropping | (e) distortion | (f) error image |

Fig. 7. Attacked images and corresponding extracted watermarks. (a) blurring attack, (b) sharpening attack, (c) scaling attack, (d) cropping attack, (e) distortion attack. (f) error image.

Compare to other schemes, we don't need all the users who has the key image, it's easier to apply in the real life. The threshold k in extracting can adjust by any situation. The comparison with some papers is shown in Table 1.

Table 1. Comparison of our scheme with others

	Ours	[4]	[8]
Robustness	✓	✓	✓
Blindness	✓	✓	✓
Security	✓	✓	✓
Multiple users	✓	✓	✗
Multiple images	✓	✓	✗
Adjust threshold	✓	✗	✗

5   Conclusion

In this paper, we proposed a watermarking scheme, which has robustness and security properties. We use the polynomial-based secret sharing scheme to improve the user participate management. This scheme can reduce the number of user who has key to recover watermark that much easier to verify the copyright. Additionally, also can deal with multiple images and multiple users, not limit in one to one relationship.

Acknowledgment. This work was partially supported National Science Council, Taiwan, R.O.C., under contract by NSC 101-2221-E-415-013 and NSC 101-2221-E-415-022.

References

1. Hou, Y.C., Chen, P.M.: An asymmetric watermarking scheme based on visual cryptography. In: Proceedings of WCCC-ICSP, 5th Int. Conf. on Signal Processing, pp. 992–995 (2000)
2. Hsu, C.S., Hou, Y.C.: Copyright protection scheme for digital images using visual cryptography and sampling methods. Optical Engineering 44(7), 077003.1–077003.10 (2005)
3. Lou, D.C., Tso, H.K., Liu, J.L.: A copyright protection scheme for digital images using visual cryptography technique. Computer Standards & Interfaces 29(1), 1530–1541 (2006)
4. Liu, F., Wu, C.K.: Robust visual cryptography-based watermarking scheme for multiple cover images and multiple owners. Information Security (IET) 5(2), 121–128 (2011)
5. Shamir, A.: How to Share a Secret. Communications of the ACM 22(11), 612–613 (1979)
6. Tai, G.-C., Chang, L.-W.: Visual cryptography for digital watermarking in still images. In: Aizawa, K., Nakamura, Y., Satoh, S. (eds.) PCM 2004, Part II. LNCS, vol. 3332, pp. 50–57. Springer, Heidelberg (2004)
7. Voyatzis, G., Pitas, I.: Applications of toralautomorphisms in image watermarking. In: Proceedings of Image Processing, vol. 2, pp. 237–240. Thessaloniki Univ., Greece (1996)
8. Wang, M.S., Chen, W.C.: A hybrid DWT-SVD copyright protection scheme based on k-means clustering and visual cryptography. Computer Standards & Interfaces 31(4), 757–762 (2009)

A Tailor-Made Encryption Scheme for High-Dynamic Range Images

Kai-Siang Lin[1], Tzung-Her Chen[1,*], Chih-Hung Lin[2], and Sheng-Shiang Chang[1]

[1] Department of Computer Science and Information Engineering,
National Chiayi University,
No. 300 University Rd., Chia-Yi City 60004, Taiwan
[2] Graduate Institute of Mathematics and Science Education,
National Chiayi University,
No. 85 Wenlong Vil., Minxiong Township, Chiayi County 62103, Taiwan
thchen@mail.ncyu.edu.tw

Abstract. Multimedia information, especially images, has been consumed in people's daily life. Due to the restriction of consumer electronics, people capture, share, and see images of low-dynamic range (LDR) on digital platform. LDR images are represented with size of 8, 10 and even 12 bits for each color channel of RGB format. Compared with range of LDR, the human visual system can distinguish more colors at a given brightness. Herein, high-dynamic range (HDR) imaging, representing each color channel with size of 32 bits, has become a welcome revolution. With the development of technology and pursuit of quality, HDR imaging is a new trend to research in academia and industry. Naturally, it will come with the issues of security to protect HDR images. However, HDR images compared with common LDR images are new encoding format, it needs specific method for processing. In this paper, the tailor-made encryption scheme is proposed to encode LogLuv format HDR images for guaranteeing confidentiality while format-compliance is achieved.

Keywords: high dynamic range, LogLUV, image encryption.

1 Introduction

Multimedia information, especially images, has been consumed in people's daily life. With the rapid development and convenience of networks, we are surfing, obtaining and transferring varied multimedia via the Internet. Consequently, there are security issues worthy to concern about such as confidentiality, authentication, etc.

Due to the restriction of consumer electronics, the monitor or projector can only display images of low-dynamic range (LDR). LDR images are represented with size of 8, 10 and even 12 bits for each color channel of RGB format. It is difficult to cover the gamut of human's vision because the human visual system can distinguish more colors in the real world than the display of LDR images.

---

[*] Corresponding author.

J.-S. Pan et al. (eds.), *Genetic and Evolutionary Computing*,
Advances in Intelligent Systems and Computing 238,
DOI: 10.1007/978-3-319-01796-9_19, © Springer International Publishing Switzerland 2014

Fortunately, high-dynamic range (HDR) imaging formats [1, 5, 11, 12] have already been developed. The HDR format is more accurate than LDR to display such that the darkest area as shadow and brightest area as sunlight in HDR images are possible. In HDR imaging, the raw format is 32-bits for each color channel R, G, B and totally 96-bits. The size of raw HDR format is comparatively huge, therefore we are used to adopt other fewer bits HDR format such as *LogLuv* [5], *RGBE* [11, 12], or *OpenEXR* [1] instead.

Nowadays, the hardware of monitor and projector cannot still display HDR images. To the end of displaying HDR images on current consumer electronics, HDR images are needed to transform by tone-mapping operator (TMO) [2, 3, 8, 10]. The TMO is a process that transforming HDR images into LDR images. In such a way, we can see the images in LDR with reserving the detail of dark and bright areas. While HDR imaging has become a new trend in academia and industry, it comes with the issues of security including confidentiality. In recent years, there is more and more attention drawn for image encryption [4, 6, 8, 13]. However, these schemes for encryption are suitable only for LDR images, not applied on HDR images directly. In this paper, the HDR tailor-made encryption scheme is proposed to encode LogLuv images for guaranteeing confidentiality while format-compliance is achieved.

The rest of this paper is organized as follows. The LogLuv HDR format is introduced in the next section. The proposed encryption scheme for HDR images is described in Section 3. The experimental results are demonstrated in Section 4. Finally, Section 5 gives the conclusions.

2   Related Work

For LogLuv format HDR images, the LogLuv format is an official part of the Tagged Image File Format (TIFF) specification [14]. In TIFF image file structure, it is mainly consist of an image file header and certain of image file dictionary (IFD) as shown in Fig. 1.

Each tag in IFD records varied information defined by TIFF, including width, length, format such as LogLuv, compression and some other information. Some specific tags record the offset of the specific data blocks. The data blocks are used to record the values of compressed LogLuv format for a HDR image. The stored values are particularly compressed by SGILog [17] which is similar to run-length encoding (RLE).

After HDR images generated by capture devices, each pixel value in HDR images is recorded with the size of 96 bits and finally stored in 32-bits LogLuv format to decrease space consuming. 96-bits format and LogLuv format are shown in Fig. 2 and Fig. 3, respectively.

The process from 96-bits floating point format to 32-bits LogLuv format is shown in Fig. 4. The RGB color system is converted sequentially into CIE XYZ color system, CIE (x,y) color system, CIE (u', v') color system and, finally, LogLuv. The transform function in each color system is shown below.

RGB color system to CIE XYZ color system:

$$\begin{bmatrix} X \\ Y \\ Z \end{bmatrix} = \begin{bmatrix} 0.497 & 0.339 & 0.164 \\ 0.256 & 0.678 & 0.066 \\ 0.023 & 0.113 & 0.864 \end{bmatrix} \begin{bmatrix} R \\ G \\ B \end{bmatrix}$$

CIE XYZ color system to CIE (x,y) color system:

$$x = X/(X + Y + Z)$$

$$y = Y/(X + Y + Z)$$

CIE (x, y) color system to CIE (u', v') color system:

$$u' = 4 * x/(-2 * x + 12 * y + 3)$$

$$v' = 9 * y/(-2 * x + 12 * y + 3)$$

CIE (u', v') color system to LogLuv format:

$$Le = \lfloor 256(\log_2(Y) + 64) \rfloor$$

$$Ue = \lfloor 410u' \rfloor$$

$$Ve = \lfloor 410v' \rfloor$$

Through the transform function mentioned above, we obtain Le, Ue and Ve values. However, before storing those Le, Ue and Ve values in TIFF image file, there is a lossless compression process like run-length encoding called 'SGILog'.There is a string value 'AAAABCDEDDDGG ', for instance. The string value after SIGLog encoding is converted to '82A4BCDE81D80G' in which '82', '4', '81', '80' are regarded as length and 'A', 'BCDE', 'D', 'G' as run. In length of '82', '4', '81', '80', all values are represented by hexadecimal. If length l_{hex} is smaller than 80_{hex}, it means there are l_{hex} different symbols sequentially. If length l_{hex} is equal to or larger than 80_{hex}, it means there are $(l - 7E)_{hex}$ the same symbols sequentially.

After SGILog compression process, the data is stored in TIFF image file. The compression process is shown in Fig. 5. After the LogLuv format color values are compressed by SGILog compression as RLE, the compressed LogLuve format color values are stored into specific data block.

In order to encrypt HDR images without destroying the original TIFF image file structure, the proposed encryption scheme is tailor-made for confidentiality and simultaneously achieving format-compliant. From those information mentioned above about LogLuv format and TIFF, we will demonstrate the proposed scheme in the next section.

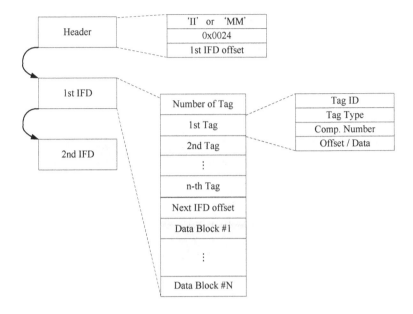

Fig. 1. TIFF image file structure

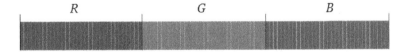

Fig. 2. 96-bits floating point format: 32bits for *R*, *G* and *B*

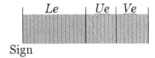

Fig. 3. 32-bits LogLuv format: 16-bits for *Le* and 8-bits for *Ue* and *Ve*

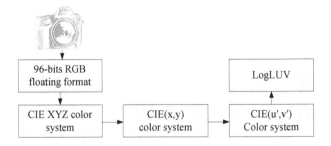

Fig. 4. The processes from 96-bits floating point format to 32-bits LogLuv format

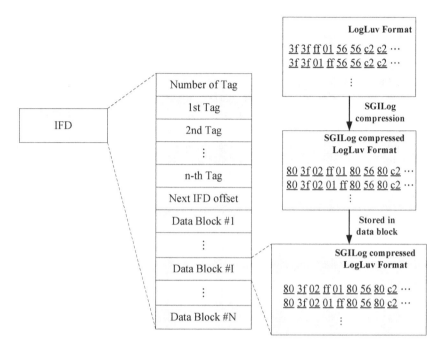

Fig. 5. Compressing and storing process

3   Proposed Scheme

Different from the tradition encryption such as DES, 3-DES, and AES [10] that generating a secure enough file but unreadable in normal decoding toolkits, we aim at generating a secure image file and simultaneously meet format-compliance that people can see the encrypted image as usual but look like noise.   In this section, we will propose tailor-made encryption to meet format-compliant and provide the confidentiality of LogLuv HDR images.

To the end of format-compliance, the LogLuv format values which are compressed in TIFF file should be firstly obtained from data block and recovered those values as uncompressed. After obtaining the uncompressed LogLuv format values, the values are disturbed by the well-defined stream cipher [10]. After that, the encrypted LogLuv format values are re-compressed by SGILog encoding and then re-stored to the data block. Some definitions are given below and shown the overview of encryption in Fig. 6.

P: An image of LogLuv format HDR
I: Compressed values of LogLuv format
I': Compressed and encrypted values of LogLuv format
$f_{SGILog}(\cdot)$: A function of SGILog compression encoding
$f_{SGILog}^{-1}(\cdot)$: A function of SGILog compression decoding

F: Uncompressed LogLuv format

F': Compressed LogLuv format by SGILog encoding.

As following, we will give an account for the encryption process of Fig. 6. Assuming there are $m \times n$ uncompressed LogLuv format values and generally the total number $m \times n$ is as the same size as HDR image P. F(i,j) is denoted each pixel of HDR image P and $F_{Le}(i,j)$, $F_{Ue}(i,j)$ and $F_{Ve}(i,j)$ represent *Le*, *Ue* and *Ve* of $F(i,j)$ where $i \in \{0, 1, \dots, m-1\}$ and $j \in \{0, 1, \dots, n-1\}$.

In the encryption process, after obtaining uncompressed LogLuv values, we focus on disturbing the values by stream cipher and the flowchart of encryption are shown in Fig. 7.

Assume that a stream consisted of *Le*, *Ue* and *Ve* of pixel values sequentially, i.e., $F_{Le}(0,0)F_{Ue}(0,0)F_{Ve}(0,0)F_{Le}(0,1)F_{Ue}(0,1)F_{Ve}(0,1) \dots F_{Le}(m-1, n-1)F_{Ue}(m-1, n-1)F_{Ve}(m-1, n-1)$.

The following operations should be done.

Step 1: To generate the random bit stream used for stream cipher, give the logistic chaotic map [13]

$$x_{n+1} = 4x_n(1 - x_n) \tag{1}$$

where x_0 is a given initial condition. Note that for randomness, after thousands of iterations the chaotic function is stable and, thus, the generated random numbers are used.

Step 2: After $x_1, x_2, \dots, x_{m \times n}$ are generated from Eq. (1), x_k as a section key ($1 \leq k \leq m \times n$) is derived.

The values should be restricted by a specific size for encryption in this step. Let e_{Le} and e_{UeVe} be integer numbers larger than $\lfloor \log_{10} 65536 \rfloor + 1$ and $\lfloor \log_{10} 256 \rfloor + 1$. And let h_k^{Le}, and h_k^{UeVe} ($1 \leq k \leq m \times n$)be the values used to limit the length of numbers generated by Eq. (2) and Eq. (3).

$$h_k^{Le} = mod(x_k \times 10^{e_{Le}}, 65536) \tag{2}$$

$$h_k^{UeVe} = mod(x_k \times 10^{e_{UeVe}}, 256) \tag{3}$$

Step 3: After Step 2, the pixel values for *Le*, *Ue* and *Ve* are encrypted here. All values of *Le*(i,j), *Ue*(i,j) and $Ve(i,j)$ are regarded as bit stream and so are h_k^{Le} and h_k^{UeVe}($k = i \times n + j + 1$). The F_x ($x = Le, Ue$ and Ve) is denoted the bit stream of *Le*, *Ue* and *Ve*. $h_k^x(x = Le, UeVe)$ is denoted the bit stream generated from Eq. (2) and Eq. (3). $F_x'(x = Le, Ue$ and $Ve)$ is denoted the bit stream after XOR-operation of F_x . The stream cipher operations are done by the following where the operator is bitwise excusive OR.

$$
\begin{aligned}
F_{Le}' &= F_{Le}(i,j) \oplus h_k^{Le} \\
F_{Ue}' &= F_{Ue}(i,j) \oplus h_k^{UeVe} \\
F_{Ve}' &= F_{Ve}(i,j) \oplus h_k^{UeVe}
\end{aligned}
\tag{4}
$$

Among Eq. (4), $k = i \times n + j + 1$, $i \in \{0, 1, \dots , M - 1\}$ and $j \in \{0, 1, \dots , N - 1\}$.

Then all the disturbed LogLuv format pixel values are re-compressed by SGILog encoding again and re-stored to data block. In such a way, the encrypted HDR image of LogLuv is format-compliant and, thus, viewed as noise-like.

In the decryption phase, the operations are similar to those of encryption. Assume the recipient has the initial condition x_0, do **Steps 1-3** by utilizing the encrypted Le, Ue and Ve to calculate with h^{Le}, h^{Ue} and h^{Ve} by XOR-operation. The encrypted HDR image will be recovered.

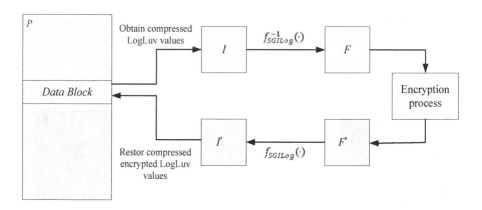

Fig. 6. Overview of the proposed encryption processes

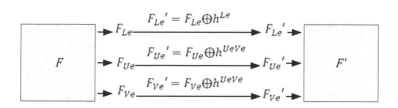

Fig. 7. The main operations of encryption/decryption process

4 Experimental Results and Discussion

For our tailor-made encryption scheme, the experimental results and discussion are shown as following. We will demonstrate the experimental results by achieving noise-like images. In our experimental process, we are using the free C++ IDE, 'Dev-C++,' for coding on Intel core i3 CPU at 3.30GHz. The experimental results are displayable by Photoshop. Herein, our experimental result images are captured by Photoshop. Table 1 lists our tested HDR images and relevant data. There are three experimental results shown in Fig. 8-10 and each serial number (a), (b) and (c) is represented for original image, encrypted image and decrypted image, respectively. By the encrypted image, we can see the HDR image via Photoshop software, it represents the encrypted HDR image meets format-compliant. And after decryption with the same key, we will recover to the original HDR image. The execution time for encryption and decryption are show in Table 2.

Table 1. List of the HDR image in the test database

Image	Source	Size
Dani_cathedral	[15]	1024 × 767
Nave	[15]	480 × 720
Platonics	[16]	1024 × 768

(a): The HDR image 'Dani_cathedral'   (b): Encrypted HDR image   (c): Decrypted HDR image

Fig. 8. The encryption and decryption results for the HDR image 'Dani_cathedral' of ref [15]

(a): The HDR image 'Nave'   (b): Encrypted HDR image   (c): Decrypted HDR image

Fig. 9. The encryption and decryption results for the HDR image 'Nave' of ref [15]

(a): The HDR image 'Platonics'	(b): Encrypted HDR image	(c): Decrypted HDR image

Fig. 10. The encryption and decryption results for the HDR image 'Platonics' of ref [16]

Table 2. The encoding time and decoding time for the proposed scheme

Image	Size	En. time (ms)	De. time (ms)	HW	Prog. tool	Display tool
Dani_ cathedral	1024 × 767	34	32	Core i3 3.3GHz	Dev- C++	Photoshop
Nave	480 × 720	14	16	Core i3 3.3GHz	Dev- C++	Photoshop
Platonics	1024 × 768	36	35	Core i3 3.3GHz	Dev- C++	Photoshop

In our experimental results, we achieve security and format-compliant and show the relevant experiment data. The HDR images are security for looking noise-like after encryption and are recovered as the same as original images after decryption. By those experimental results captured by Photoshop, simultaneously, the encryption meets the format-compliant. Finally, the relevant experiment data are shown in Table 2.

5   Conclusions

To display widely range for illumination and colors, the HDR images are essential and weighty for human in the future. In this paper, we propose tailor-made encryption scheme for HDR images. And as our best knowledge, this paper is the first attempt to provide the confidentiality of LogLuv HDR images. After encryption, the HDR images are visually meaningless but still format-compliant. The experimental results tells that the proposed scheme does work.

Acknowledgment. This work was partially supported National Science Council, Taiwan, R.O.C., under contract by NSC 101-2221-E-415-013 and NSC 101-2221-E-415-022.

References

1. Bogart, R., Kainz, F., Hess, D.: The OpenEXR File Format. In: Siggraph 2003 Technical Sketch (2003), http://www.openexr.com
2. Durand, F., Dorsey, J.: Fast bilateral filtering for the display of high-dynamicrange images. In: Proceedings of ACM Siggraph 2002, pp. 257–266 (2002)
3. Drago, F., Myszkowski, K., Annen, T., Chiba, N.: Adaptive logarithmic mapping for displaying high contrast scenes. Computer Graphics Forum 22(3), 419–426 (2003)
4. Jin, J.: An image encryption based on elementary cellular automata. Optics and Lasers in Engineering 50(12), 1836–1843 (2012)
5. Larson, G.W.: LogLuv encoding for full-gamut, high-dynamic range images. Journal of Graphics Tools 3(1), 15–31 (1998)
6. Liao, X., Lai, S., Zhou, Q.: A novel image encryption algorithm based on self-adaptive wave transmission. Signal Processing 90(9), 2714–2722 (2010)
7. Mantiuk, R., Myszkowski, K., Seidel, H.P.: A perceptual framework for contrast processing of high dynamic range images. ACM Transactions on Applied Perception 3(3), 286–308 (2006)
8. Pareek, N.K., Patidar, V., Sud, K.K.: Diffusion–substitution based gray image encryption scheme. Digital Signal Processing 23(3), 894–901 (2013)
9. Reinhard, E., Pattanaik, S., Ward, G., Debevec, P.: High Dynamic Range Imaging: Acquisition, Display, and Image-Based Lighting (2005)
10. Stallings, W.: Cryptography and Network Security: Principles and Practice, 4th edn. (2006)
11. Ward, G.: Real Pixels. In: Graphic Gems II, ch. 11.5, pp. 80–83 (1991)
12. Ward-Larson, G., Shakespeare, R.A.: Rendering with Radiance. Morgan Kaufmann, San Francisco (1988)
13. Zhou, Q., Liao, X.: Collision-based flexible image encryption algorithm. Journal of System and Software 85(2), 400–407 (2012)
14. TIFF Revision 6.0: Specification for revision 6.0, in PDF http://partners.adobe.com/public/developer/tiff/index.html
15. HDR image database, http://www.anyhere.com/gward/hdrenc/pages/originals.html
16. HDR image database, http://www.artoolkit.org/Gallery/General/
17. Example LogLuv images and other information, http://www.anyhere.com/gward/pixformat/tiffluv.html

NBA All-Star Prediction Using Twitter Sentiment Analysis

Yi-Jen Su[*] and Yue-Qun Chen

Department of Computer Science and Information Engineering, Shu-Te University
{iansu,s11639113}@stu.edu.tw

Abstract. As Web 2.0 services become more popular, social network analysis related research have received more attention. Typically, most Internet users contact others through a variety of social media, such as Facebook or Twitter. This research explores human behavior by conducting opinion mining on Twitter to predict the final voting results of NBA All-Star 2013. The term-feature model is proposed to filter out noise for enhancing the quality of the tweet corpus. Tweenator, an emotion detector, assists to decide whether the emotion tag for each gathered article is positive or negative. Two factors are counted in this research: the number of tweets and the ratio of positive tweets for each candidate player. According to experimental result, the positive tweets has direct ratio with the number of votes in the NBA All-Star Game, a result suggesting that sentiment analysis is an effective tool for predicting human voting outcomes.

Keywords: Sentiment Analysis, Opinion Mining, Social Network Analysis.

1   Introduction

In recent years, multifarious social media services have brought a revolutionary change to people's communicative behavior on the Internet. For example, there are more than 200 million tweets published every day on Twitter, one of the most popular micro-blogging services. Twitter users are accustomed to posting their subjective judgments on events any place any time, and those who endorse or disapprove of the ideas respond right away. As these tweets tend to be spontaneous reactions, this type of emotional expression is very different from that derived from questionnaires, which often restrict ideas and bias people toward opinions.

Sentiment analysis or Opinion Mining analyzes emotions exhibited in interaction content by Natural Language Processing (NLP). Basically sentiment analysis adopts distinct technologies to determine the *polarity* of varying-sized objects, including the Document Level, the Sentence Level, and the Phrase Level [1]. Because all tweets on twitter are limited in length to a maximum of 140 characters with time stamp [2], users need to be concise and focused. Unfortunately, this length restriction causes

---

[*] Corresponding author.

J.-S. Pan et al. (eds.), *Genetic and Evolutionary Computing*,   193
Advances in Intelligent Systems and Computing 238,
DOI: 10.1007/978-3-319-01796-9_20, © Springer International Publishing Switzerland 2014

difficulty in determining content or emotion features. This research, therefore, aims to find an efficient method to identify users' emotions at the sentence level.

Voicing opinions online has become a popular way for people to participate in community life. People seem to especially like to express their opinions when there is an important social event going on, like the presidential election, a natural disaster or a sports event. Once they post emotional comments on specific events in a social media service, it becomes an invaluable opinion source to study the preferences of the majority of people. People's preference opinions can be collected and, then, based on influence of these opinions on behavioral patterns, predict the results of the events. National Basketball Association (NBA), a professional basketball organization, always attracts lots of sports fans. Some of them like to discuss and share game results with others in Internet public spaces, e.g. Twitter. This research collects discussion contents as an important information source to predict the final voting results of NBA All-Star. Two major factors, the number of tweets and the ratio of positive tweets, are counted to determine if an all-star candidate is well-liked.

There are two major reasons to support tracing the distribution of preferences by collecting tweets from the Twitter platform: representativeness and popularity. The first reason is that the NBA was held in the USA and that most fans of NBA players come from the USA. The other is Twitter is one of the most popular social media services in America. In addition, Twitter provides API functions that allow social science researchers to easily retrieve communication data for opinion mining.

To enhance the quality of the tweet corpus, a novel term-feature model is proposed to filter out noise tweets. Then Tweenator [3], an emotion detector, is adopted to set emotion tags on gathered articles to label them either positive or negative. The experiment checks the total number of tweets for each NBA player to measure his popularity. The positive tweet ratio is also an important index to the support percentage of a player. Finally, the experimental result is compared to the final voting results of NBA All-Star 2013 to verify the effectiveness of the prediction method using data accessed from Twitter APIs.

2   Literature Review

2.1   Microblogging

Microblogging provides a variety of social media services to users for broadcasting their views in real-time. These messages often contain spontaneous emotions and personal expressions. Most of these messages are limited in length to 200 characters or less and, at the document level, present a style very different from that of traditional blogs. Twitter is one of the most popular microblogging services in the world that produces more than 200 million tweets every day. In general, based on the content of messages (or tweets), a post or status update on Twitter can be divided into various categories: empty meaning messages account for 40.55%, conversation 37.55%, retweets 8.7%, self-promotion 5.85%, small advertisements 3.75%, and news 3.6%.

2.2   Sentiment Analysis

Sentiment Analysis uses such technologies as NLP, machine learning, and data mining, to discover emotions from large amounts of online messages posted on some specific topics. The term *sentiment* has been employed in prediction of final results since 2001 by Das and Chen [4] when analyzing consumers' preferences and behaviors. Consumers' comments are collected to measure their preferences for specific commercial products in [5]. After 2007, *sentiment analysis* has been applied extensively in other fields to predict human behaviors. For example, Snyder [6] used it to estimate a restaurant's popularity by customers' comments on food materials and the dining atmosphere. Generally, the sentiment polarities of comments are divided into three categories: positive, negative and neutral emotions. Wang [7] classifies emotion polarities into six classes: anger, surprise, sadness, disgust, fear, and joy.

2.3   Rank Correlation Coefficient

Rank Correlation Coefficient (RCC), also named Spearman Rank Correlation, was proposed by Spearman [8] in 1904. The metric method reflects the degree of similarity between two rankings. An increasing rank correlation coefficient implies increasing agreement between rankings. The RCC value, r, is given as Equation (1),

$$r = 1 - \frac{6 \sum d_i^2}{n(n^2 - 1)} \tag{1}$$

In Equation (1), a sample of size n represents the total number of observed items, and the difference d_i between the ranks of each observation on the two variables is calculated. When the value of r is higher than 0.5, it suggests that there is a high positive correlation between the observed items. In other words, the lower the r value, the lower the correlation between the observed items. When r equals 0, there is no relationship between the two variables. Once r is a negative value, it means a negative correlation between the two observed variables.

Table 1. An example of player ranking by the number of votes and related tweets

Player	No. of Votes	No. of Tweets	Vote Rank	Tweet Rank
Kobe Bryant	1500	2000	1	1
LeBron James	1200	1600	2	2
Ray Allen	1000	1000	3	4
Jeremy Lin	900	1300	4	3

There are four players in Table (1). The order between players is decided firstly by the number of votes, and secondly by the number of tweets. Based on Equation (1), n represents 4 players and $\sum d_i^2$ is 2. The value of r equals 0.8, which means that these two ranking methods have higher correlation.

3   Research Method

The NBA All-Star prediction is proposed as shown in Fig.1. First, in the data collection step, the candidates' lists (50 players) are compiled from the official website of NBA All-Star. Then, the Twitter API functions, Streaming API and Search API, are used to access and gather all tweets related to these candidates on Twitter.

In the corpus construction step, both the full-name (FN) and partial-name (PN) of the players are chosen to filter out noise tweets. When the FN of the candidates is used to retrieve tweets, only a small number of tweets with high relatedness can be gathered for candidate players. The frequently-occurring terms in FN tweets form a term-feature model, which assists in identifying meaningful terms and in filtering out large amounts of poor quality tweets in the PN retrieving process.

Fig. 1. The research processes for NBA All-Star prediction

In the step of sentiment tag detection, all tweets collected by the FN method are passed on to TreeTagger, a well-developed technology since proposed in 1994, to check morphological features such as parts of speech (POS). A sentence can be divided into independent terms. Each term, with its POS identified and frequency marked, is then integrated into the corpus after removing stop words.

Using the PN and the term-feature model to retrieve tweets can significantly reduce noise. All qualified tweets are tagged by Tweenator as shown in Fig. 2. A web spider is adopted to collect tweets and pass them on to Tweenator to determine if the tweets convey positive or negative emotions.

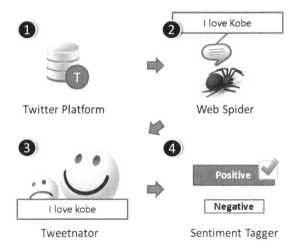

Fig. 2. The sentiment tag setting processes

4   Experimental Result

The tweet dataset of All-Star players are collected from the Twitter platform by APIs during 2013-01-14~2013-01-26. There are in total 121,589 tweets collected by the FN of the candidate players, which in turn form a term-feature model. The goal of the term-feature model is to effectively eliminate noise when retrieving related PN tweets. In Fig. 3, there is no related tweet queried by the full name of the index No. 45, 46, 47, and 48 players. Neither is the amount of tweets related to other candidates sufficient to support the prediction process. A total of 345,523 tweets eventually pass the term-feature model, which is a significant reduction from the original 448,011 tweets collected by the PN of players.

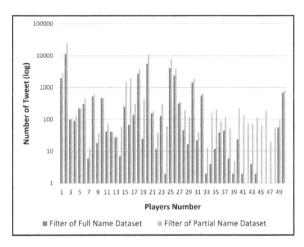

Fig. 3. The number of related PN and FN tweets for NBA All-Star players

In Fig 4, the number of votes for each NBA All-Star candidate is directly proportional to T_p, the number of related tweets collected by PN that pass the term-feature model evaluation. In other words, once an NBA player owns higher All-Star votes, he is much more likely to be discussed on Twitter with high frequency. In the experiment, the value of RCC is 0.58, which suggests that the relatedness is high. The research also uses sentiment analysis to explore the relationship between positive emotion tweets and real votes for each NBA player.

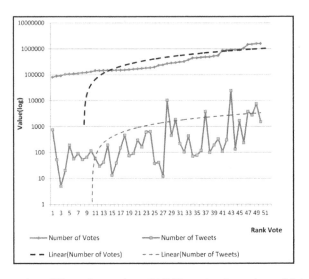

Fig. 4. The number of T_p vs. the number of All-Star votes for each candidate player

Following the selection rules of the NBA All-Star Game, all players are divided into five positions on offense, Point Guard (PG), Shooting Guard (SG), Center(C), Power Forward (PF), and Small Forward (SF). The player that has the most votes at each offense position will be the starting lineup in the NBA-Star Game. Then the head coaches of the two districts, Eastern and Western, will choose the other seven bench players based on their professional skills. The result does not always match the voting result.

Based on the comparison results of Table 2, the highest vote owner of each offense position almost invariably has the largest number of discussion tweets and the highest Positive Rate (PR) of related tweets, except the position of SF. Therefore, using sentiment analysis to predict the NBA All-Star lineup players can reach the accuracy of 80%. On the other hand, only 40% of accuracy can be reached when the number of related tweets is adopted to make prediction.

As shown in Table 3, the statistical results of the Western candidate list attain 80% accuracy using the PR prediction method, which is almost the same as the final voting result of the Eastern district. There is only one starting player whose offense position is Center and his PR ranking is different from the ranking of the votes. If based solely on the number of discussion tweets to predict the final voting result, the accuracy is

Table 2. Comparison of No. of tweets, No. of votes, and positive emotions for Eastern players

	Eastern				
	Player	No. of Tweets	Avg. of Positive	No. of Votes	Position
PG	Rajon Rondo	5543	0.42	924180	Pioneer
	Kyrie Irving	6055	0.37	445730	Reserve
	Jrue Holiday	949	0.35	103146	Reserve
SG	Dwyane Wade	4936	0.53	1052310	Pioneer
	Paul George	4809	0.44	80060	Reserve
C	Kevin Garnett	2384	0.59	553222	Pioneer
	Chris Bosh	7056	0.37	528014	Reserve
	Tyson Chandler	7522	0.34	467968	Reserve
	Joakim Noah	2652	0.52	230796	Reserve
	Brook Lopez	3155	0.46	108978	Reserve
PF	LeBron James	13722	0.48	1583646	Pioneer
SF	Carmelo Anthony	16543	0.43	1460950	Pioneer
	Luol Deng	6295	0.27	130744	Reserve

Table 3. Comparison of No. of tweets, No. of votes, and positive emotions for Western players

	Western				
	Player	No. of Tweets	Avg. of Positive	No. of Votes	Position
PG	Chris Paul	11367	0.52	929155	Pioneer
	Russell Westbrook	11080	0.34	376411	Reserve
	Tony Parker	4751	0.45	176168	Reserve
SG	Kobe Bryant	14555	0.6	1591437	Pioneer
	James Harden	10273	0.42	485986	Reserve
C	Dwight Howard	41874	0.34	922070	Pioneer
	Tim Duncan	5913	0.49	492373	Reserve
PF	Blake Griffin	3573	0.66	863832	Pioneer
	David Lee	4294	0.55	165875	Reserve
	LaMarcus Aldridge	2945	0.43	160197	Reserve
	Zach Randolph	1584	0.5	146980	Reserve
SF	Kevin Durant	11732	0.55	1504047	Pioneer

60%. How the coaches choose the bench players is only remotely related to the PR ranking. The phenomenon is not hard to understand: sometimes the experts' ideas may not necessarily match the expectations of the general public.

In order to examine the connection between the number of related tweets and the PR, the Rank Correlation Coefficient (RCC) is used to measure the degree of similarity, as shown in Table 4. In the Eastern district, a comparison of the number of tweets and the number of positive tweets for each candidate indicates that, only three offense locations, SG, PF and SF, have significant positive correlation with the average 0.53, while PG and C have low positive correlation. In addition, whereas the average correlation value between PR and the vote ranking is 0.59, the value for the position of SF is -0.1, which means that the correlation between the two attributes is inversely proportional and that almost no relationship exists between them.

Table 4. The RCC value of No of tweets, PR, weighted ranking in Eastern and Western districts

District	Eastern			Western		
RCC ＼ Position	Tweets	Positive	Tweets+ Positive (Weight)	Tweets	Positive	Tweets+ Positive (Weight)
PG	0.10	0.83	0.30	0.43	0.36	0.43
SG	0.60	0.90	0.80	1.00	1.00	1.00
C	0.04	0.32	0.82	1.00	0.40	1.00
PF	1.00	1.00	1.00	0.43	0.69	0.74
SF	0.90	-0.10	0.90	1.00	0.50	1.00

All in all, predicting the NBA All-Star final voting results by using the number of related tweets and PR achieves higher accuracy. The Western district shows the same result as the Eastern district and the result even has a higher correlation value. The value of RCC similarity computing on the number of related tweets and PR in both districts is higher than 0.5. This research suggests a higher accuracy when applying sentiment analysis technology in estimating voting results.

5   Conclusion

This research adopts sentiment analysis to identify various emotions when users publish their views regarding NBA All-Star candidate players on Twitter. Two useful factors, the number of related tweets and the positive rate of tweets for each player, are used to predict the final voting results. In order to promote the quality of the dataset, a term-feature model is applied to filter out noise tweets collected by the PN of players.

The experimental result shows that there is a positive correlation between the number of related tweets and the number of All-Star votes. Once an NBA player has a higher mentioned rate on related tweets and PR, the player will have more votes. Adding weights to these two factors significantly enhances prediction accuracy, a result proving that the proposed sentiment analysis method is effective.

Future research could increase more emotion tags to improve emotion identification. The same technology could also be applied to more research domains.

References

1. Wilson, T., Wiebe, J., Hoffmann, P.: Recognizing contextual polarity in phrase-level sentiment analysis. In: The Conference on Human Language Technology and Empirical Methods in Natural Language Processing, pp. 347–354. Association for Computational Linguistics, Vancouver (2005)
2. Tumasjan, A., Sprenger, T.O., Sandner, P.G., Welpe, I.M.: Predicting Elections with Twitter: What 140 Characters Reveal about Political Sentiment. In: 4th International AAAI Conference on Weblogs and Social Media, pp. 178–185 (2010)

3. Saif, H., He, Y., Alani, H.: Alleviating data sparsity for twitter sentiment analysis. In: 2nd Workshop on Making Sense of Microposts with WWW 2012, Lyon, France, pp. 2–9 (2012)
4. Das, S., Chen, M.: Yahoo! for Amazon: Extracting market sentiment from stock message boards. In: 8th Asia Pacific Finance Association Annual Conference, vol. 35, p. 43 (2001)
5. Pang, B., Lee, L., Vaithyanathan, S.: Thumbs up?: sentiment classification using machine learning techniques. In: ACL 2002 Conference on Empirical Methods in Natural Language Processing, vol. 10, pp. 79–86. Association for Computational Linguistics, New Jersey (2002)
6. Snyder, B., Barzilay, R.: Multiple aspect ranking using the good grief algorithm. In: Joint Human Language Technology/North American Chapter of the ACL Conference (HLT-NAACL), New York, USA, pp. 300–307 (2007)
7. Wang, C.Y.: Sentiment Detection of Micro-blogging Short Texts via Contextual, Social and Responsive Information. Master Thesis, Electrical Engineering and Computer Science, National Taiwan University, Taipei, Taiwan (2011)
8. Spearman, C.: The Proof and Measurement of Association between Two Things. The American Journal of Psychology 15, 72–101 (1904)

Information Hiding Based on Binary Encoding Methods and Crossover Mechanism of Genetic Algorithms

Kuang Tsan Lin[1,*] and Pei Hua Lin[2]

[1] Department of Mechanical and Computer-Aided Engineering, St. John's University
499, Section 4, Tam King Road, Tamsui, New Taipei City 25135, Taiwan
ktlin@mail.sju.edu.tw
[2] Department of Computer Science and Information Engineering, National Central University
300, Jhongda Rd., Jhongli City, Taoyuan County 32001, Taiwan
pepelpink@hotmail.com

Abstract. Information hiding for digital images based on a binary encoding method and a crossover mechanism of genetic algorithm is presented in this paper. First, a crossover operation technique is used to change the pixel values of a covert image to form a crossover-operated matrix by using a specific crossover technique. Then, the crossover-operated matrix is encoded into a host image to form an overt image by using a specific encoding rule. The overt image contains eight groups of binary codes, i.e. identification codes, dimension codes, graylevel codes, crossover operating time codes, crossover technique codes, parent organism length codes, starting and ending position codes, and information codes. The parameters are used to encode and hide the covert image. According to the simulation results, the proposed method does well, larger image scrambling degree for the scrambled matrix.

Keywords: Information hiding, Binary encoding, Crossover mechanism of genetic algorithm.

1   Introduction

Information scrambling techniques have been applied to information hiding, which are enhancing the information security. Researchers have proposed image scrambling approaches differently, such as Arnold transformations [1-2], p-Fibonacci trans-formations [3]. Using these methods can do well but the image decoding processes must have correct parameters about transformations or lesser image scrambling degree and some distortion might be found for the image decoding processes.

Some other researches have proposed to combine information scrambling techniques and other image encrypting methods to increase image security. Hennelly and Sheridan [4] and Zhao et al. [5] combined an image scrambling technique and a fractional Fourier transform to hide a covert image. Meng et al. [6] used an image scrambling technique and an iterative Fresnel transform to hide a covert image. There

---

* Corresponding author.

J.-S. Pan et al. (eds.), *Genetic and Evolutionary Computing*,
Advances in Intelligent Systems and Computing 238,
DOI: 10.1007/978-3-319-01796-9_21, © Springer International Publishing Switzerland 2014

were pretty good robustness for all the hybrid methods, but they might contain some distortion in reconstructed covert images during covert-image reconstruction.

This paper proposes a method that combines a binary encoding method [7] and a crossover mechanism of genetic algorithm to hide a covert image in a host image to form an overt image. This method can be applied to equilateral or non-equilateral images, and the overt images look nearly the same as their corresponding host images. And, the method is good security, and free of distortion for the decoded covert images, and larger image scrambling degree for the scrambled image.

2   Crossover Mechanism of Genetic Algorithm for Encoding Images

Let C be a $r \times c$ covert image to be crossover-operated and let Z be a $r \times c$ matrix formed from the crossover mechanism of genetic algorithm of C. The processes for deriving Z from C are shown below. First, transform the pixels of C into form a pixel string A with $r \times c$ elements according to a specified order (from the first row to the last row and from the first column to the last column for the same row). Second, use a decimal-to-binary operation process from a pixel string A with $r \times c$ elements to generate a binary string B with $r \times c \times g$ elements (g is the graylevel of the covert image). Third, assign a parent organism length s and transform the binary string B with $r \times c \times g$ to a $(r \times c \times g/s) \times s$ binary matrix D. Fourth, determine a crossover-operated binary matrix $(r \times c \times g/s) \times s$ E by using a specific crossover mechanism of genetic algorithm from the $(r \times c \times g/s) \times s$ binary matrix D. Fifth, transform the crossover-operated binary matrix E (i.e. children organism) into form a crossover-operated binary string F according to a specified order (from the first row to the last row and from the first column to the last column for the same row). Sixth, determine the crossover-operated matrix Z with $r \times c$ by using a binary-to-decimal operation process from the binary string $1 \times (r \times c \times g)$ F. An example for the crossover mechanism of genetic algorithm processes from a 3×4 covert image C to a 3×4 matrix Z is depicted in Fig. 1.

The reconstruction processes from a $r \times c$ crossover-operated matrix Z^* to a $r \times c$ decoded covert matrix C^* are all identical to that the crossover mechanism of genetic algorithm processes from a $r \times c$ covert matrix C to a $r \times c$ crossover-operated matrix Z. So we do not figure it again.

3   Binary Encoding Methods for Hiding Crossover-Operated Matrices

Assume H is an $M \times N$ host image used to hide a crossover-operated matrix Z to form an $M \times N$ overt image H^*, where the crossover-operated matrix Z is processed a crossover mechanism of genetic algorithm from a covert image C. All of the pixels in H^* are classified into eight groups, i.e. identification codes, dimension codes, graylevel codes, crossover operating time codes, crossover technique codes, parent organism length codes, starting and ending position codes and information codes.

The identification codes are used to justice whether the codes in H^* is encoded with the proposed encoding method or not; the dimension codes are used to denote the dimensions of the covert image C; the graylevel codes are used to denote the graylevel of the covert image C; the crossover operating time codes are used to denote the times of crossover mechanism of genetic algorithm process repeated; the crossover technique codes are used to denote the technique of crossover mechanism of genetic algorithm like one-point technique, two-point technique; the parent organism length codes are used to denote the length of parent organism string during operating crossover mechanism of genetic algorithm; the starting and ending position codes are used to denote the starting point and the ending point of parent organism string during operating crossover mechanism of genetic algorithm. The information codes are used to hide the crossover-operated matrix Z and they are encoded at the second row to the last row of H^*. In addition, the other seven groups of binary codes are encoded at the first row of H^* with the orders specified by the designer.

$$C = \begin{bmatrix} 9 & 10 & 1 & 6 \\ 8 & 4 & 2 & 5 \\ 7 & 11 & 12 & 3 \end{bmatrix} \quad A = \begin{bmatrix} 9 & 10 & 1 & 6 & 8 & 4 & 2 & 5 & 7 & 11 & 12 & 3 \end{bmatrix}$$

(a)   (b)

$B = [1\ 0\ 0\ 1\ 1\ 0\ 1\ 0\ 0\ 0\ 0\ 1\ 0\ 1\ 1\ 0\ 1\ 0\ 0\ 0\ 0\ 1\ 0\ 0$
$0\ 0\ 1\ 0\ 0\ 1\ 0\ 1\ 0\ 1\ 1\ 1\ 1\ 0\ 1\ 1\ 1\ 1\ 0\ 0\ 0\ 0\ 1\ 1]$

(c)

$$D = \begin{bmatrix} 1 & 0 & 0 & 1 & 1 & 0 \\ 1 & 0 & 0 & 0 & 0 & 1 \\ 0 & 1 & 1 & 0 & 1 & 0 \\ 0 & 0 & 0 & 1 & 0 & 0 \\ 0 & 0 & 1 & 0 & 0 & 1 \\ 0 & 1 & 0 & 1 & 1 & 1 \\ 1 & 0 & 1 & 1 & 1 & 1 \\ 0 & 0 & 0 & 0 & 1 & 1 \end{bmatrix} \quad E = \begin{bmatrix} 1 & 0 & 1 & 0 & 0 & 1 \\ 1 & 0 & 0 & 1 & 1 & 1 \\ 0 & 1 & 1 & 1 & 1 & 1 \\ 0 & 0 & 0 & 0 & 1 & 1 \\ 0 & 0 & 0 & 1 & 1 & 0 \\ 0 & 1 & 0 & 0 & 0 & 1 \\ 1 & 0 & 1 & 0 & 1 & 0 \\ 0 & 0 & 0 & 1 & 0 & 0 \end{bmatrix} \quad Z = \begin{bmatrix} 10 & 6 & 7 & 7 \\ 12 & 3 & 1 & 9 \\ 1 & 10 & 8 & 4 \end{bmatrix}$$

(d)   (e)   (g)

$F = [1\ 0\ 1\ 0\ 0\ 1\ 1\ 0\ 0\ 1\ 1\ 1\ 0\ 1\ 1\ 1\ 1\ 1\ 0\ 0\ 0\ 0\ 1\ 1$
$0\ 0\ 0\ 1\ 1\ 0\ 0\ 1\ 0\ 0\ 0\ 1\ 1\ 0\ 1\ 0\ 1\ 0\ 0\ 0\ 0\ 1\ 0\ 0]$

(f)

Fig. 1. Example for crossover mechanism of genetic algorithm processes from C to Z. (a) assumed covert matrix C; (b) A; (c) B with 4 graylevel; (d) D with a parent organism length 6; (e) E (crossover with one-point technique at the third point and operate between i-th row and $(i+4)$th row, $1 \leq i \leq 4$); (f) F; (g) Z.

For the identification codes, the number of the codes must be large enough to avoid incorrect judgment and the codes are binary, e.g. 110001100011100111001111011 1101.

For the dimension codes, they are two sets of ten-digit binary codes. The first set of binary codes is used to denote the row dimension r ($r > 1$) of C and it includes $r1$ to $r10$. The relationship between r and $r1$- $r10$ is followed by Eq. (1). The second set of binary codes is used to denote the column dimension c ($c > 1$) of C and it includes $c1$ to $c10$. The relationship between c and $c1$- $c10$ is followed by Eq. (2).

$$r = \sum_{i=1}^{10} r_i \cdot 2^{i-1} + 2 , \tag{1}$$

$$c = \sum_{i=1}^{10} c_i \cdot 2^{i-1} + 2 . \tag{2}$$

For the graylevel codes, they are eight-digit binary codes $g1$ to $g8$, and they are used to denote the g value of C. The relationship between g and $g1$- $g8$ is similar to Eq. (3).

$$g = \sum_{i=1}^{8} g_i \cdot 2^{i-1} + 1 . \tag{3}$$

For the crossover operating time codes, they are ten-digit binary codes $t1$ to $t10$, and they are used to denote the t times of crossover operation repeated. The relationship between t and $t1$- $t10$ is similar to Eq. (4).

$$t = \sum_{i=1}^{10} t_i \cdot 2^{i-1} . \tag{4}$$

For the crossover technique codes, they are ten-digit binary codes $p1$ to $p10$, and they are used to denote the pattern p of crossover operation. The relationship between p and $p1$- $p10$ is similar to Eq. (5).

$$p = \sum_{i=1}^{10} p_i \cdot 2^{i-1} . \tag{5}$$

For the parent organism length codes, they are twenty-digit binary codes $s1$ to $s20$, and they are used to denote the length s of parent organism. The relationship between s and $s1$- $s20$ is similar to Eq. (6).

$$s = \sum_{i=1}^{20} s_i \cdot 2^{i-1} + 3 . \tag{6}$$

For the starting and ending position codes, they are two sets of twenty-digit binary codes. The first set of binary codes is used to denote the starting point $h1$ of parent organism string and it includes $h11$ to $h120$. The relationship between $h1$ and $h11$- $h120$ is followed by Eq. (7). The second set of binary codes is used to denote the ending point $h2$ of parent organism string and it includes $h21$ to $h220$. The relationship between $h2$ and $h21$- $h220$ is followed by Eq. (8).

$$h1 = \sum_{i=1}^{20} h1_i \cdot 2^{i-1} + 5 .$$
(7)

$$h2 = \sum_{i=1}^{20} h2_i \cdot 2^{i-1} + 7 .$$
(8)

For the information codes, they are used to encrypt Z and the encoding processes to encrypt Z in the $M \times N$ host image H to form an $M \times N$ overt image H^* are simply illustrated below.

(1) Create a binary array R with N elements. Some of the elements of R contain identification codes, dimension codes, graylevel codes, crossover operating time codes, crossover technique codes, parent organism length codes, starting and ending position codes. The other elements of R are not available and they are all set to be 0.

(2) Transform the elements of Z to form the elements of a crossover-operated string G followed by

$$G((r'-1) \times c + c') = Z(r',c') ,$$
(9)

where $1 \le r' \le r$ and $1 \le c' \le c$.

(3) After $G(k) = \sum_{i=0}^{g-1} a_i(k) \cdot 2^{i-1}$ are known, transform G into a binary data string K according to

$$K(i + h \times (k-1)) = a_i(k) .$$
(10)

(4) Use the elements of K to form the elements of a $(M-1) \times N$ data matrix S. Since the array data number L of K may be smaller than $(M-1) \times N$, there are $(M-1) \times N - L$ dummy elements in S not formed from the elements of K. As a result, the values of dummy elements are all set to be 0.

(5) Constitute the row array R with N elements and the $(M-1) \times N$ matrix S to form a $M \times N$ binary matrix T. The first row of T is duplicated from R, while other rows of T are duplicated from S in succession.

(6) Modify the element $H(u,v)$ of the host image H to form the element $H'(u,v)$ of a modified matrix H' followed by

$$H'(u,v) = 2 \times floor(H(u,v)/2),$$
(11)

where the function floor(x) modulates the value of x to the nearest integer x_n (< x), and every $H'(u,v)$ is an even integer.

(7) An overt image H^* is formed by summing the corresponding elements of matrices T and H', i.e.

$$H^*(u,v) = T(u,v) + H'(u,v) .$$
(12)

Figure 2 depicts an assumed host matrix H and an assumed binary matrix T, and the resulted modulated matrix H' and the resulted overt matrix H^*.

$$H = \begin{bmatrix} 12 & 13 & 14 \\ 0 & 1 & 1 \\ 83 & 84 & 85 \end{bmatrix} \quad T = \begin{bmatrix} 1 & 1 & 1 \\ 0 & 1 & 0 \\ 0 & 0 & 0 \end{bmatrix} \quad H' = \begin{bmatrix} 12 & 12 & 14 \\ 0 & 0 & 0 \\ 82 & 84 & 84 \end{bmatrix} \quad H* = \begin{bmatrix} 13 & 13 & 15 \\ 0 & 1 & 0 \\ 82 & 84 & 84 \end{bmatrix}$$

$$\text{(a)} \qquad\qquad \text{(b)} \qquad\qquad \text{(c)} \qquad\qquad \text{(d)}$$

Fig. 2. (a) Assumed host matrix H; (b) assumed binary matrix T ; (c) resulted modulated matrix H' ; (f) resulted overt image $H*$

For a $r \times c$ image Z scrambled by the crossover mechanism of genetic algorithm from a $r \times c$ image C, the definition of the image scrambling degree δ is [8]

$$\delta = \frac{\sum_{i=1}^{r}\sum_{j=1}^{c}[W_{-1,0}(i,j) + W_{1,0}(i,j) + W_{0,-1}(i,j) + W_{0,1}(i,j)]}{255^2 \times r \times c}, \tag{13a}$$

where

$$W_{m,n} = \left| [Z(i+m, j+n) - Z(i,j)]^2 - [Z(i+m, j+n) - Z(i,j)]^2 \right|. \tag{13b}$$

A larger δ value indicates that C and Z are more different.

The *PSNR* (peak signal to noise ratio) values of the two images H and $H*$ is used to check image quality. The definition of PSNR is [9]

$$PSNR = 10 \times \log\left(\frac{M \times N}{MSE}\right), \tag{14a}$$

where

$$MSE = \frac{1}{M \times N}\sum_{i=1}^{M}\sum_{j=1}^{N}[H(i,j) - H*(i,j)]^2 . \tag{14b}$$

Basically, if the *PSNR* is larger than 30, it will hardly distinguish the difference between H and $H*$ for naked eyes; that is the image $H*$ looks almost the same as H [10].

4   Simulations

Figure 3 shows a 120×256 256-graylevel image as the covert image C. Fig. 4 shows a 512×512 256-graylevel picture used as the host image H.

The simulation about the covert image C in Fig. 3 is introduced below. Because the dimension of H is 512×512, the dimension of the row array R is 1×512. The 1st to 32nd elements of R are used to be the identification codes, and they are designated as 11000110001110011110011110111101. The 33rd to 52nd elements of R are used to be the covert image dimension codes. Since the size of C is 120×256, the two sets of dimension codes for r and c are 0001110110 and 0011111110, respectively. The 53rd to 60th elements of R are used to be the graylevel codes. Since the number of the gray values is 256 (=2^8), g=8, the codes are 00000111. The 61st to 70th elements of R are used to the crossover operating time codes t. Since t is set to be 2 here, the codes are 0000000010. The 71st to 80th elements of R are used to the crossover technique codes p. Since p is 1 for selecting the one-point technique of the crossover mechanism of genetic algorithm operation, the codes are 0000000001. The 81st to 120th elements of R are used to the parent organism length codes s for the first crossover operating. Since s is equal to 27, the codes are 0000000000000011000. The 121st to 140th elements of R are used to the parent organism length codes s for the second crossover operating. Since s is equal to 67, the codes are 0000000000001000000. The 141st to 180th elements of R are used to the starting and ending position codes $h1$ and $h2$ for the first crossover operating. Since $h1$ and $h2$ are equal to 16 and 27 respectively, the codes are 0000000000000001011 and 0000000000000010100. The 181st to 200th elements of R are used to the starting and ending position codes $h1$ and $h2$ for the second crossover operating. Since $h1$ and $h2$ are equal to 16 and 67 respectively, the codes are 0000000000000001011 and 0000000000000111100.

Fig. 3. A covert image for test

Fig. 4. A host image for encoding to form an overt image

First we crossover-operate the 120×256 covert image C into a 120×256 matrix Z by using the proposed crossover operation technique. The crossover-operated matrix Z is depicted in Fig. 5(a). Then we transform Z into a crossover-operated string G

with 30720 elements. Then, transform G into a binary-data string K with 245760 elements. Subsequently, copy the elements of K to form the elements of a 511×512 data matrix S. The 245761st to 261632nd elements of K are all set to be 0.

Second, we change K into a 511×512 matrix S, and we combine the 1×512 binary row array R and the 511×512 matrix S to form a 512×512 binary matrix T. The matrix T is depicted in Fig. 5(b). Furthermore, the host image H is modulated to form a modified image H'. Then, the corresponding elements of the matrices T and H' are summed to form an overt image H^*. The overt image H^* looks nearly the same as the host image H in Fig. 4.

The *PSNR* value between H and H^* is equal to 51.2 for the case of the covert binary image. Therefore, the two images H and H^* look almost identical for both the case. Moreover, for the case the *PSNR* value between the original covert image C and the decoded covert image C^* is infinity, i.e. there is no distortion during the covert image decoding.

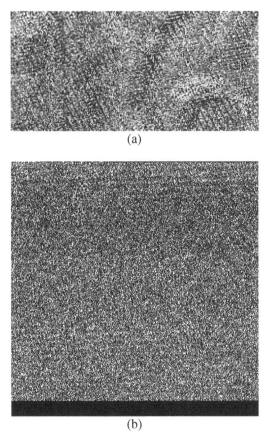

(a)

(b)

Fig. 5. (a) The crossover-operated matrix Z; (b) matrix T; (c) the overt matrix H^* for encoding the covert image in Fig. 3

5   Discussions

The proposed combine the binary encoding method and the crossover mechanism of genetic algorithm technique are demonstrated only for clear and simple in this paper, but changing the definitions of parameters or changing specified pixel positions for encoding parameters can obtain a better security for the proposed method.

We compute the image scrambling degree percentages δ_p ($=100\% \times \delta / \delta_{max}$) of the scrambled images for different algorithm. In Fig. 6(a) shows δ_p corresponding to the covert image in Fig. 3 by the proposed method, δ_p are always larger than 90% for $t \geq 5$ (the case with t=71 has δ_{max} 0.6684).

In Fig. 6(b) shows δ_p by using the p-Fibonacci transformation, δ_p are not stable and they are larger than 90% for some discrete t values (the case with t=28 has δ_{max} 0.3705). In Fig. 6(c) shows δ_p by using the Arnold transformation, δ_p are also unstable and they are larger than 90% for some discrete t values (the case with t=39 has δ_{max} 0.3961). On the other hand, δ_p based on the proposed method is very stable and it is always larger than 90% for $t \geq 5$. And the minimum δ ($\delta_{min} = 0.4234$) of the proposed method are completely larger than the maximum δ of the p-Fibonacci transformation and the Arnold transformation. Therefore, the proposed method does better than the p-Fibonacci transformation and the Arnold transformation for using $t \geq 5$.

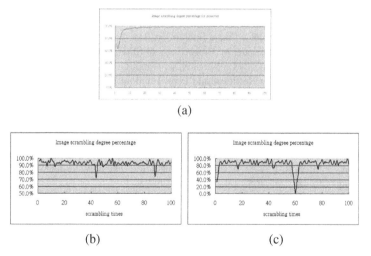

(a)

(b)   (c)

Fig. 6. Image scrambling degree percentages with different t values and by using (a) the proposed method; (b) the p-Fibonacci transformation; (c) the Arnold transformation

6   Conclusions

The proposed method combines the crossover mechanism of genetic algorithm method and the binary encoding method to encode a covert image C into the host image H to form an overt image H^*. The crossover mechanism of genetic algorithm method scrambles from the covert image C to form a crossover-operated matrix Z. The crossover-operated matrix Z can provide a larger value of the image scrambling degree. The binary encoding method hides the crossover-operated matrix into the overt image H^*. Because the *PSNR* of H^* and H is larger than 50, the two images H and H^* look almost the same. And, the decoded covert image C^* can be directly extracted from the overt image H^* not needing the host image H. Furthermore, because the *PSNR* of C and C^* is infinity, C is identical to C^*, i.e. there is no distortion for the decoding processes of the covert image. Lastly, the reconstruction of a decoded covert image from an overt image not being authorized can be difficult, whereas the reconstruction of the decoded covert image can be easy for authorized users.

References

1. Yang, Y.L., Cai, N., Ni, G.Q.: Digital Image Scrambling Technology Based on the Symmetry of Arnold Transform. Journal of Beijing Institute of Technology 15, 216–221 (2006)
2. Li, B., Xu, J.W.: Period of Arnold Transformation and Its Application in Image Scrambling. Journal of Central South University of Technology 12, 278–283 (2005)
3. Zhou, Y., Agaian, S., Joyner, V.M., Panetta, K.: Two Fibonacci p-code based image scrambling algorithms. In: Proceedings of SPIE, vol. 6812, p. 681215 (2008)
4. Hennelly, B., Sheridan, J.T.: Optical Image Encryption by Random Shifting in Fraction Fourier Domains. Optics Letters 28, 269–277 (2003)
5. Zhao, J., Lu, H., Fan, Q.: Color Image Encryption Based on Fractional Fourier Transforms and Pixel Scrambling Technique. In: Proceedings of SPIE, vol. 6279, p. 62793B (2007)
6. Meng, X.F., Cai, L.Z., Yang, X.L., Shen, X.X., Dong, G.Y.: Information Security System by Iterative Multiple-phase Retrieval and Pixel Random Permutation. Applied Optics 45, 3289–3295 (2006)
7. Lin, K.T.: Digital Information Encrypted in an Image Using Binary Encoding. Optics Communications 281, 3447–3453 (2008)
8. Yu, X.Y., Ren, H., Li, E.S., Zhang, X.D.: A New Measurement Method of Image Encryption. Journal of Physics: Conference Series 48, 408–413 (2006)
9. Kutter, M., Petitcolas, F.A.P.: A Fair Benchmark for Image Watermarking Systems. In: Proceedings of SPIE, vol. 3657, pp. 226–231 (1999)
10. Shih, T.K., Lu, L.C., Chang, R.C.: An Automatic Image in Paint Tool. In: Proceedings of the Eleventh ACM International Conference on Multimedia, pp. 102–103 (2003)

A Modified Method for Constructing Minimum Size Homogeneous Wireless Sensor Networks with Relay Nodes to Fully Cover Critical Square Grids

Bing-Hong Liu, Yue-Xian Lin, Wei-Sheng Wang, and Chih-Yuan Lien

Department of Electronic Engineering,
National Kaohsiung University of Applied Sciences,
415, Chien Kung Rd., Kaohsiung 80778, Taiwan
{bhliu,cylien}@kuas.edu.tw, kyps503002@yahoo.com.tw, dj3662002@hotmail.com

Abstract. Sensor deployment is an important issue to provide QoS guarantees in the wireless sensor network. Recently, CRITICAL-SQUARE-GRID COVERAGE problem, which is the problem of constructing the minimum size wireless sensor network with full coverage of critical square grids in the sensor field, has been presented and addressed by many researchers. However, this problem does not take the relay nodes into consideration, where the relay nodes may exist in the sensing field and can provide data transmission. In this paper, our problem is to construct a minimum size homogeneous wireless sensor network to fully cover critical square grids when the relay nodes exist in the sensing field. A modified algorithm is proposed by considering the existence of relay nodes. The simulation results show that our method has better performance than others.

Keywords: Wireless sensor network, sensor deployment, coverage problem, relay nodes.

1   Introduction

A wireless sensor network is consisted of many sensors each having the capabilities of sensing objects, processing data, and transmitting messages. The data sensed and generated by sensors are transmitted by multi-hop routing protocols to some specific sensor, such as a sink, to gathering data. Recently, many applications [1,9,10] have been developed by wireless sensor networks, such as environmental monitoring, medical monitoring, collection of data related to the pollution of the environment, and etc. In many applications [1,9], sensors are deployed to construct wireless sensor networks to provide QoS guarantees. Therefore, coverage problem is an important issue in the wireless sensor network. In this paper, we undertake the development of deployment policy for constructing wireless sensor networks.

The full coverage problem [12,13] is a typical coverage problem, which is focused on efficiently deploying sensors such that a complete sensing field can

J.-S. Pan et al. (eds.), *Genetic and Evolutionary Computing,*
Advances in Intelligent Systems and Computing 238,
DOI: 10.1007/978-3-319-01796-9_22, © Springer International Publishing Switzerland 2014

be fully covered. In [13], Zhang and Hou study how to efficiently wake up fewer sensors in a large wireless sensor network to form a connected wireless sensor network with full coverage of the sensor field. In [12], Yun et al. study and present deployment patterns for wireless sensor networks to provide full coverage and k-connectivity with $k \leq 6$.

Another deployment issue, called barrier coverage problem [8], studies deploying sensors to construct a wireless sensor network, which acts as a barrier for detecting the intruders attempting to across the barrier. In [11], Silvestri considers the barrier coverage problem with the assistance of mobile sensors. An efficient algorithm is proposed by Silvestri to move mobile sensors to cover uncovered space such that a barrier is completely constructed.

In some applications, a number of interesting points in the sensing field have to be monitored periodically in order to ensure QoS guarantees. The security patrol is an example, where the security patrol has to move to check important areas/targets periodically. In [3], Du et al. propose an efficient algorithm to schedule fewer mobile sensors to periodically sense some specific locations while the speed of mobile sensors are considered.

Recently, a new coverage problem [5], termed CRITICAL-SQUARE-GRID COVERAGE problem, has been presented, where the problem is to construct a minimum size connected wireless sensor network to fully cover critical square grids distributed in a sensing field. In addition, Ke, Liu, and Tsai [6] propose an approximation algorithm for CRITICAL-SQUARE-GRID COVERAGE problem. However, there may exist a number of relay nodes in a sensing field, implying that messages can be transmitted through the relay nodes. Therefore, we can use the relay nodes to connect sensors to minimize the number of deployed sensors, which is not considered in the research [6]. In this paper, we take the existence of the relay nodes into consideration and propose an efficient algorithm to construct a homogeneous wireless sensor network while minimizing the number of deployed sensors. The remainder of this paper is organized as follows. The network model and problem definition are presented in Section 2. In Section 3, our algorithm is proposed. In Section 4, the performance of our method and others are evaluated. Finally, we conclude this paper in Section 5.

2   Network Model and Problem Definition

In a homogeneous wireless sensor network, every sensor has the same sensing range, R_s, and the same transmission range, R_t. Every sensor may detect events within its sensing range and send messages to sensors within its transmission range. In this paper, the binary sensor model [4] is assumed to be the sensors' sensing model, where an event is detected by a sensor if the event is within the sensor's sensing range; otherwise, the event cannot be detected by the sensor. In addition, the unit disk graph model [2] is assumed to be the sensor's communication model, where two sensors can communicate with each other if one of the sensors is within the transmission range of the other sensor. When given a sensor field, the field is divided into grids of squares with length ℓ. Let the center of each grid be called grid point. The sensors are allowed to be deployed to grid points. In the field, the grid that is critical and has to be fully covered by a sensor is called a critical grid. Let a relay node be a node that cannot sense events but can forward messages from sensors/relay nodes to sensors/relay nodes. In the presence of relay nodes, two or more disjointed sensors can be connected by the relay nodes in order to minimize the number of sensors required to form a connected wireless sensor network. Given R_s, R_t, a sensor field, the critical grids, and relay nodes, our problem is to find a minimum size homogeneous wireless sensor network to fully cover all critical grids.

Take Fig. 1 as an example of constructing a minimum size homogeneous wireless sensor network to fully cover all critical grids. Fig. 1(a) shows a given sensor field, where the sensor field is divided into 9 square grids and has 3 critical grids shown in green. Given sensor's sensing range $R_s = \frac{\sqrt{2}}{2}\ell$, sensor/relay node's transmission range $R_t = \ell$, the location of relay nodes, Fig. 1(b) shows a connected homogeneous wireless sensor network fully covering all critical grids, where the relay node is located on the grid point $(2,2)$ and three sensors are, respectively, deployed on grid points $(1,2)$, $(2,1)$, and $(2,3)$. Note that, in the presence of the relay node located on the grid point $(2,2)$, only three sensors are required to construct a connected wireless sensor network that fully covers three critical grids in the sensor field.

3   Our Method

The idea of our method is similar to Steiner-Tree-Based Critical Grid Covering Algorithm (STBCGCA) [6], which is an approximation algorithm of constructing a minimum size wireless sensor network to fully cover critical grids in a sensor field without considering relay nodes. In our method, when given the instance of our problem, including R_s, R_t, a sensor field, critical grids, and relay nodes, we can first construct an undirected weighted graph $G(V, E, w)$ by the instance. Then, a node-weighted Steiner tree is constructed by the graph $G(V, E, w)$. Finally, the deployment policy is determined by the node-weighted Steiner tree.

In the first step, an undirected weighted graph $G(V_G, E_G, w_G)$ is constructed when R_s, R_t, a sensor field, critical grids, and relay nodes are given. Let V_1 be the set of nodes $v_{i,j}$ for all grid points (i, j) in the sensor field. Let V_2 be the set of nodes $v'_{x,y}$ for all grid points (x, y) on which the relay nodes are located. And let V_3 be the set of nodes $t_{z,w}$ for all grid points (z, w) in the critical grids. Then $V_G = V_1 \cup V_2 \cup V_3$. Let E_1 denote the set of edges $(v_{i,j}, v_{p,q})$ for all grid points (i, j) and (p, q) in the sensor field such that the distance between (i, j) and (p, q) is less than or equal to R_t. Let E_2 denote the set of edges $(v_{i,j}, v'_{x,y})$ for all grid points (i, j) and (x, y) in the sensor field such that the distance between (i, j) and (x, y) is less than or equal to R_t. And let E_3 denote the set of edges $(v_{i,j}, t_{z,w})$ for all grid points (i, j) and (z, w) such that the critical grid with grid point (z, w) can be fully covered by the sensor deployed on grid point (i, j). Then $E_G = E_1 \cup E_2 \cup E_3$. Let $w_G(v) = 1$ for all $v \in V_1$ and $w_G(v) = 0$ for all $v \in V_2 \cup V_3$. Also let $w_G((u, v)) = 0$ for all $(u, v) \in E_1 \cup E_2$ and $w_G((u, v)) = 4\lceil \frac{R_s}{\lfloor \frac{R_t}{\ell} \rfloor \ell} \rceil + 1$ for all $(u, v) \in E_3$.

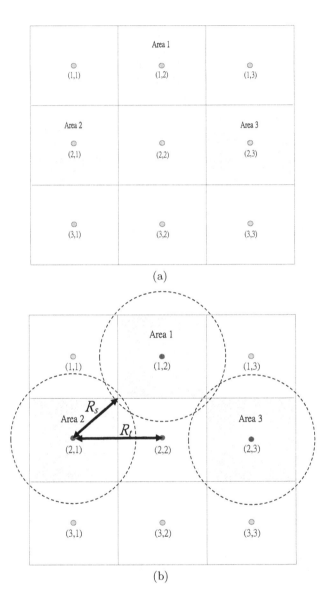

Fig. 1. Example of the problem of constructing a minimum size homogeneous wireless sensor network to fully cover all critical grids. (a) The sensor field that is divided into 9 square grids with length ℓ, where each grid has a gray grid point used to deploy sensors. (b) A homogeneous wireless sensor network formed by 3 sensors and 1 relay node, where the transmission range of each of the sensors and relay node is ℓ, the sensing range of each sensor is $\frac{\sqrt{2}}{2}\ell$, the sensors are deployed on grid points $(1,2)$, $(2,1)$, and $(2,3)$, respectively, shown in red, and the relay node is located on the grid point $(2,2)$ shown in blue.

Take Fig. 2 for example. When given the instance, including $R_s = \frac{\sqrt{2}}{2}\ell$, $R_t = \ell$, a sensor field, the critical grids, and the relay nodes, as shown in Fig. 1, an undirected weighted graph $G(V_G, E_G, w_G)$ can then be constructed by the first step of our method, as shown in Fig. 2(a). In the first step, $V_1 = \{v_{1,1}, v_{1,2}, v_{1,3}, v_{2,1}, v_{2,2}, v_{2,3}, v_{3,1}, v_{3,2}, v_{3,3}\}$, $V_2 = \{v'_{2,2}\}$, and $V_3 = \{t_{1,2}, t_{2,1}, t_{2,3}\}$. In addition, $E_1 = \{(v_{1,1}, v_{1,2}), (v_{1,1}, v_{2,1}), (v_{1,2}, v_{1,3}), (v_{1,2}, v_{2,2}), (v_{1,3}, v_{2,3}), (v_{2,1}, v_{2,2}), (v_{2,1}, v_{3,1}), (v_{2,2}, v_{2,3}), (v_{2,2}, v_{3,2}), (v_{2,3}, v_{3,3}), (v_{3,1}, v_{3,2}), (v_{3,2}, v_{3,3})\}$, $E_2 = \{(v_{1,2}, v'_{2,2}), (v_{2,1}, v'_{2,2}), (v_{2,2}, v'_{2,2}), (v_{2,3}, v'_{2,2}), (v_{3,2}, v'_{2,2})\}$, and $E_3 = \{(v_{1,2}, t_{1,2}), (v_{2,1}, t_{2,1}), (v_{2,3}, t_{2,3})\}$. Furthermore, $w_G(v) = 1$ for all $v \in V_1$ and $w_G(v) = 0$ for all $v \in V_2 \cup V_3$. $w_G((u,v)) = 0$ for all $(u,v) \in E_1 \cup E_2$ and $w_G((u,v)) = 5$ for all $(u,v) \in E_3$.

In the second step, we assume that the nodes in V_3 are terminal nodes and the other nodes in V_G are non-terminal nodes. We apply Klein and Ravi's algorithm

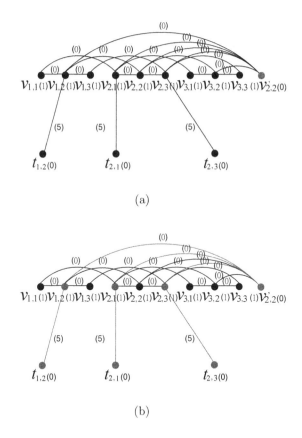

(a)

(b)

Fig. 2. Example of MINIMUM NODE-WEIGHTED STEINER TREE. (a) A graph G with nonnegative weighted nodes and edges, whose weights are shown in parentheses, where $t_{1,2}$, $t_{2,1}$, and $t_{2,3}$ are terminal nodes, and $V'_{2,2}$ is relay node. (b) A node-weighted Steiner tree ST(V_ST, E_ST), where $V_ST = v_{1,2}, v_{2,1}, v'_{2,2}, v_{2,3}, t_{1,2}, t_{2,1}, t_{2,3}$, $E_ST = (v_{1,2}, t_{1,2}), (v_{1,2}, v'_{2,2}), (v_{2,1}, v'_{2,2}), (v_{2,1}, t_{2,1}), (v_{2,3}, v'_{2,2}), (v_{2,3}, t_{2,3})$.

[7] in $G(V_G, E_G, w_G)$ to construct a node-weighted Steiner tree $T(V_T, E_T)$. Based on $G(V_G, E_G, w_G)$ shown in Fig. 2(a), we can construct $T(V_T, E_T)$ shown in Fig. 2(b), where $V_T = \{v_{1,2}, v_{2,1}, v_{2,3}, t_{1,2}, t_{2,1}, t_{2,3}, v'_{2,2}\}$ and $E_T = \{(t_{1,2}, v_{1,2}), (v_{1,2}, v'_{2,2}), (t_{2,1}, v_{2,1}), (v_{2,1}, v'_{2,2}), (t_{2,3}, v_{2,3}), (v_{2,3}, v'_{2,2})\}$.

In the third step, we obtain the locations of grid points for deployment by the node-weighted Steiner tree $T(V_T, E_T)$. Let $U = V_1 \cap V_T$. The locations for deployment are grid points (g, h) for all $v_{g,h} \in U$. Take Fig. 2 for example. By Fig. 2(a), we have $V_1 = \{v_{1,1}, v_{1,2}, v_{1,3}, v_{2,1}, v_{2,2}, v_{2,3}, v_{3,1}, v_{3,2}, v_{3,3}\}$. By Fig. 2(b), we have $V_T = \{v_{1,2}, v_{2,1}, v_{2,3}, t_{1,2}, t_{2,1}, t_{2,3}, v'_{2,2}\}$. We then have $U = \{v_{1,2}, v_{2,1}, v_{2,3}\}$. Therefore, the locations for deployment are grid points $(1, 2)$, $(2, 1)$, and $(2, 3)$.

4   Simulation Result

Here we use computer simulation to generate the instances of the deployment problem and evaluate the performance. In the simulation, a sensor field was divided into 20×20 grids each having grid length 1. The critical grids were randomly selected from the grids of the sensor field. In addition, the relay nodes with $R_t = 2$ were also randomly deployed on the grids of the sensor field. In order to form a homogeneous wireless sensor network, sensors with $R_s = \frac{3}{\sqrt{2}}$ and $R_t = 2$ were used to be deployed. In the simulation, the performance of our method and STBCGCA [6] were studied under different number of critical grids and different number of relay nodes. In the following simulation results, the empirical data were obtained by averaging the data of 100 sensor fields.

Fig. 3 shows the simulation results in terms of the average number of deployed sensors in sensor fields with 150 relay nodes and 30 to 150 critical grids. It is clear that our method needs less sensors than STBCGCA. This is because in

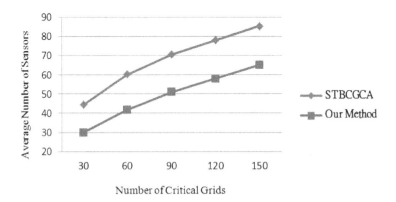

Fig. 3. Average numbers of sensors deployed to construct connected homogeneous wireless sensor networks by our method and STBCGCA in sensor fields under different number of critical grids.

our method relay nodes are considered to be used to connect sensors, which thus reduces the number of deployed sensors. In addition, it is noted that the more critical grids in the sensor field, the more sensors required to be deployed by our method and STBCGCA. This is because the sensing range $R_s = \frac{3}{\sqrt{2}}$, one sensor can fully cover only one critical grid, and thus more sensors are required to cover critical grids.

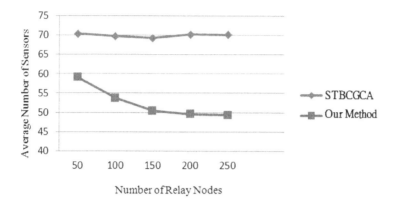

Fig. 4. Average numbers of sensors deployed to construct connected homogeneous wireless sensor networks by our method and STBCGCA in sensor fields under different number of relay nodes.

Fig. 4 shows the simulation results in terms of the average number of deployed sensors in sensor fields with 90 critical grids and 50 to 250 relay nodes. As the same observations in the previous simulation results, our method needs less sensors than STBCGCA. In addition, it is clear that the more relay nodes in the sensor field, the less sensors required to be deployed by our method. This is because more relay nodes rather than sensors are used to connect sensors.

5   Conclusion

Here we address the problem of constructing a minimum size homogeneous wireless sensor network fully covering critical square grids of the sensor field in the presence of relay nodes. An algorithm modified from STBCGCA [6] is developed by considering the existence of the relay nodes. In the simulation, the results show that our method has better performance than STBCGCA under different number of critical grids and different number of relay nodes.

Acknowledgment. This work was supported by the National Science Council under the Grant NSC 101-2221-E-151-001.

References

1. Akshay, N., Kumar, M.P., Harish, B., Dhanorkar, S.: An efficient approach for sensor deployments in wireless sensor network. In: International Conference on Emerging Trends in Robotics and Communication Technologies, pp. 350–355 (2010)
2. Clark, B.N., Colbourn, C.J., Johnson, D.S.: Unit disk graphs. Discrete Math. 86(1-3), 165–177 (1990)
3. Du, J., Li, Y., Liu, H., Sha, K.: On sweep coverage with minimum mobile sensors. In: IEEE 16th International Conference on Parallel and Distributed Systems, pp. 283–290 (2010)
4. Heo, N., Varshney, P.K.: Energy-efficient deployment of intelligent mobile sensor networks. IEEE Trans. Systems, Man, and Cybernetics-Part A: Systems and Humans 35(1), 78–92 (2005)
5. Ke, W.C., Liu, B.H., Tsai, M.J.: The critical-square-grid coverage problem in wireless sensor networks is NP-Complete. Computer Networks 55(9), 2209–2220 (2011)
6. Ke, W.C., Liu, B.H., Tsai, M.J.: Efficient algorithm for constructing minimum size wireless sensor networks to fully cover critical square grids. IEEE Trans. Wireless Communications 10(4), 1154–1164 (2011)
7. Klein, P.N., Ravi, R.: A nearly best-possible approximation algorithm for node-weighted steiner trees. J. Algorithms 19(1), 104–115 (1995)
8. Kumar, S., Lai, T.H., Arora, A.: Barrier coverage with wireless sensors. In: Proceedings of ACM MobiCom, Cologne, Germany (2005)
9. Lei, Y., Zhang, Y., Zhao, Y.: The research of coverage problems in wireless sensor network. In: International Conference on Wireless Networks and Information Systems, pp. 31–34 (2009)
10. Mittal, R., Bhatia, M.P.S.: Wireless sensor networks for monitoring the environmental activities. In: IEEE International Conference on Computational Intelligence and Computing Research, pp. 1–5 (2010)
11. Silvestri, S.: Mobibar: Barrier coverage with mobile sensors. In: IEEE Global Telecommunications Conference, pp. 1–6 (2011)
12. Yun, Z., Bai, X., Xuan, D., Lai, T.H., Jia, W.: Optimal deployment patterns for full coverage and k-connectivity (k ≤ 6) wireless sensor networks. ACM/IEEE Transactions on Networking 18(3), 934–947 (2010)
13. Zhang, H., Hou, J.C.: Maintaining sensing coverage and connectivity in large sensor networks. Ad Hoc & Sensor Wireless Networks 1(1-2), 89–124 (2005)

Relative Location Estimation over Wireless Sensor Networks with Principal Component Analysis Technique

Shao-I Chu[1], Chih-Yuan Lien[1], Wei-Cheng Lin[2], Yu-Jung Huang[3],
Chung-Long Pan[2], and Po-Ying Chen[4]

[1] Department of Electronic Engineering,
National Kaohsiung University of Applied Sciences, Taiwan
[2] Department of Electrical Engineering, I-Shou University, Taiwan
[3] Department of Electronic Engineering,
I-Shou University, Taiwan
[4] Department of Information Engineering,
I-Shou University, Taiwan
erwinchu@kuas.edu.tw

Abstract. This paper presents the maximum likelihood (ML)-based approaches for relative location estimation over the correlated wireless sensor networks (WSNs), which innovatively exploit the principal component analysis (PCA) and probabilistic PCA to transform all received signal strength (RSS) measurements into useful information. Simulation results reveal that the proposed approaches remarkably outperform the other existing schemes when a high correlation exists or a strong noise power occurs. Taking the path-loss exponent into consideration, it is observed that the higher the path-loss exponent, the lower the location estimation error. These results show that the proposed approach is suitable for the practical correlated wireless channels.

Keywords: received signal strength (RSS), maximum likelihood (ML), principal component analysis (PCA), correlation.

1   Introduction

Location estimation or tracking of target is one of the most interesting applications in wireless sensor networks (WSNs) [1-2]. Localization can be applied to a variety of applications such as pet-tracking, patient-caring and so on. There have been several location estimation techniques [3-5] by using the measurements such as time of arrival (TOA), time difference of arrival (TDOA), angle of arrival (AOA), and received signal strength (RSS). There exist tradeoffs between the localization accuracy and the implementation complexity. The RSS-based technique is a simple solution when the low-cost and easy-to-implement issues are taken into account. Moreover, the current wireless communication standards, such as IEEE 802.11 and IEEE 802.15.4, support the function of measuring RSS values in their protocols.

The commonly used techniques for relative location estimation are classified into three types: the minimum mean square error (MMSE) estimation [6], the maximum

J.-S. Pan et al. (eds.), *Genetic and Evolutionary Computing*,
Advances in Intelligent Systems and Computing 238,
DOI: 10.1007/978-3-319-01796-9_23, © Springer International Publishing Switzerland 2014

likelihood (ML) location estimation [5] and maximum a posteriori probability (MAP) estimation. In the MMSE scheme, the estimated position of an unknown node is obtained by minimizing the total squared errors over all differences between the measured distances and the estimated distances. The ML approach [5] exploits the log-normal channel model to estimate the target position. When the statistics of the measurement error are known, the ML estimator is asymptotically optimal. For the planned deployment of the wireless sensor nodes, some information such as the size and shape of a room and the arrangement of the anchor nodes may be known as a priori. Consequently, the MAP algorithm may be an option by utilizing the prior information. Similar to the MAP concept, Chang *et al.* [7-8] recently proposed a probability-based maximum likelihood (PML) scheme to significantly improve the estimation accuracy. However, the correlation effects on the RSS values among all anchor nodes were not addressed.

Instead of solving the optimization problem for location estimation, the fingerprinting architecture was proposed. The basic design of fingerprint includes two phases: off-line and online. In the off-line phase, the RSS values are collected at sampling locations to establish the database (radio map) for the environment. In the online phase, the physical location of the unknown node can be estimated by comparing the measured RSS with the stored RSS values in the database. More recently, Fang *et al.* [10-11] exploited the technique of conventional principal component analysis (PCA) and the hybrid projection approach to reduce the dimension of RSS values in the stored database. Although the idea of fingerprint is quite simple, there are plenty of measurements needed to construct the database for online RSS comparison.

In this paper, we investigate the relative location estimation problem over the WSNs, where the channel correlation exists. Instead of picking the several strongest RSS values, the proposed estimation procedure condenses all measurements into the useful information of a lower dimension via PCA or probabilistic PCA techniques. By such a linear transformation, the ML algorithm is therefore applied to estimate the coordinate of the unknown node or object. Simulation results show that the proposed approaches are significantly superior to other existing schemes in a wireless environment with a high correlation or a strong noise power. In the cases of a high path-loss exponent, the estimation error is remarkably improved. Such results demonstrate that the proposed schemes achieve a good performance on the estimation accuracy.

2   Wireless Channel Model and Well-Known Localization Algorithms

2.1   Lognormal-Distributed Wireless Channel

Consider a wireless sensor network, including N anchor (reference) nodes and an unknown (target) node. The coordinate location of anchor node i is $\mathbf{z}_i = (x_i, y_i)$, $i = 1$, $2,\dots, N$, and the location of the unknown (target) node is $\mathbf{z}_i = (x_0, y_0)$. Let d_i be the Euclidean distance from anchor node i to the unknown node. That is,

$$d_i \equiv \|\mathbf{z}_i - \mathbf{z}_0\| = \sqrt{(x_i - x_0)^2 + (y_i - y_0)^2}, i = 1, 2, \cdots, N. \tag{1}$$

The RSS value measured from anchor node i can be modeled as follows [9]:

$$r_i = r_0 - 10\eta \log_{10} d_i + W_i, i = 1, 2, \cdots, N, \tag{2}$$

where r_0 is the received signal strength at 1 m and η is the path-loss exponent. W_1, W_2, ... and W_N are jointly Gaussian distributed random variables with zero mean and covariance matrix Σ_W. Note that η and Σ_W can be calibrated from the environment.

Let $\tilde{d}_i$ be the estimated distance from the unknown node to anchor node i. Based on (2), $\tilde{d}_i$ can be estimated by

$$\tilde{d}_i = 10^{(r_0 - r_i)/(10\eta)}, i = 1, 2, \cdots, N. \tag{3}$$

In practice, we may not utilize all RSS values from N anchor nodes for location estimation. Instead, only K strongest RSS values ($K \leq N$) are selected because of computational burden on the device. To convey the following discussions, let A be the set of anchor nodes whose RSS values are among the K strongest ones. Some well-known localization algorithms are presented as follows.

2.2 Minimum Mean Square Error Estimation (MMSE)

The minimum mean square error estimation [6] is commonly used, which aims to minimize the square error over all differences between the measured distances and the estimated distances. That is,

$$\tilde{\mathbf{z}}_0 = \arg\min_{\mathbf{z}_0} \sum_{i \in A} (d_i - \tilde{d}_i)^2, \tag{4}$$

where $\tilde{d}_i$ can be calculated by (3).

2.3 Maximum Likelihood Estimation (ML)

By taking the probabilistic concept into account, Patwari et al. [5] proposed Bias-reduced Maximum Likelihood (BML) to better estimate node position by using the log-normally distributed distance measure in (2). In [5], the correlation effects of RRS values among the anchor nodes were not discussed. Therefore, the generalized maximum likelihood estimation is described as follows:

$$\tilde{\mathbf{z}}_0 = \arg\min_{\mathbf{z}_0} p(\mathbf{r}_A | \mathbf{d}_A), \tag{5}$$

where $p(\mathbf{r}_A | \mathbf{d}_A) = (2\pi)^{-K/2} |\Sigma_A|^{-1/2} \exp\left\{-\frac{1}{2}(\mathbf{r}_A - \boldsymbol{\mu}_A)'(\Sigma_A^{-1})(\mathbf{r}_A - \boldsymbol{\mu}_A)\right\}$. $\mathbf{r}_A = (r_{i_1} \cdots r_{i_K})'$, $\mathbf{d}_A = (d_{i_1} \cdots d_{i_K})'$ and $\boldsymbol{\mu}_A = (r_0 - 10\eta \log d_{i_1} \cdots r_0 - 10\eta \log d_{i_K})'$, where $\{i_1, i_2, \cdots, i_K\} \in A$. $|\Sigma_A|$ denotes the determinant of the covariance matrix Σ_A.

2.4   Maximum A Posteriori Estimation (MAP)

The objective of MAP is to maximize the conditional probability $p(\mathbf{d}_A|\mathbf{r}_A)$. Since $p(\mathbf{d}_A|\mathbf{r}_A) = p(\mathbf{r}_A|\mathbf{d}_A)p(\mathbf{d}_A)/p(\mathbf{r}_A)$ and $p(\mathbf{r}_A)$ is a constant. Therefore, the location estimation by MAP is expressed as

$$\tilde{\mathbf{z}}_0 = \arg\max_{\mathbf{z}_0} p(\mathbf{r}_A \mid \mathbf{d}_A)p(\mathbf{d}_A). \tag{6}$$

where $p(\mathbf{d}_A) = \prod_{i \in A} p_i(d_i)$. For the planned deployment, the probability density function (PDF) of d_i for anchor node i, $p_i(d_i)$, serves as a priori. However, $p_i(d_i)$ actually depends on the size and shape of a room to be monitored, and the arrangement of the anchor nodes. Note that the concept of MAP is equivalent to that of the probability–based maximum likelihood estimation (PML) proposed in [7-8], where they assume all channel are independent and do not take the channel correlation into consideration.

3   Proposed PCA-Based Maximum Likelihood Localization

In this section, we intend to determine the transformation matrix $\mathbf{\Phi}$ instead of selecting the strongest anchor nodes. The transformation matrix is used to transform the original RSS vector $\mathbf{r}$ into a new variable $\mathbf{y}$. That is,

$$\mathbf{y} = \mathbf{\Phi r}, \tag{7}$$

where $\mathbf{r} = (r_1 \quad \cdots \quad r_N)'$, $\mathbf{y} = (y_1 \quad \cdots \quad y_K)'$ and $\mathbf{\Phi} \in \Re^{K*N}$. The PCA and probabilistic PCA are applied to reduce the dimension of RSS values in this paper.

3.1   Principal Component Analysis

The key idea of PCA [13] is to minimize the mean square error, i.e., $E(\| \mathbf{r} - \mathbf{\Phi}^T \mathbf{y} \|^2)$. The transformation matrix $\mathbf{\Phi}$ can be obtained by solving of an eigenvalue-eigenvector problem for a positive-semidefinite symmetric matrix $\mathbf{S}$. $\mathbf{S}$ is defined as

$$\mathbf{S} = \frac{1}{N_s} \sum_{i=1}^{N_s} (\mathbf{r}_i - \bar{\mathbf{r}})(\mathbf{r}_i - \bar{\mathbf{r}})^T, \tag{8}$$

where N_s is the number of training samples, $\mathbf{r}_i$ is the RSS vector for the ith sample and $\bar{\mathbf{r}}$ is the sample mean of all RSS samples. For the matrix $\mathbf{S}$, the eigenvalues and their corresponding eigenvectors are denoted as $\{\lambda_1,...., \lambda_N\}$ and $\{\mathbf{v}_1,....\mathbf{v}_N\}$, respectively. They satisfy the following relationship.

$$\mathbf{S}\mathbf{v}_i = \lambda_i \mathbf{v}_i, i = 1,2,\cdots,N. \tag{9}$$

$$\lambda_i \geq \lambda_j, \forall i > j. \tag{10}$$

The optimized $\mathbf{\Phi}_{PCA}$ [13] by the PCA technique has the form.

$$\mathbf{\Phi} = \mathbf{\Phi}_{PCA} = [\mathbf{v}_1 \quad \cdots \quad \mathbf{v}_K]^T. \tag{11}$$

3.2   Probabilistic Principle Component Analysis (PPCA)

Given the desirable dimension of RSS values (K), the probabilistic principal component analysis [12] can be viewed as a maximum likelihood procedure based on the isotropic Gaussian noise model, as mentioned in factor analysis. Therefore, the maximum likelihood estimator of Φ_{PPCA} can be computed by the closed-form expression in [12] or the iterative expectation maximization (EM) algorithm. See [12] for the details. Note that Φ_{PCA} and Φ_{PPCA} can be obtained from the baseline and calibration experiments in the real environment.

3.3   Proposed Location Estimation Procedure

The proposed location estimation procedure is composed of two phases. The first phase intends to get a good initial point for the ML optimization solver by applying the MMSE procedure; the second phase aims at calculating an optimal location coordinate based on ML criterion incorporated with PCA or PPCA.

***Phase I*:** Input: $\mathbf{r}$, Output: $\tilde{\mathbf{z}}_0^{MMSE}$

1. Set an arbitrary initial point for searching.
2. Execute Levenberg–Marquardt algorithm [14] for the optimization problem in (4).

***Phase II*:** Input: $\mathbf{r}$, $\tilde{\mathbf{z}}_0^{MMSE}$, Output: $\tilde{\mathbf{z}}_0$

1. Set $\Phi = \Phi_{PCA}$ or $\Phi = \Phi_{PPCA}$.
2. Compute $\mathbf{y} = \Phi\mathbf{r}$.
3. Execute Levenberg–Marquardt algorithm with the initial point $\tilde{\mathbf{z}}_0^{MMSE}$ to solve the optimization problem as follows:

$$\tilde{\mathbf{z}}_0 = \arg\max_{\mathbf{z}_0} p(\mathbf{y} \mid \mathbf{d}), \tag{12}$$

where $p(\mathbf{y} \mid \mathbf{d}) = (2\pi)^{-K/2} \mid \Phi\Sigma\Phi^T \mid^{-1/2} e^{-\frac{1}{2}(\mathbf{y}-\Phi\mu)'(\Phi\Sigma\Phi^T)^{-1}(\mathbf{y}-\Phi\mu)}$.

Here, ML location estimation algorithms, which are combined with PCA and PPCA techniques, are called ML-PCA and ML-PPCA, respectively.

4   Performance Evaluation and Comparison

This section evaluates and compares the estimation accuracy of the proposed algorithms (ML-PCA and ML-PPCA) with the MMSE, ML and MAP techniques. Consider an indoor environment of a 50m*20m corridor, which includes six anchor nodes as shown in Fig. 1. In this scenario, the probability density functions (PDF) of d_i serve as the prior information and can be derived via the cumulative probability function (CDF). See Appendix for the details. For each anchor node, the RSS perceived at the unknown node is randomly generated based on the radio channel model in (2), where the RSS value at 1 m, r0, is set to -55dB. For easy discussion, the covariance matrix of the channel has the form $\Sigma_W = \{\sigma_{ij}\} \in \Re^{N*N}$, where $\sigma_{ij} = \rho\sigma_2$ if $i \neq j$, and $\sigma_{ij} = \sigma^2$ if $i = j$. ρ means the correlation coefficient and $0 \leq \rho \leq 1$. In the baseline and calibration experiments, we collect 3000 RSS samples to calculate the transformation matrix Φ_{PCA} and Φ_{PPCA}. In addition, 2000 simulations were conducted to evaluate the average distance error as the performance metric, which is the Euclidean distance between the estimated location and the true coordinates.

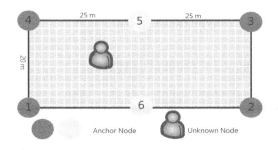

Fig. 1. A 50m*20m corridor with six anchor nodes

4.1   Effects of Noise Power

This simulation experiment investigates the noise effect on the estimation performance. In this simulation, the path-loss exponent is $\eta = 3.5$ and the correlation coefficient is $\rho = 0.8$. For MMSE, ML and MAP algorithms, they exploit the five strongest RRS values for localization. In Fig. 2, it is observed that the estimation accuracy of all approaches deteriorates with the increase of σ. Comparison of all localization algorithms indicates that ML-PCA and ML-PPCA significantly outperform MMSE, ML and even MAP in the cases of large σ. ML-PCA has a better performance when $\sigma = 8$. However, the performances of ML, MAP, ML-PCA and ML-PPCA are almost the same when the noise power is relatively weak.

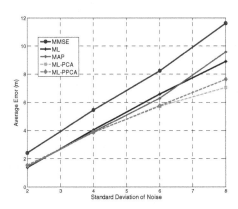

Fig. 2. Average estimation error with respect to noise power

4.2   Effects of Channel Correlation

How the channel correlation affects the estimation error is the aim of this simulation study. The path-loss exponent is $\eta = 3.5$ and the standard deviation of the noise is $\sigma = 6$. As the previous simulation setting, the five strongest RRS values are utilized for localization. Fig. 3 depicts the average estimation errors with respect to ρ when different estimation methods are applied. It reveals that the proposed algorithms do not give any benefits from the reduction of RSS dimension when the channels have a low correlation. As the correlation coefficient increases, ML-PCA and ML-PPCA surpass other approaches. The reason comes from the fact that PCA converts a set of observations of possibly correlated variables into a set of values of uncorrelated variables. Via such a transformation, the new measurements provide better localization accuracy. Such results demonstrate the superiority of the proposed scheme over a correlated wireless sensor network.

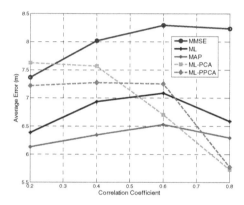

Fig. 3. Average estimation error with respect to correlation coefficient

4.3   Effects of Selected Anchor Nodes

The objective of the experiment intends to analyze the effect of the K strongest RSS values from anchor nodes on the estimation accuracy. Again, we set $\eta = 3.5$, $\sigma = 6$ and $\rho = 0.8$. We evaluate the estimation performances when $K = 3, 4, 5$ and 6, respectively. In Fig. 4, when $K = 3$, the proposed algorithms have worse performance. In the cases of $K = 4$ and 5, ML-PCA and ML-PPCA have good estimation accuracy. As all RSS values are adopted for location estimation, i.e., $K = 6$, the performances of ML, ML-PCA and ML-PPCA are obviously the same. In addition, MAP outperforms other schemes in the case of $K = 3$, while its performance gets worse with the increase of K.

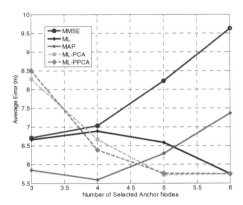

Fig. 4. Average estimation error with respect to anchor nodes

4.4   Effects of Path-Loss Exponent

This numerical experiment aims to investigate the effect of the path-loss exponent η on the estimation performance. The higher the value of η, the more the signal strength is attenuated. The simulation environment is set as follows: $\sigma = 6$, $\rho = 0.8$ and $K=5$. Fig 5 shows that the average estimation error decreases with the increase of η. ML-PCA and ML-PPCA are always superior to other schemes. As compared to MAP, the proposed algorithms still have the outstanding performance even if MAP has the prior information for location estimation based on the geometry of the coverage area.

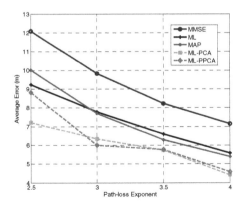

Fig. 5. Average estimation error for with repect to path-loss exponents

5   Conclusions

The PCA-based ML approaches for relative location estimation have been proposed such that the all RSS information is effectively utilized. Simulation results showed that the proposed schemes significantly outperform other existing schemes in the cases of a high correlation and a strong noise power. It revealed the superiority of ML-PCA and ML-PPCA in the estimation accuracy. However, in practice, the integration of ML with PCA or PPCA may lead to additional computations that require more power consumption and performance overhead. Therefore, the evaluation needs to be further investigated in the future.

Appendix

By evaluating cumulative density functions (CDFs), the prior probability density of d_i for anchor node i in Fig.1 is derived as follows:

For $i = 1, 2, 3$ and 4,

$$p_i(d_i) = \begin{cases} \dfrac{\pi d_i}{2wl}, & d_i \le w \\ \dfrac{d_i(\pi/2 - \cos^{-1}(w/d_i))}{wl}, & w < d_i \le l \\ \dfrac{d_i(\pi/2 - 2\cos^{-1}(w/d_i))}{wl}, & l < d_i \le \sqrt{2}l \\ 0, & d_i > \sqrt{2}l \end{cases} \tag{13}$$

For $i = 5$ and 6,

$$p_i(d_i) = \begin{cases} \dfrac{\pi d_i}{wl}, & d_i \le w \\ \dfrac{2d_i(\pi/2 - \cos^{-1}(w/d_i))}{wl}, & w < d_i \le l/2 \\ \dfrac{2d_i(\pi/2 - \cos^{-1}(w/d_i) - \cos^{-1}(l/2d_i))}{wl}, & l/2 < d_i \le \sqrt{w^2 + (l/2)^2} \\ 0, & d_i > \sqrt{w^2 + (l/2)^2} \end{cases} \tag{14}$$

References

1. Sendonaris, A., Erkipand, E., Aazhang, B.: User cooperation diversity—Part I: System description. IEEE Trans. Commun. 51(11), 1927–1938 (2003)
2. Liu, H., Darabi, H., Banerjee, P., Liu, J.: Survey of wireless indoor positioning techniques and systems. IEEE Trans. Syst., Man, Cybern. C, Appl. Rev. 37(6), 1067–1080 (2007)
3. Vossiek, M., Wiebking, L., Gulden, P., Wieghardt, J., Hoffmann, C., Heide, P.: Wireless local positioning. IEEE Microw. Mag. 4(4), 77–86 (2003)
4. Al-Jazzar, S., Ghogho, M.: A joint TOA/AOA constrained minimization method for locating wireless devices in non-line-of-sight environment. In: Proc. IEEE VTC-Fall (2007)
5. Al-Jazzar, S., Caffery, J.J.: ML and Bayesian TOA location estimators for NLOS environments. In: Proc. IEEE VTC-Fall (2002)
6. Patwari, N., Hero, A.O., Perkins, M., Correaland, N.S., O'Dea, R.J.: Relative location estimation in wireless sensor networks. IEEETrans. Signal Processing 51(8), 2137–2148 (2003)

7. Li, Z., Trappe, W., Zhang, Y., Nath, B.: Robust statistical methods for securing wireless localization. In: Proc. Symposium on Information Processing in Sensor Networks (2005)
8. Chang, C.-H., Liao, W.: Revisiting relative location estimation in wireless sensor networks. In: Proc. IEEE ICC (2009)
9. Chang, C.-H., Liao, W.: A probabilistic model for relative location estimation in wireless sensor networks. IEEE Commun. Lett. 13(12), 893–895 (2009)
10. Chris, S., Jan, M.R., Koen, L.: Robust positioning algorithms for distributed ad-hoc wireless sensor networks. In: Proc. USENIX Annual Technical Conference (2002)
11. Fang, S., Lin, T.: Principal component localization in indoor WLAN environments. IEEE Trans. Mobile Comput. (published online)
12. Fang, S., Wang, C.-H.: A dynamic hybrid projection approach for improved Wi-Fi location fingerprinting. IEEE Trans. Veh. Technol. 60(3), 1037–1044 (2011)
13. Tipping, M.E., Bishop, C.M.: Probabilistic principal component analysis. Journal of the Royal Statistical Society, Series B 61(3), 611–622
14. Duda, R., Hart, P., Stork, D.: Pattern Classification. John Wiely& Sons (2000)
15. Levenberg, K.: A method for the solution of certain non-linear problems in least squares. The Quarterly of Applied Mathematics 2, 164–168 (1944)

Hierarchical Particle Swarm Optimization Algorithm of IPSVR Problem

Shang-Kuan Chen[1,*], Gen-Han Wu[2], Yen-Wu Ti[3], Ran-Zan Wang[4], Wen-Pinn Fang[1], and Chian-Jhu Lu[2]

[1] Department of Computer Science and Information Engineering, Yuanpei University
[2] Graduate Institute of Logistics Management, National Dong Hwa University
[3] Department of Computer Science & Information Engineering,
Hwa Hsia Institute of Technology
[4] Department of Computer Science and Engineering, Yuan Ze University
skchen@mail.ypu.edu.tw

Abstract. Production scheduling (PS) and vehicle routing (VR) are integrated to solve a production issue with timing requirement. The issue of multiple production chains and multiple vehicles is considered to produce productions and deliver them to customers. In the integrated production scheduling and vehicle routing (IPSVR) problem, each order is normally defined by its dependent setup time and processing time for the producing process and by its delivering time and time window for the delivering process. In this paper, a hierarchical particle swarm optimization algorithm is proposed for solving the IPSVR problem and reaching the minimum tardiness time.

Keywords: production scheduling, vehicle routing, hierarchical particle swarm optimization.

1 Introduction

For the progressively competing environment, more and more production companies need to change their production line to meet the requirement of the customers. Scheduling the orders in the production line becomes an important issue for effectively utilizing production line. However, the customers usually restrict a due time and a deadline for their orders. When their order is accomplished over a due time but meets deadline, the customers will ask penalty for the late order. Moreover, if the order doesn't meet the deadline, the customer will require more indemnifications from the company and the company will damage its reputation and loss the customer. Therefore, before accepting the orders, the company must evaluate which order should be taken and which order should be given up in order to avoid past the deadline. This kind of production scheduling problem is called order acceptance and scheduling (OAS) which is addressed in [1-3].

---

[*] Corresponding author.

J.-S. Pan et al. (eds.), *Genetic and Evolutionary Computing*,
Advances in Intelligent Systems and Computing 238,
DOI: 10.1007/978-3-319-01796-9_24, © Springer International Publishing Switzerland 2014

Normally, in a supply chain, the problem of production scheduling [4-6] ignores the back-end delivery and the problem of vehicle routing [7-10] does not consider the issue of the front-end production scheduling. Therefore, the result optimizes individually instead of optimizes integrally. In order to obtain the entire optimal solution, the problems of just-in-time production and that of just-in-time delivery are integrated into one problem.

Some important issues, such as identical parallel machines, orders, sequence-dependent setup times, identical vehicles, customers, time windows are considered. A mixed integer programming model is developed to integrate these two sub-problems and an optimization algorithm minimizes the total tardiness time.

Due to the extremely complexity of this integrated problem, only the small-sized problems are tested for confirming our algorithm with the ability of obtaining optimal solution. We adopt the commercial software LINGO 11.0 that is based on branch & bound method to find the optimal solution.

The rest of this paper is organized as follows. Section 2 describes the proposed mathematical model. The proposed method is shown on Section 3. Experimental results are demonstrated in Section 4. Finally, brief conclusions are given in Section 5.

2   The Proposed Mathematical Model

An integer linear programming formulation is adopted for the mathematical model of the IPSVR problem. The parameters are defined as follows:

pt_i = the processing time of order i
st_{0i} = the setup time of order i producing first
st_{ij} = the setup time of order j producing after order i
tr_{0i} = the transporting time from factory to customer i
tr_{ij} = the transporting time from customer i to customer j
a_i = the lower bound of time window of order i
b_i = the upper bound of time window of order i
w_i = the weight of order i
u_i = size of order i
cap_k = capacity of vehicle k
M = a large number
n = number of orders

Two sets of binary variables, za, xa, are defined to handle machine production and two sets of binary variables, zb, xb, are to handle vehicle transportation.

$$za_i^m = \begin{cases} 1 & \text{if order } i \text{ is produced by machine } m \\ 0 & \text{otherwise} \end{cases}$$
$$xa_{ij}^m = \begin{cases} 1 & \text{if machine m produces order } i, \text{then produces order } j \\ 0 & \text{otherwise} \end{cases}$$
$$zb_i^k = \begin{cases} 1 & \text{if order } i \text{ is transported by vehicle } k \\ 0 & \text{otherwise} \end{cases}$$
$$xb_{ij}^k = \begin{cases} 1 & \text{if vehicle } k \text{ transports order } i, \text{then transports order } j \\ 0 & \text{otherwise} \end{cases}$$

The continuous variables, $T1_i$, $T2_i$, $T3_i^k$, T_i, are defined to handle the production finish time of order i, the arrival time of order i, the departure time of vehicle k carry with order i, the delay of order i, respectively.

Objective function:

$$\min \sum_{i=1}^{n} w_i T_i$$

s.t.

$$\sum_m za_i^m = 1 \qquad \forall i \tag{1}$$

$$xa_{ij}^m \leq za_i^m \qquad i \neq j, \forall i, j, m \tag{2}$$

$$xa_{ij}^m \leq za_j^m \qquad i \neq j, \forall i, j, m \tag{3}$$

$$T1_i \geq st_{oi} za_i^m + pt_i za_i^m \qquad \forall i, m \tag{4}$$

$$T1_j - T1_i + M(3 - xa_{ij}^m - za_i^m - za_j^m) \geq st_{ij} + pt_j$$
$$i \neq j, \forall i, j, m \tag{5}$$

$$T1_i - T1_j + M(2 + xa_{ij}^m - za_i^m - za_j^m) \geq st_{ji} + pt_i$$
$$i \neq j, \forall i, j, m \tag{6}$$

$$T1_i \geq 0 \qquad \forall i \tag{7}$$

$$st_{oi} \geq 0 \qquad \forall i \tag{8}$$

$$st_{ij} \geq 0 \qquad i \neq j, \forall i, j \tag{9}$$

$$pt_i \geq 0 \qquad \forall i \tag{10}$$

$$\sum_k zb_i^k = 1 \qquad \forall i \tag{11}$$

$$xb_{ij}^k \leq zb_i^k \qquad i \neq j, \forall i, j, k \tag{12}$$

$$xb_{ij}^k \leq zb_j^k \qquad i \neq j, \forall i, j, k \tag{13}$$

$$T3_i^k \leq M \cdot zb_i^k \qquad \forall i, k \tag{14}$$

$$T3_i^k - T1_j + M(2 - zb_i^k - zb_j^k) \geq 0 \qquad \forall i, j, k \tag{15}$$

$$T3_i^k - T3_j^k \leq M(2 - zb_i^k - zb_j^k) \qquad \forall i, j, k \tag{16}$$

$$T2_i \geq T3_i^k + tr_{oi} zb_i^k$$
$$\forall i, k \tag{17}$$

$$T2_j - T2_i + M(3 - xb_{ij}^k - zb_i^k - zb_j^k) \geq tr_{ij}$$
$$i \neq j, \forall i, j, k \tag{18}$$

$$T2_i - T2_j + M(2 + xb_{ij}^k - zb_i^k - zb_j^k) \geq tr_{ji}$$
$$i \neq j, \forall i, j, k \tag{19}$$

$$\sum_i u_i \, zb_i^k \leq cap_k \qquad \forall k \qquad (20)$$

$$T2_i \geq a_i \qquad \forall i \qquad (21)$$

$$T_i \geq T2_i - b_i \qquad \forall i \qquad (22)$$

$$T_i \geq 0 \qquad \forall i \qquad (23)$$

$$T2_i \geq 0 \qquad \forall i \qquad (24)$$

$$tr_{oi} \geq 0 \qquad \forall i \qquad (25)$$

$$tr_{ij} \geq 0 \qquad i \neq j, \forall i, j \qquad (26)$$

$$T3_i^k \geq 0 \qquad \forall i, k \qquad (27)$$

3   The Proposed Method

In this section, a particle swarm optimization and variable neighborhood search method is adopted for solving the IPSVR problem. The mathematical model having shown in previous section rules the variables of the IPSVR problem. We design a hierarchical variable neighborhood search (VNS) methods both in machine production and vehicle transportation. The adopted VNS method includes order exchanging on machine, machine correcting, order inserting on machine, and machine exchanging (machine level), and order exchanging on machine, machine correcting, order inserting on machine, and machine exchanging (vehicle level). First, we randomly generate p particles by a set of random numbers for initial solutions. For each particle, it will search better solutions by VNS. If there is no better solution can be found by VNS, The particles will change their directions by their personal best solution and global best solution. The searching process will routine until it meets a terminal condition. The algorithm is summarized as follows.

Step 1: Generate k machine particles by their location set $xm_1, xm_2, \ldots, xm_k$ and their velocity set $vm_1, vm_2, \ldots, vm_k$. The $xm_i = \{xm_{i1}, xm_{i2}\}$ and $vm_i = \{vm_{i1}, vm_{i2}\}$, where xm_{i1} means the production sequence, xm_{i2} means machine set for production.

Step 2: Set parameters: the inertia weight "w", the weight of personal best solution "c_1", the weight of global best solution "c_2", random parameters r_1 and r_2 generated by uniform distribution $U(0, 1)$.

Step 3. Update personal best solution pbm_{ij} and global best solution gbm_j.

Step 4: Generate k vehicle particles by their location set $xv_1, xv_2, \ldots, xv_k$ and their velocity set $vv_1, vv_2, \ldots, vv_k$. The $xv_i = \{xv_{i1}, xv_{i2}\}$ and $vv_i = \{vv_{i1}, vv_{i2}\}$, where xv_{i1} means the routing sequence, vv_{i2} means vehicle set for production.

Step 5: For each particle, do the VNS processes for finding personal best solution pbv_{ij}.

Step 6: From personal best solutions, decide global best solution gbv_j.

Step 7: Update velocities and locations about vehicle routing by the following equations.

$$vv_{ij} = wvv_{ij} + c_1r_1(pbv_{ij} - xv_{ij}) + c_2r_2(gbv_j - xv_{ij})$$
$$xv_{ij} = xv_{ij} + vv_{ij}$$

Step 8: If it does not meet the local terminal condition, then go to Step 5.

Step 9: Update velocities and locations about machine production by the following equations.

$$vm_{ij} = wvm_{ij} + c_1r_1(pbm_{ij} - xm_{ij}) + c_2r_2(gbm_j - xm_{ij})$$
$$xm_{ij} = xm_{ij} + vm_{ij}.$$

Step 10: If it meets the terminal condition, the algorithm ends; otherwise, go to Step 3Experimental Results

We use LINGO 11.0, the optimization software, to get optimum solution. The parameters are given by random numbers with given lower bound and upper bound.

Table 1. The parameters

Parameters	(lower bound, upper bound)
Processing time	(25, 50)
Weight of order	(1, 2)
Setup time(first)	(10, 20)
Setup time(dependent)	(8, 18)
Size of order	(1,3)
Capacity of vehicle	(16, 21)
Lower bound of time window	(65, 99)
Upper bound of time window	(95, 180)
Transporting time (first)	(18, 45)
Transporting time (dependent)	(20, 50)

There is a set of parameters in this experiment for 5 orders. The parameters are generated as follows:

Processing time
49   44   44   43   43

Setup time (first)

13	14	15	12	20

Setup time (dependent)

0	9	11	15	13
15	0	18	13	17
11	9	0	17	8
11	10	9	0	10
11	10	8	14	0

Lower bound of time window

88	74	82	88	99

Upper bound of time window

122	105	115	176	115

Transporting time (from factory)

23	31	41	29	23

Transporting time (dependent)

0	38	42	39	23
36	0	37	26	47
36	26	0	40	22
27	20	43	0	43
35	24	22	22	0

Weight of order

2	2	1	2	2

If the number of machine equals to 3 and that of vehicle equals to 2, then the best solution of the proposed IPSVR problem based on the above parameters is shown as follows.

The best schedule (machine):
Machine 1   (order 5 ==> order 2)
Machine 2   (order 3 ==> order 4)
Machine 3   (order 1)
The best schedule (vehicle):
Vehicle 1   (order 1 ==> order 5 ==> order 3)
Vehicle 2   (order 2 ==> order 4)
The minimum delay: 108

The minimum tardiness time obtained by the proposed method is the same as the one obtained by LINGO 11.0. We summarized the experimental results in Table 2 and the corresponding parameters are shown as follows.

Processing time

49	44	44	43	43	32

Setup time (first)

14	15	12	20	17	13

Setup time (dependent)

0	15	18	13	17	11
9	0	17	8	11	10
9	10	0	11	10	8
14	10	13	0	17	12
9	14	16	14	0	9
13	14	10	17	13	0

Lower bound of time window

82	88	99	74	67	71

Upper bound of time window

176	115	111	124	155	142

Transporting time (from factory)

23	36	19	24	18	39

Transporting time (dependent)

0	43	35	24	22	22
37	0	49	28	36	44
49	46	0	24	29	20
40	34	28	0	30	22
26	48	43	43	0	37
40	26	44	21	35	0

Weight of order

1	1	1	1	2	2

Table 2. The experimental results

Minimal tardiness time	(No. of orders, No. of machines, No. of vehicles)	
	(6, 1, 1)	(6, 2, 2)
The optimal solution of LINGO 11	2004	262
The solution of the proposed method	2004	262

The following table shows that our method is efficient.

Table 3. The experimental results

Time spanning (seconds)	(No. of orders, No. of machines, No. of vehicles)	
	(5, 3, 2)	(6, 1, 1)
The optimal solution of LINGO 11	99	515
The solution of the proposed method	1.201	1.435

4   Conclusions

A particle swarm optimization algorithm of integrating production scheduling and vehicle routing is proposed for effectively solving the optimum problem. A mathematical model of the integrated production scheduling and vehicle routing problem is also generated for obtaining the optimum solution by LINGO 11. The object function is to obtain minimum tardiness time. The proposed method can obtain the optimal solution that is also evaluated from the optimization software LINGO 11.

References

1. Rom, W., Slotnick, S.A.: Order acceptance using genetic algorithms. Computers and Operations Research 36, 1758–1767 (2009)
2. Slotnick, S.A., Morton, T.E.: Order acceptance with weighted tardiness. Computers and Operations Research 34(10), 3029–3042 (2007)
3. Oğuz, C., Salman, F.S., Yalçın, Z.B.: Order acceptance and scheduling decisions in make-to-order systems. International Journal of Production Economics 125, 200–211 (2010)
4. Monkman, S.K., Morrice, D.J., Bard, J.F.: A production scheduling heuristic for an electronics manufacturer with sequence-dependent setup costs. European Journal of Operational Research 187, 1100–1114 (2008)
5. Zhang, L., Lu, L., Yuan, J.: Single machine scheduling with release dates and rejection. European Journal of Operational Research 198, 975–978 (2009)
6. Lin, S.W., Lee, Z.J., Ying, K.C., Lu, C.C.: Minimization of maximum lateness on parallel machines with sequence-dependent setup times and job release dates. Computer and Operations Research 38, 809–815 (2011)
7. Chiang, W.C., Russell, R.A.: A meta-heuristic for the vehicle-routing problem with soft time windows. Journal of the Operational Research Society 55, 1298–1310 (2004)
8. Feillet, D.: A tutorial on column generation and branch and price for vehicle routing problems. 4 OR-A Quarterly Journal of Operations Research 8, 407–424 (2010)
9. Liberatore, F., Righini, G., Salani, M.: A column generation algorithm for the vehicle routing problem with soft time windows. 4 OR-A Quarterly Journal of Operations Research 9, 49–82 (2011)
10. Belfiore, P., Yoshizaki, H.T.Y.: Scatter search for a real-life heterogeneous fleet vehicle routing problem with time windows and split deliveries in Brazil. European Journal of Operational Research 199, 750–758 (2009)

An Approach to Mobile Multimedia Digital Rights Management Based on Android

Zhen Wang, Zhiyong Zhang, Yanan Chang, and Meiyu Xu

Electronics Information Engineering College, Henan University of Science and Technology,
Luoyang 471023, P.R. of China
z.zhang@ieee.org

Abstract. For digital rights management of mobile multimedia in mobile terminals, an Android based Digital Rights Management (DRM) approach to implementing mobile audio and video media usage control was proposed. The solution adopts 3DES encryption and decryption algorithm for protecting multimedia contents security as a whole. The usage of control and display for protected contents were completed in mobile terminal according to the acquired digital license. A prototype confirms that the solution has the features of high security and faster encryption speed, can be helpful to protect the copyright of digital multimedia contents on the highlighted Android of mobile platforms.

Keywords: Digital Rights Management, Mobile Multimedia, Encryption, Usage Control, Android.

1   Introduction

As the technologies for communication network and information dissemination rapidly develop, 3G mobile similarly enjoys increasing popularity while 4G is on its way to contributing to these advancements. Mobile terminals have become a primary electronic equipment and carrier for internet applications. In the realm of 4G networking, wireless data transmission will be as fast and as convenient as cable internet. Moreover, the convenience of mobility and hand holding is an advantage of smart phones over PC terminals. Data from the study of International Data Corporation (IDC) indicate that smart phones with an Android system account for at least half of the market shares of smart phones throughout the world. The Android system is a smart phone system preferred by many users at present. As digital contents (e.g., e-books, digital images, multimedia audio and videos, etc.) are easily copied and distributed without any damage or omission, valuable digital content products protected by the intellectual property law can also be copied by batch without permission and be distributed [1], spread, and abused through various communication network carriers. Therefore, undesirable outcomes and significant losses are incurred, affecting economic, social, and cultural development. To address this technological problem, digital rights management (DRM) was designed. The DRM comprises a series of technologies, tools, flow, and treatment methods mainly

J.-S. Pan et al. (eds.), *Genetic and Evolutionary Computing*,
Advances in Intelligent Systems and Computing 238,
DOI: 10.1007/978-3-319-01796-9_25, © Springer International Publishing Switzerland 2014

used for protecting the contents of digital products as well as the legal rights and benefits of copyright owners and users [2].

2   Related Works

Encryption is currently a popular method for protecting digital contents. This technique encrypts common digital content documents (plain text) into ciphertexts to prevent valuable information from being illegally blocked or stolen, and to protect the copyright of digital contents. To protect the copyright of digital contents, solutions such as Windows Media DRM, which is based on the Windows Media Player by Microsoft and Helix DRM for media streaming by Real Company, were developed and applied to PCs.

Prompted by the growing significance of smart phones, personal digital assistants, and other mobile equipment, the new trend in the field of DRM research and development focuses on mobile terminals. A series of smart phones based on the Windows Mobile system of Microsoft [3], as well as smart phones and mobile equipment by Apple Inc., have been developed for and applied to commercial trade. The former supports Windows Media DRM solution and handles documents with Windows media video (WMV) and Windows media audio (WMA) formats. The latter utilizes the iTunes developed by Apple Inc. to secure the encrypted digital contents, the copyright of which is protected.

Bhatt et al. [4] proposed an individual DRM system based on the peer-to-peer model for the Motorola E680i smart phone to protect users' individual documents, such as photos and recorded videos. The native Android platform protects digital contents and applications through OMA DRM 1.0 solution [5]. Given its inherent vulnerability, OMA DRM 1.0 solution cannot effectively protect the contents in the equipment. Shuo Zhang, Zhao-Feng Ma et al. [6] proposed a dynamic decryption and playing solution for MP3 documents. This solution encrypts MP3 documents frame by frame according to MP3 document structure. Therefore, it can realize the dynamic decryption and playing of ciphertext during document playing, without creating any temporary documents in the mobile terminal.

The abovementioned systems and solutions install the DRM system in several kinds of common smart phone systems. However, the DRM system that can effectively protect audio and video contents is not installed in most popular Android smart phones.

3   Audio and Video Protection Solution for the Android Platform

To solve the problems on the audio and video digital copyright protection of mobile terminals, the originally designed mobile DRM (MDRM) system is installed and realized by considering the Android system, which currently has the most market shares, as the object platform and the OMA DRM v2.0 [7] standard. The system architecture chart is shown in Figure 1.

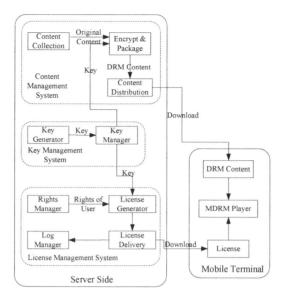

Fig. 1. System architecture of the MDRM system

The MDRM system performs a cycle that starts from content encryption, package, and issuance to the distribution of permits by the content provider and the decryption and content use right control by the users. It prevents contents from being abused and shared without permission by separating protected digital contents from the applicable permit, distributing the same, and controlling the authorized use. This step is done to achieve the safe use control and digital copyright protection of contents. This model consists of the server and the mobile terminal. For this MDRM system, the decryption and playing, including the use control of multimedia audio and video contents based on Android platform, are the emphases of the study.

3.1   Server Side

The server side is composed of the management systems for content, key, and license to perform the encryption and for the issuance of multimedia audios and videos, generation and management of content encryption key, and generation and issuance of multi-media videos and audios.

Content Management System
The triple data encryption standard (3DES) cryptographic algorithm is used for the content management system to encrypt and pack video and audio documents. It is a transitional cryptographic algorithm from DES to advanced encryption standard (AES), and it uses three 56-digit keys to process data thrice.

For audio and video documents requiring encryption for protection, the encryption and package program of the content management system reads the data from the source document through a module with fixed size, initiate the 3DES encryption program to encrypt the read data, and input the encrypted data into the new document. The process is repeated until all the data in the original document are encrypted. The main codes for the encryption program are as follows:

```
/* len refers to the length of the source document; buffer_size refers to the data
   module with a fixed size read every time; fileIn refers to the source document; and
   fileOut refers to the protected document encrypted */
   long j=len/buffer_size;
   for(i=0; i<=j; i++)
   {memset(buffer,'\0',buffer_size);
    fread(buffer,1,buffer_size,fileIn);
    3DesEncrypt(key,buffer,buffer,buffer_size);
    fwrite(buffer,1,buffer_size,fileOut1);}
```

Key Management System

One of the tasks of the key management system is to generate a random character string key for each of the original content to be encrypted and packed, and to supply the generated key to the content management system for encrypting and packing the original document. Another task of this system is to manage effectively these generated keys and return to the encrypting key of the applicable protected content after receiving the request from the license management system in order to generate a license.

License Management System

The license management system generates and distributes the license of the protected contents to legally authorized users. The license is composed of the decryption key of the protected contents and the right of the user to the digital contents. As a prototype system, the license management system defines the right of the user to the protected contents as the authorized times of playing the protected contents in the mobile equipment.

3.2   Mobile Terminal

The mobile terminal, that is, the client side, plays multimedia audio and video through the MDRM player installed in the Android platform. The MDRM Agent module in the system runtime library identifies, decrypts, and controls the use of protected contents. Figure 2 illustrates the implementation of this module in the Android platform.

In the original system of the Android platform, when the superior multimedia application contacts the MediaPlayer class of the application framework layer, the MediaPlayer class directly initiates the multimedia modules in the system runtime for processing. In the designed Android player, MDRM Agent modules are integrated in the system runtime. When the user utilizes the MDRM Player to play audio and

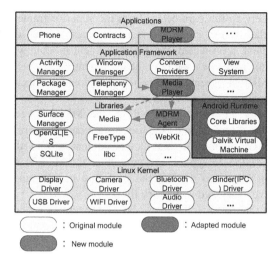

Fig. 2. Android system architecture

video, the MediaPlayer class in the application framework layer first contacts the MDRM Agent module, which then processes the parameters from the MediaPlayer class and initiates the operation of the multimedia module. Figure 3 presents the details of the parameter transmission course.

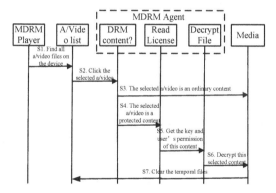

Fig. 3. Information processing by the MDRM player

Step S1. The user opens the MDRM player application, which scans the memory space of the equipment after searching for audio and video documents with the supported format. After scanning, the audio and video documents are shown to the user in list form. For audio document, the list presents not only the document name but also its protection status and the right of the user to it (playing times).

Step S2. The user selects a favorite content to play according to the information indicated in the document list of the audio and video.

Step S3. After receiving the parameters from the superior framework, the MDRM Agent module in the system runtime layer evaluates if the selected content is protected. If the content is protected, the MDRM Agent module will directly initiate the multimedia module to resolve and play the document.

Step S4. If the MDRM Agent assesses that the content selected by the user is protected, it will search for the correct permit.

Step S5. After detecting the correct permit, the MDRM Agent reads its sub-modules to validate the information on the user's right and decryption key.

Step S6. If the right of the user to the protected content is valid (the playing times is higher than zero time), the decryption sub-module of the MDRM Agent will analyze the key acquired in Step 6 to decrypt the protected content and will initiate the multimedia to play the decrypted temporary file.

Step S7. After the multimedia module is played, the temporary document created is cleared and Step S2 is repeated.

The system neither forces the user to place protected contents in a specific area in the equipment nor limits the protected contents to the download sources through the specific browser installed in the equipment. Instead, the user may store the protected contents in other areas in the equipment. Moreover, the source of protected contents may be downloaded by the user through the same or other equipment, or from a PC through Wi-Fi, USB, Bluetooth, and so on provided that the protected contents are complete and damage-free. However, to manage the right permit of the protected contents, it should be saved in a fixed folder in the equipment.

4   Performance Assessment

To test the system performance of the design, the content decryption and encryption speeds of the MDRM system and system safety are respectively tested and analyzed.

4.1   Speed Tests of Decryption and Encryption

The test environment for the encryption package program was a common PC with i3-2130 CPU. The test environment of the MDRM player program was the Android simulator run by a PC-based UBUNTU system virtual computer. Block sizes were set to 1024000, 102400, and 10240 to test the two programs.

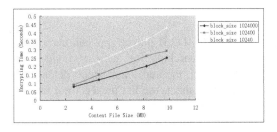

Fig. 4. Test of the encryption package program

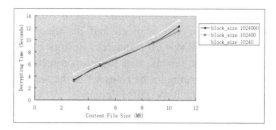

Fig. 5. Test of the decryption program

Figures 4 and 5 present the test results of the encryption package program and the decryption program, respectively, in different block sizes. The block size of 1024000 had the best performance in the encryption package program with an average speed of 38 m/s, and the 102400 block size had the best performance in the decryption program with an average speed of 0.906 m/s . Therefore, in different hardware environments, the scales of the block size of the same encryption and decryption programs vary in terms of the highest speed. Block size scale should be adjusted according to actual conditions during deployment to enhance decryption.

4.2   System Safety Analysis

In the designed MDRM Agent module, the directory of the decrypted temporary document of the protected contents is specified as a folder under the "data" of the system directory. Considering safety, common users of the Android system do not have the right to access and directly manipulate the document under the "data" directory [8]. The temporary decrypted document of protected contents in the Android system is also deleted immediately after being played. Accordingly, the designed system can finish protecting the encrypted contents in the mobile terminal. In this study, temporary document refers to the document created when protected contents are played.

Table 1. Comparison among solutions

	Our system	Literature2	Literature3	Literature 4
Encryption algorithm	3DES	RC4	Unspecified	AES
Save contents in the equipment	Encrypted contents	Encrypted contents	Original directory Specific location	Encrypted contents
Temporary document	with	with	without	Without
Object platform	Android	Mobilinux	Android	Windows mobile

5   Conclusion

For audio and video digital copyright protection in mobile terminals, the author of this paper selected the Android system, which currently has the most market shares, as the object platform. The source codes and compiling rules of Android 2.3 were analyzed. The designed prototype system was realized and installed based on the Android platform according to OMA DRM v2.0 standards. Results confirm that the MDRM player, one of the system components, can present protected contents by playing within the users' right and according to the rules set in the server-side of the Android platform, thus complying with the basic DRM demands.

Acknowledgments. The work was sponsored by National Natural Science Foundation of China Grant No. 61003234, Plan for Scientific Innovation Talent of Henan Province Grant No. 134100510006, Program for Science & Technology Innovation Talents in Universities of Henan Province Grant No.2011HASTIT015, and Key Program for Basic Research of The Education Department of Henan Province Grant No.13A520240. We would also like to thank the reviewers for their valuable comments, questions, and suggestions.

References

1. Security, Z.: Trust and Risk in Digital Rights Management Ecosystem. Science Press, China (2012)
2. Zhang, Z.Y.: Digital Rights Management Ecosystem and its Usage Controls: A Survey. International Journal of Digital Content Technology & Its Applications 5(3), 255–272 (2011)
3. Toma, C., Boja, C.: Survey of Mobile Digital Rights Management Platforms. Journal of Mobile, Embedded and Distributed Systems 1(1), 32–42 (2009)
4. Bhatt, S., Sion, R., Carbunar, B.: A Personal Mobile DRM Manager for Smartphones. Computers & Security 28(6), 327–340 (2009)
5. Chuang, C.Y., Wang, Y.C., Lin, Y.B.: Digital Right Management and Software Protection on Android Phones. In: IEEE Vehicular Technology Conference, Taipei, Taiwan, pp. 1–5 (2010)
6. Zhang, S., Ma, Z.F., Lu, X.F., Yang, Y.X., Niu, X.X.: Design and Implementation of Music Content Dynamic Encryption and License Authorization System. Computer Science 38(12), 43–48 (2011)
7. Open Mobile Alliance™, OMA DRM Requirements Candidate Version 2.0, OMA-RD-DRM- V2_0-20040715-C
8. Liu, C.P., Fan, M.Y., Wang, D.W., Zheng, X.L., Gong, Y.F.: Light-weight access control oriented tow and Android. Application Research of Computers 27(7), 2611–2613 (2010)

A Path-Combination Based Routing Scheme for Cognitive Radio Networks

Li Zi, Zhao Hongyang, and Pei Qingqi[*]

State Key Laboratory of Integrated Service Network, Xidian University, Xi'an 710071, China
qqpei@mail.xidian.edu.cn

Abstract. Cognitive radio networks (CRNs) has been popular in the field of wireless network. In CRNs, researches focus mainly on the single hop scenes rather than multi-hop networks. Studies for cognitive routing are also scarce. A new routing scheme is proposed based on path combination that communicating pairs share the links for all nodes. The cognitive networks are divided into two sub-networks: data network and control network. Each of the two network forms has their own missions. Spectrum resources are divided into conflict units and carefully allocated to links. The concept of price and spectrum blocks is used to judge the shared links. This scheme is good in performance because of the reduction of network load. With the combination of paths, spectrum resources will have a very high utilization rate.

Keywords: cognitive, routing, path, spectrum.

1 Introduction

The concept of cognitive radio was first proposed in 1999 by Dr. Joseph Mitola [1, 2], and expounded further in his doctoral thesis [3]. According to the Federal Communications Commission, the definition of cognitive radio is described as based on interaction with the environment and the dynamic change of radio transmitter parameters [4]. With the rapid development of wireless communication industry, the contradiction between the limited spectrum resources and the growing demand for wireless applications becomes more and more prominent. At the same time, in the allocated spectrum for a legitimate user, from the angle of time and space, and there is a spectrum utilization inequality problem [5, 6]. In this premise, the FCC puts forward a new spectrum management and allocation strategy [7, 8] as a new mode of wireless communication. Optimizing the use of spectrum has become very important and dynamic spectrum access technology has attracted wide attention. Cognitive radio technology is considered to be an important technical means of implementation DSA and plays a very important role [8].

Most of the researches for CRNs are based on the single hop to deal with questions about the physical and medium access control layer, including spectrum sensing, spectrum decision and spectrum sharing technologies [9, 10].Researchers have only

---

[*] Corresponding author.

J.-S. Pan et al. (eds.), *Genetic and Evolutionary Computing,*
Advances in Intelligent Systems and Computing 238,
DOI: 10.1007/978-3-319-01796-9_26, © Springer International Publishing Switzerland 2014

recently realized the multiple hops CRNs has potential to open up a new frontier, simply because multi-hops network configuration meets the needs of a wide range of communication applications. In order to fully release the potential of multi-hop cognitive radio networks, we have to face new challenges and difficulties. In particular, in cognitive scenarios with unique characteristics, efficient routing scheme should be integrated into the studies of existing bottom (physical layer or Mac layer).

Routing is used for communication of the source and purpose of transfer information of users to establish a path. Traditional wireless networks work in a fixed band. Therefore, there is no need to consider after getting the path spectrum allocation. Cognitive wireless network of the cognitive users to use the free authorized spectrum, their use of spectrum is uncertain, and set a path for every link on the path to the distribution of spectrum resources, since the traditional wireless network routing methods cannot be directly applied to CNRs, we need a new cognitive wireless network routing method.

According to the routing optimization object, M. Cesana et al [11] classified the existing cognitive wireless network routing schemes. The scheme optimizes network throughput, transmission delay, energy consumption, routing quality, path stability and disturbance to the authorized users etc. The problem is there is no plan to optimize the demand for spectrum resource, when the user sends a message concentrated in one area of the network, and the user receives the message concentrated in another area, when each pair of user interaction data sending and receiving quantity is not big, these methods lead to different transceiver users on the use of the routing not overlap or less overlap each other, thus increasing the spectrum resource requirements.

Hao Chen et al [12] put forward a central routing method. This method comes together all the sending and receiving users, using the maximum flow algorithm to get maximum flow distribution, and as a new network topology, distribution path and the transmission time. A weakness of this method is that the entire network routing work shall be borne by a center, when routing demand increases in the network, the burden of the center will increase rapidly, making routing delay increase.

This paper is organized as follows. In the second part, the paper introduces the network construction. In the third part, the paper introduces the routing scheme based on path-combination. In the fourth part, the paper introduces the routing scheme simulation and performance. Finally, in the fifth part the paper gives the conclusion.

2   Network Model

A CR network consists of a sink node and several CR users. Logically Speaking, CR network can be divided into two parts, data network and control network. A data network is a multi-hop and fully distributed network, which bears the interactions and sessions among CR users. Sink node does not participate in the data transmission in data network as is shown in Fig 1. On the other hand, control network contains those basic transmissions in order to maintain the normal operations of CR network. Sink node is the center of control network, and every CR users communicates directly with it. Moreover, adjacent users can communicate with each other, but just in one hop, as is shown in Fig 2.

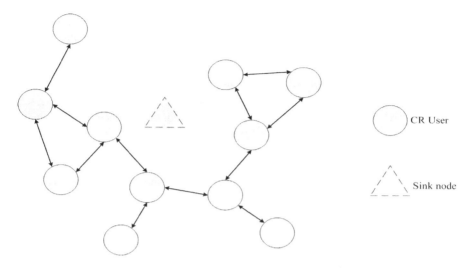

Fig. 1. Control network construction

Every CR users has two radios which work for data network (radio1) and control network (radio2) respectively. The two radios work in different spectrum band to be guaranteed to work at the same time and without interference.

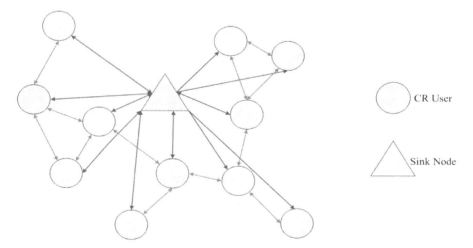

Fig. 2. Data network construction

2.1 Obtain Sensing Results

Each radio1of CR user carries out Quite Period Sensing - Adjacent CR Users Discovering-Communicating/ Sleeping/ Auxiliary Sensing. As is shown in Fig 4, All CR users execute these steps synchronously.

Quite Period Sensing: radio1 in all CR users stop transmission on data network, sense the existence of primary users and spectrums holes, then store the result.

Adjacent CR users discovering: CR users discover each other via radio1, every user store the node listing.

Communicating: send and receive packets of various kinds of business, sessions and so on.

Sleeping: when no packet needs to be sent or received, that is to say, radio1 can be turned off in order to reduce energy consumption.

Auxiliary sensing: when no packet needs to be sent or received, that is to say, CR users can perform the specific sensing task that allocated by sink node, and at this stage, CR users can achieve information related to channel quality, Adjacent CR users' actions, channel load, etc.

Finally, all the information is collected and summarized, which is named sensing results.

2.2   Gather Sensing Results

Sink node and CR users achieve this goal in radio2. Each CR user carries out polling- waiting/adjacent users' interaction- Receiving network topology- waiting/adjacent users' interaction circularly on radio2. Polling and receiving network topology must be asynchronous with sink node.

Polling: the CR user which is polled should report sensing result for sink node and receive control instructions accordingly.

Receiving network topology: each CR user receives CR network topology from sink node simultaneously. The CR network topology is used to describe CR users and their Interconnection performance of CR network

Sink node receives sensing result from CR user X that is polled, and sends control instructions accordingly to user X. In this process, radio2 keeps monitoring state, when the polling massage is sent to user X, radio2 begin to communicate with sink node.

After all sensing results of every CR user has been found, sink node starts gathering and processing the information, forming the CR network topology. Then the network topology information is spread via broadcasting in order to ensure all CR users will receive it.

3   A Path-Combination Based Routing Scheme for Cognitive Radio Networks

The task of route is getting a path and determining the spectrum allocation of each hop. There are two steps to accomplish this task, namely path generation and spectrum allocation. Path generation is that the given source users s and destination users create the path p connected s to t. Spectrum allocation is that allocate spectrum resource to p which is created by the connection from s to t unified.

3.1   Path Generation

Path generation is accomplished by the source CR user to create one path p. Assumed that the source user is s, and destination user is t. for a link, link e=(u, v), the definitions are as follows: The price of Link e is a parameter released and maintained by the center node;

- The distance from Link e to source user is defined as $ds = \| u - s \| + \| v - s \|$;
- The distance from Link e to destination user is defined as $dt(e) = \| u - t \| + \| v - t \|$;
- The synthesized distance from Link e to source uses and destination user is : $d(e) = \min\{ds(e), dt(e)\}$;
- For users s and t, $d_{max} = \max_e d(e)$;
- The distance coefficient of Link e is $rd(e) = \dfrac{\log(d(e)+1)}{\log(d_{max}+1)}$. It means that when

  the Link is close to source user or destination user, the distance coefficient is low and the distance coefficient of distance d (e) is increased rapidly and approaching 1 gradually.
- On one path, the price of Link e to source or destination user is $rp(e) = price(e) \times rd(e)$. It means that when the Link e is closed to source or

  destination, the discount is high. Source or destination user must pay the price for maintaining its surrounding Links. So the price is low when it uses the surrounding links.
- Indicator (e) represents the interference Binary tag. $indicator(e) = 1$ represents

  interference from u or v to primary user, while $indicator(e) = 0$ was defined no interference.

Sink node define prices for every link related to the load of link and the number of available spectrum channels.

Every price has the same initial value with no actions for CR users. Price in used and unsaturated link is lower than that in unused link, that is to say, we encourage building paths via using the unsaturated Link; when one link has been used, and prices of those links within the scope of the interference have a slight increase; when one link has been used and is almost saturated, price increases to prevent the overload.

For used channel, the price of a link increases from the lowest value with the saturation level rising, finally reaches a steady value. For unused channel, the price of a link increases from the second lowest value to the second highest value with the number of undisturbed spectrum channels decrease.

Price has two forms: price1 and price2.

Price1 is applied when the link is used. Because of the denying of overload, price1 is defined in (1).

$$price1 = \frac{\lambda}{1 - sat} + (Iv1 - \lambda) \qquad (1)$$

where λ is the attenuation parameter; $sat \in [0,1)$ is the saturation; Iv1 is initial value of price1 when sat=0.

Price2 is applied when the link is unused and defined in (2).

$$price2 = BI \times PPB + Iv2 \qquad (2)$$

where BI is the number of interference bands in the link; BBP is the price of each band; Iv2 is the initial value of price1, when sat=0. Iv2 > Iv1.

The source node of a path regards the price as the edge weight of a topology calculates the shortest path from source node to destination node. The performance parameter of a path p is defined in (3).

$$sp(p) = \sum_{e \in p} (rp(e) + hp + indicator(e) \times MAX) \qquad (3)$$

where MAX is a constant.

The sp (p) of a link which can infer primary user is greater than MAX. Therefore, the path is abandoned. The source node can obtain a path within the conditions via the theory maximum spanning tree on condition that the source node has known the network topology.

The purpose of this rule for building paths is to make the shared links of several l paths as much as possible. Because the shared parts allocate spectrum recourse just for once, the number of those links which interferences each other decreases. Therefore, the number of available spectrum channels that are used in CR networks decreases as well. When a shared link is almost saturated, the price will rise and source node will choose those unused links to build a path.

3.2   Spectrum Allocation

Let B define the set of available spectrum channels, T define the set of available time slots. Let i, j, p, q, u, v represent CR users, s represent source user and t represent destination user.

Suppose the transmission distance R_{Tr} and interference distance R_{If} are all the same initially.

The conflicting set of time slots can be defined in (4).

$$SI_e = \{ f(i,j) \mid u = i \text{ or } u = j \text{ or } v = i \text{ or } v = j, f \neq e \} \qquad (4)$$

where $SI_e \in E$

The conflicting set of bands can be defined in (5).

$$BI_e = \{g(i,j) \mid d_{u,i} \le R_{IF} \text{ or } d_{u,j} \le R_{IF} \text{ or } d_{v,i} \le R_{IF} \text{ or } d_{v,j} \le R_{IF}, \; g \ne e\} \quad (5)$$

where $BI_e \in E$

Each link e has an indicator $x_{e,\tau}^b$, $x_{e,\tau}^b > 0$ presents that e uses band b in time slot . $x_{e,\tau}^b = 0$ resents that e use no band in time slot . $x_{e,\tau}^b < 0$ presents that e has no rights to use bands in time slot . $\mid x_{e,\tau}^b \mid$ is the number of parameters which leads to the unavailable band.

Define $Count^b = \displaystyle\sum_{g \in E, \tau \in T} \frac{x_{g,\tau}^b + \left| x_{g,\tau}^b \right|}{2}, b \in B$, is the total number that $x_{g,\tau}^b \ge 0$.

When link g wants to use spectrum recourses, we firstly check $x_{e,\tau}^{b1}$. If $x_{e,\tau}^{b1} = 0$ can be found, g uses the spectrum recourses. When more than one band makes $Count^b > 0$, g chooses the band that has smaller $Count^b$. When the above method is unable to achieve, check $x_{g,\tau}^{b2}, b_2 \in \{b \mid b \in B, Count^b = 0\}, \tau \in T$, choose a series of new spectrum recourses. If there is no satisfaction, the sink node refuses the request for spectrum recourses of a link.

4   Simulation and Performance

In order to prove our method, simulation circumstance is defined as follows:

- Area dimension: length 600m, wide 200m
- Number of CR users: 60.
- Communication distance: 100m
- Interference distance: 150m

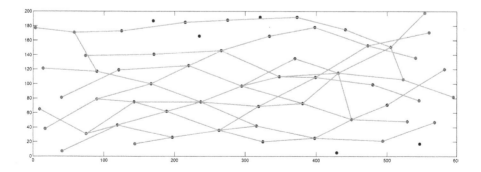

Fig. 3. Paths of traditional scheme

Each link contains communication requests at most five paths; one is reserved for the ends of a link, other four forward requests can be contained for a link.

Fig 3 is the traditional method for building paths, while Fig 4 is our method. Obviously, the number of our paths using our method is simple. That is to say, our method makes CR nodes choose paths that are highly shared and reduce the number of links that is need to assign the spectrum resources. Thus decrease the demand for spectrum resources in cognitive radio networks, and raise the efficiency of utilization of spectrum resources.

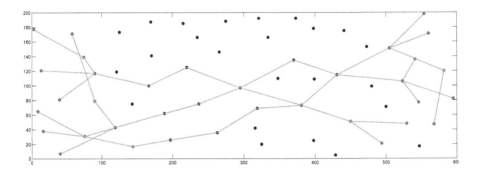

Fig. 4. Paths of our scheme

Through the analysis of the specifications, it is illustrated that the Path-combination based Routing Scheme has great advantages in performance, as is shown in Fig 5, 6.

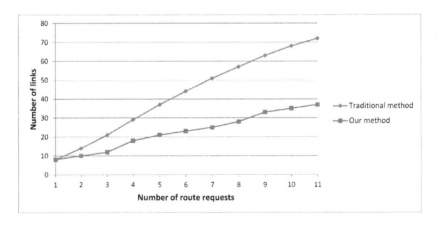

Fig. 5. Simulation about links and route requests

Fig 5 makes it clear that when number of communication pairs (source and destination nodes) increases from 1 to 11, the change trend of the quantity of links has an apparent growth. We can see the more the communication pairs are, the greater the difference is. It is excellent news for the crowded networks and the increasingly scarce spectrum.

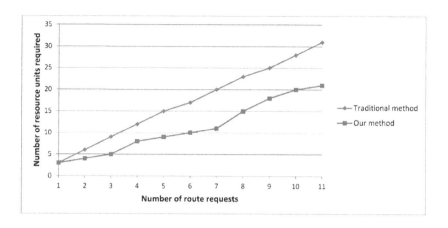

Fig. 6. Simulation about the number of resource units and route requests

Fig 6 shows the demand of the spectrum varies from communication pairs. It is a more intuitive explanation to highlight the advantages of our scheme. Trends of the lines in the picture are roughly the same. The line reflecting our scheme is smoother and the resource units represents the spectrum resources at other aspects. The number of bands is the quality of spectrum channels used in nodes via the paths. In this section, numerical results are presented to illustrate the superiority. A simulation of routing path is given and some explanations have been provided. It is thought our scheme is good in saving spectrum recourses and simplifying routing paths.

5   Conclusion

Routing in CRNs consists of path generation and resource allocation. Amount of resource required is determined by conflicts in resource allocation. Less links used can bring about fewer conflicts in resource allocation. In this paper, we use a combination based method to generate routing paths. Then we divide resource into nonlict units and carefully allocate these units to the links.

Then we divide resources into nonclict units and carefully allocate these units to the links. Simulation results demonstrate that our routing scheme can effectively reduce the links used by the network and requires less spectrum bands. So it can be used in CRNs to improve spectrum utilization.

Acknowledgement. This work is supported by the National Natural Science Foundation of China under Grant No. 61172068, the Program for New Century Excellent Talents in University (Grant No.NCET-11-0691), and the Fundamental Research Funds for the Central Universities (Grant No.K50511010003).

References

1. Mitola, J., Maguire, G.: Cognitive radio: making software radio more personal. Personal Communications 6(4), 13–18 (1999)
2. Mitola, J.: Cognitive radio for flexible mobile multimedia communications. In: 1999 IEEE International Workshop on Mobile Multimedia Communications (MoMuC 1999), pp. 3–10. IEEE (1999)
3. Mitola, J.: Cognitive Radio: An Integrated Agent Architecture for Software Defined Radio. Doctor of Technology, Royal Inst. Technol., Stockholm, Swede, pp. 271–350 (2000)
4. FCC Notice of Proposed Rule Making and Order. ET docket, no. 03-322 (2003)
5. FCC Spectrum Policy Task Force Report. ET docket, no. 02-135 (2002)
6. FCC Notice of Inquiry and Notice of Proposed Rulemaking. ET docket, no. 03-237 (2003)
7. Staple, G., Werbach, K.: The end of spectrum scarcity. Spectrum 41(3), 48–52 (2004)
8. Chen, R., Park, J., Hou, Y.T., Reed, J.H.: Toward secure distributed spectrum sensing in cognitve radio networks. IEEE Communications Magazine 46(4), 50–55 (2008)
9. Cormio, C., Chowdhury, K.R.: A survey on MAC protocols for cognitive radio networks. Ad Hoc Networks 7(7), 1315–1329 (2009)
10. Haykin, S., Reed, J.H., Li, G.Y., Shafi, M.: Scanning the issue. Proceedings of the IEEE 97(5), 784–786 (2009)
11. Matteo, C., Cuomo, F., Ekici, E.: Routing in cognitive radio networks: Challenges and solutions. Ad Hoc Networks 9(3), 228–248 (2011)
12. Chen, H., Du, Q., Ren, P.: A Joint Routing and Time-Slot Assignment Algorithm for Multi-Hop Cognitive Radio Networks with Primary-User Protection. International Journal of Computing Communication 7(3), 403–416 (2012)

Digital Rights Management and Access Control in Multimedia Social Networks

Enqiang Liu[1], Zengliang Liu[1], and Fei Shao[2]

[1] University of Science and Technology Beijing, Beijing, China
[2] Xidian University, Xi'an, China
en815@163.com

Abstract. The emerging multimedia social networks (MSN) services have significantly improved and enriched consumers' experiences on multimedia digital contents, and further spurred the misuse and malicious dissemination of digital media (rights) among users. Thus, Digital Rights Management (DRM) issue is becoming more prominent. The theories and methods for security, trustworthiness, controllability of DRM systems available do not meet the needs of the novel Internet application, which is an urgent problem to solve. This paper analyzed traditional access control technologies and social network properties, and made a survey on related home and abroad research progresses on DRM security technologies, multimedia social networks access control and media contents copyrights protection. Finally, several main research directions were addressed on the key technologies of DRM, from three layers of cloud media services, media networks admission and user social networks, respectively, that is media content access control, media terminal trusted admission and media rights secure distribution.

Keywords: multimedia social networks, content security, access control, digital rights management.

1   Introduction

Digital rights management (DRM) is an open problem, a challenge to the healthy development of the digital content industry. Since the middle 1990s, the research and applications of DRM have experienced offline use, Internet online, content distribution networks, and peer-to-peer network phases [1, 2]. In recent years, technologies, such as the multimedia cloud computing technology that integrates the server hosting computing mode and the client computing mode [3, 4], and other social media network services, such as Facebook, Twitter, and micro-blogging, have emerged. According to the data released by iResearch, an authoritative information technology consulting firm in China, more than 1.2 billion users around the world have used social networking web sites at least once a month by December 201; global social network users are expected to maintain double-digit growth in 2014; and the social elements have become the basic applications in the global Internet. In other

J.-S. Pan et al. (eds.), *Genetic and Evolutionary Computing,*
Advances in Intelligent Systems and Computing 238,
DOI: 10.1007/978-3-319-01796-9_27, © Springer International Publishing Switzerland 2014

words, social media sharing has become the main driving force for the development of social networks.

Multimedia social networks (MSN), such as YouTube, SongTaste, Tudou, Youku, are social networking platforms that are organized by users' social relationships; and are mainly used for using, sharing, and disseminating digital media content. The basic characteristic of this emerging Internet application is the integration of the cloud server-based centralized access mode (online) and the cloud terminal-based distributed access mode (offline). MSNs provide digital media users with more convenient and efficient, and higher quality multimedia service for them to gain a richer experience of the digital content. Meanwhile, media social networks also share the essential characteristics of small-world networks, that is, easy to spread. As a result, to the large number of copyrighted media content, the infringement of their copyright and digital rights have become increasingly serious during the process of access, use, share, and dissemination, which has brought an unprecedented negative impact to the digital content/media industry, and has also made the DRM problem stands out even more [5, 6].

2   Digital Rights Management and Content Distribution

2.1   Security Technologies for Digital Rights Management

The research of DRM has two major technical paths: the preventive DRM technology and the reactive DRM technology. The preventive DRM technology is mainly based on the theory of cryptography and the usage control technology. It explores the safe storage, distribution, and usage control of digital content [7], as well as the fair use of digital content in accordance with the copyright law requirements. Among them, digital content protection mainly involves safe encryption and decryption, provable security for cryptographic protocols [8], and identity-based domain key distribution protocols [9]; whereas the usage control of digital rights covers the language description of the digital rights, usage control technologies, the transparent access based on the file system layer and content semantics [10], a trusted computing environment for end-users, and the trusted execution of the content use policy. For this reason, Gong-Xuan Zhang and his colleagues [11] proposed the TCM-based access control model. The reactive DRM technology, on the other hand, mainly focuses on users' violations of digital copyrights, including tracking violations through digital watermarks and distinguishing copyright of digital content [12].

2.2   Distribution and Dissemination of the Digital Content (Right)

The existing shared digital rights are mainly limited to domains with ordinary authority levels, such as family domains or personal entertainment domains. In this approach, the distributed digital rights and content were tied to the devices and end-users through the use of a secondary distribution method and a strict usage control security policy, thus to ensure the legal use of the digital content within the authorized domains. In terms of the implementation of the shared digital rights, because OMA RI usually bound the content, license, and equipment (user) together when authorizing an end-user license, it, therefore, puts a strong limit on the shared use of the digital content. Bhatt et al [13] realize implemented a personal DRM prototype system in the Motorola E680i smart mobile terminal. In this personal DRM system, the end-users can set the digital license independently and transfer licenses between devices freely, so as to protect the personal digital content. In addition, [14] proposed a secondary distribution plan and the relevant security protocol based the content using time between devices. This research was a useful attempt to share the time elements in the digital rights; and it also broadened the view of research on the shared digital rights. Xue Feng and Zhi Tang proposed a DRM license sharing plan based on Ergodic Encryption and the license sharing mechanism of the machine authentication technology to reduce the cost burden of the traditional authorized domain dependent approach [15]. Win et al. [16] proposed a safe and interoperable distribution mechanism that supports multiple authorized domains, which made the safe and effective content sharing among domains become possible.

3   Access Control and Copyright Protection in Multimedia Social Network

3.1   Features of Traditional Access Control and Social Networking

Access control refers to using end-users' identities and defining groups to limit their access to certain information, and to restrict their use of some control functions. The traditional access control is based on the specified access control policy to decide whether an access request is allowed. It cannot be well positioned to meet the fine-grained access control requirements (e.g., different access control parameters for accessing different parts of an image). In addition, there exist a large number of users and concurrent accesses in MSN. The traditional access control mechanisms cannot adapt to this new scale of needs [17]. Mainly as follows:

1. In a dynamic, highly decentralized environment, such as collaborative groups, the centralized that is responsible for executing the access control may become the bottleneck of the entire system.
2. The traditional access control mechanism applies centralized access control architecture. It completely relies on the system administrators to manage users' data and the relevant access control policy, which may cause problems in user privacy and data security.
3. Although the traditional access control mechanism can also achieve fine-grained access control requirements by using the access control lists, it requires specification of which user can access the data, which increases system overhead and management complexity.

In social networks, the new applications, there are a lot of collaborative users and concurrent data accesses, which brings new requirements for access control. Mainly as follows:

1. Support the user relationship-based access control, and take the depth and credibility of a user relationship into consideration;
2. Resources and data that allows fine-grained access;
3. To lift the restriction on collaboration, must be flexible about force ruling out unauthorized users from accessing data;
4. The access control model should be dynamic, so as to allow the running policy to change with the environment changes;
5. The administrator rights assignment and data ownership issues;
6. The reasonable range of performance and resource overhead.

3.2   Relationship-Based Access Control

Gates [18] described a security example of a new relationship-based access control that can satisfy the requirements of Web 2.0. Hart et al. [19] proposed a content and relationship-based access control system. This proposal suggested using relational information to represent the authorized party in a web based social network (WBSN), which meets the key requirement for protect WBSN resources. However, this system has its several short comes. First, it does not satisfy the increasing needs of privacy in access control. Second, it only considers the direct relationship, whereas the node trust is not considered as a parameter when authorizing an access. In point of privacy concerns, at present, it mainly focuses on privacy protection, data mining technology, and the permission to conduct social network analysis on disclosed, potentially sensitive information.

Barbara et al [20] pointed out that the enhanced social network access control system is the first step to solve the existing security and privacy issues in online social networks. In order to resolve the current limitations, they proposed an expandable, fine-grain access control model based on semantic, web-online social networking. This model contains authorization, management and filtering; and uses web ontology language (OWL) and semantic web rule language (SWRL) for modeling. Park et al [21] proposed a user-behavior centered access control framework. They identified four core control behaviors: attributes, strategies, relationships, and sessions. Among them, sessions represent the valid users who have been recorded in the online social network (OSN). In a simplest example, a session inherits all the properties and policies of a user. However, the existing social networks cannot support this function. Park et al. suggested that OSN should have the following features. First, personalized policies, that is, OSN users have their own security and privacy policies and attributes. Second, the users and resources policies are separated. Third, support access control of those who are independent of the user relationship, and support active sessions; and take into account the enhanced control. The existing OSN services do not have these functions, and many of the latest literature on OSN access control cannot distinguish between sessions and users.

3.3   Trust-Based Access Control

Ali et al [22] applied a multi-level security approach, in which trust is the only parameter that is used to determine the security levels of users and resources. More precisely, each user is assigned a reputation value. The reputation value is a user's average trust grade that is specified by other users. However, Ali and his colleagues only considered direct trust relationship without taking account the indirect trust relationship. Kruk et al [23], then, proposed a distributed authentication management system based on the second round "friend" relationship to bring out the management of access rights and trust authorization. Wang et al [24] proposed a trust-related management framework that includes the access control policies and a mechanism that supports privacy protection. This mechanism administers the access policy on the data that contain the provable information; enhances the support to the highly complex privacy related policies; takes consideration of the purpose and obligations. Under this mechanism, the main body can perform access rights on the objects based on relationships, trusts, purposes, and obligations. This mechanism also introduced strategic operations and the concept of policy conflicts; and proposed a purpose related access control policy framework. Amit Sachan and Sabu Emmanuel [25] pointed out that the traditional access control cannot meet the fine-grained access control requirements and the large number of users. To solve this problem, they proposed an efficient bit-vector transform based access control mechanism suitable for MSNs. They converted the content related certificate into an efficient structure, and, then, verified the security, storage, and execution efficiency of the proposed mechanism rough simulations. Villegas [17] proposed a personal data access control (PDAC) scheme. PDAC computes a "trusted distance" measure between users that is composed of the hop distance on the social network and an affine distance derived from experiential data.

In a distributed system, privacy policy must be implemented on a user's private information, such as the P3P (Platform for Privacy Preferences) standard. In particular, Agrawal, etc. [26] proposed a Hippocratic database system. This proposed database system combines privacy protection in relational database systems. An important feature of this proposed database is the use of privacy metadata, in which the external-recipients and retention attributes are in the privacy-policies table, while the authorized-users attribute is in the privacy-authorizations table.

However, Agrawal and his colleagues did not discuss the hierarchical goals, nor did they discuss the target prohibited, target joint, or data elements. LeFevre et al [27] proposed a method about how to administer privacy policies in a database system - the two-cell level model, whereas Ni et al [28] proposed a role-based access control model. However, none of these two models consider access management. The continuing evolving access technology has brought many challenges to access control and the model structure becomes necessary, which leads to the next generation of access management issues.

Li-Qin Tian and Chuang Lin proposed a game-theoretic control mechanism of user behavior trust based on predictions in a trustworthy network. The proposed mechanism not only can predict behavior trust grade under the single trust attribute conditions, but also can predict trust grade under the multi-trust-attribute conditions. Li-Qin Tian and Chuang Lin proposed the whole process of the user-behavior-trust-prediction-based game control mechanism. The main idea is to increase the control and management of user behavior trust in addition to the user identity trust, including behavior trust prediction, risk analysis and decision-making control, strengthening the dynamic process on a network user's status, thus to provide a strategic basis for the implementation of the intelligent and adaptive network security mechanisms. As shown in Figure 1, in this model, when service providers receive a user's request of access, they authenticate the user's identity trust first, if the authentication fails, the access is denied, otherwise they will continue to forecast the user's behavior trust and conduct the game-control decisions. The process of behavior trust prediction-based game-control process is shown in Figure 1 [29].

The existing WBSN enforcement access control model is relatively simple. That is to say, the resource owners can define three security settings: 1) public, 2) private, 3) accessible for directly related users. This existing model assumes that all the friends

Table 1. Comparison of Several Typical Access Control Policies

Functionality and Security Mechanisms	Traditional access control models	Reference [17]	Reference [21]	Reference [24]	Reference [25]	Reference [30]
Access control mechanisms	Trusted software module	Personal data access control	User behavior	Relationship, trust, purpose, obligations	Binary vector transformation	license
Fine-grained access control	Access control lists	Guard interval	Separation of individual users and resources policy	Attributes	Binary vector transformation	Policy
Trust calculation	No	Behavior	Behavior	Behavior	Behavior	License chain
Type of trust	Without	Full trust	Full trust	Full trust	Full trust	Full trust
Depth of relationship	Direct	Direct, indirect	Direct, indirect	Direct, indirect	Direct, indirect	Direct, indirect
User privacy protection	Not protected	protected	Users and resources	Protected	Protected	Protected
Security settings in the replication resources	Not exist	exist	Not exist	Not exist	Not exist	Not exist
System overhead	Larger	Smaller	Larger	Moderate	Smaller	Larger
Applicable settings	Ordinary security domains	Social networks	Online social networks	Social networks	Multimedia social networks	Social networks

are the same without considering the relationship type between users; it does not have a comprehensive consideration of the of the relationship depth; it only allows users who have a direct relationship with the resource owners to access the data, and does not allow those who have indirectly relationship (the second time "Friends") to access the data; it allows un-authenticated user access, which is not flexible enough for the authenticated users; it does not distinguish data that focus on sharing from data that focus on privacy. Table 1 illustrates a comparative analysis of typical access control policies and models regarding their functionalities and security mechanisms.

3.4   Media Content Copyright Protection

Zhi Wang, Li-Feng Sun et al [31] presented a paper at ACM Multimedia conference. In their study, they measured reality of online social networking systems, analyzed the main features of the video mode of transmission, and proposed a new audio and video content distribution method. In the proposed method, the video content is deployed in the appropriate servers and node caches, and was assigned to the appropriate network resources based on its propagation, so as to enhance the efficiency of the network communication. In order to ensure the safe sharing of media content in social networks, He-Fei Ling et al. proposed JFE (combination of fingerprint and encryption) based on a tree structure conversion security mechanism, which combines the fingerprint technology and the encryption technology to provide multiple layers of protection for media sharing [32].

In order to improve media content copyright protection and to diminish the illegal spread of media content in social networks, Lian et al [33] proposed a content distribution and copyright authentication system based on the media index and watermarking technology. The results of the experiments confirmed that the system had strong robustness and stability. In addition, Chung et al [34] proposed a novel video matching algorithm, as well as developed an intelligent copyright protection system based on this algorithm. Confirmed by experiments, the proposed algorithm can effectively conduct video matching; and the proposed system was suitable for copyright protection for video sharing networks. With the intention of solving the problem of content security in online social networks, [35] proposed a security model based on multi-party authentication and key agreement. This proposed model can achieve user authentication between communities with a strong non-repudiation and flexibility.

Ming-Chu Li et al [36] proposed a fine-grained trust computation model. They defined a fine-grained QoS in order to achieve the calculation of recommendation trust; used Gaussian function to measure the preference similarity between peers; and verified the effectiveness and flexibility of the proposed trust model through a large number of simulation experiments. We [37] proposed a MSN trust model based on small-world theory. This model can effectively evaluate and dynamically update the value of trust between users, as well as identify malicious share users.

4   Conclusion

The current DRM research on MSNs mainly includes traditional digital content encryption, usage control, digital watermarking, and digital content distribution. By means of integrating the recently emerged cloud computing technology and its theoretical architecture, the cloud media social networks (CMSN) were developed. However, the systematic research on the three elements (safe media content access control, credible media terminal access authentication, media spread rights and risk assessment) that can impact the implementation of the digital media copyright protection has not been carried out yet. Consequently, the three elements have become the key generic technologies and methods for the cloud media DRM. This paper suggests that the future DRM research should be based on the basic attributes of cloud media social networking, combined with the practical applications of the DRM, and focus on the explorations of the key technologies from the cloud media content service layer, the media terminal network access layer, and the media users social networking layer, so as to achieve the expected cloud media content security and copyright protection. This new research approach is of fundamental theoretical significance to the achievement of the cloud media content security and copyright protection. It is also has great application prospects and practical value for the promotion of digital media content platform, its industry health, and healthy development.

Acknowledgement. The work was sponsored by National Natural Science Foundation of China Grant No.61003234, Plan For Scientific Innovation Talent of Henan Province Grant No.134100510006, program for Science & Technology Innovation Talents in Universities of Henan Province Grant No.2011HASTIT015, and Key Program for Basic Research of The Education Department of Henan Province Grant No.13A520240.

References

1. Zhang, Z.Y.: Digital Rights Management Ecosystem and its Usage Controls: A Survey. International Journal of Digital Content Technology & Its Applications 5(3), 255–272 (2011)
2. Zhang, Z.Y.: Security, Trust and Risk in Digital Rights Management Ecosystem. Science Press, China (2012)
3. Zhu, W.W., Luo, C., Wang, J.F., Li, S.P.: Multimedia cloud computing. IEEE Signal Processing Magazine 28(3), 59–69 (2011)
4. Gadea, C., Solomon, B., Ionescu, B., et al.: A collaborative cloud-based multimedia sharing platform for social networking environments. In: Proc. of International Conference on Computer Communications and Networks, Maui, HI, USA, pp. 1–6 (August 2011)
5. Diaz-Sanchez, D., Almenarez, F., Marin, A., Proserpio, D., Cabarcos, P.A.: Media cloud: An open cloud computing middleware for content management. IEEE Transactions on Consumer Electronics 57(2), 970–978 (2011)
6. Huang, T., Zhang, Z.Y., Chen, Q.L., et al.: A Method for Trusted Usage Control over Digital Contents Based on Cloud Computing. International Journal of Digital Content Technology & Its Applications 7(4), 795–802 (2013)

7. Pretschner, A., Hilty, M., Schütz, F., et al.: Usage Control Enforcement: Present and Future. IEEE Security & Privacy 6(4), 44–53 (2008)

8. Koushanfar, F.: Provably secure active IC metering techniques for piracy avoidance and digital rights management. IEEE Transactions on Information Forensics and Security 7(1), 51–63 (2012)

9. Yan, X.X., Ma, J.F., Yang, Y.F., et al.: Identity-based domain key distribution protocol in the E-document security management. Journal of Communications 33(5), 12–20 (2012)

10. Lee, S., Lee, H.R., Lee, S.K.: DRMFS: A file system layer for transparent access semantics of DRM-protected contents. The Journal of Systems and Software 85(5), 1058–1066 (2012)

11. Zhang, G., Zhu, Z., Wang, P., Song, B.: A TCM-enabled access control scheme. In: Xiang, Y., Cuzzocrea, A., Hobbs, M., Zhou, W. (eds.) ICA3PP 2011, Part II. LNCS, vol. 7017, pp. 312–320. Springer, Heidelberg (2011)

12. Thomas, T., Emmanuel, S., Subramanyam, A.V.: Joint Watermarking Scheme for Multiparty Multilevel DRM Architecture. IEEE Transactions on Information Forensics and Security 4(4), 758–767 (2009)

13. Bhatt, S., Sion, R., Carbunar, B.: A Personal Mobile DRM Manager for Smartphones. Computers & Security 28(6), 327–340 (2009)

14. Lee, S., Kim, J., Hong, S.J.: Redistributing Time-based Rights between Consumer Devices for Content Sharing in DRM System. International Journal of Information Security 8(4), 263–273 (2009)

15. Feng, X., Tang, Z., Yu, Y.Y.: An Efficient Contents Sharing Method for DRM. In: Proc. of 2009 Consumer Communications and Networking Conference. 5th IEEE Workshop on DRM, Las Vegas, NV, pp. 1–5 (2009)

16. Win, L.L., Thomas, T., Emmanuel, S.: Secure interoperable digital content distribution mechanisms in a multi-domain architecture. Multimedia Tools and Applications 60(1), 97–128 (2012)

17. Villegas, W.: A trust-based access control scheme for social networks. School of Computer Science, McGill University, Montreal (2008)

18. Gates, C.: Access Control Requirements for Web 2.0 Security and Privacy. Position paper accepted to the Workshop on Web 2.0 Security and Privacy, Oakland, California, United States (2007)

19. Hart, M., Johnson, R., Stent, A.: More content – less control: access control in the web 2.0. In: Proc. of the Web 2.0 Security and Privacy Workshop on Online Social Networks, WOSN (2007)

20. Barbara, C., Elena, F., Raymond, H., Murat, K.: Semantic web-based social network access control. Computers and Security 30(2), 108–115 (2011)

21. Park, J., Sandhu, R., Cheng, Y.: A User-Activity-Centric Framework for Access Control in Online Social Networks. IEEE Internet Computing 15(5), 62–65 (2011)

22. Ali, B., Villegas, W., Maheswaran, M.: A trust based approach for protecting user data in social networks. In: Proc. of the 2007 Conference of the Center for Advanced Studies on Collaborative Research, Richmond Hill, Ontario, Canada, pp. 288–293 (2007)

23. Kruk, S.R., Grzonkowski, S., Gzella, A., Woroniecki, T., Choi, H.-C.: D-FOAF: Distributed Identity Management with Access Rights Delegation. In: Mizoguchi, R., Shi, Z.-Z., Giunchiglia, F. (eds.) ASWC 2006. LNCS, vol. 4185, pp. 140–154. Springer, Heidelberg (2006)

24. Wang, H., Sun, L.L.: Trust-involved access control in collaborative open social networks. In: Proc. of the 2010 4th International Conference on Network and System Security (NSS), Melbourne, VIC, pp. 239–246 (2010)

25. Sachan, A., Emmanuel, S., Kankanhalli, M.: An efficient access control method for multimedia social networks. In: Proc. of the 2nd ACM SIGMM Workshop on Social Media, Firenze, Italy, pp. 33–38 (October 2010)
26. Agrawal, R., Kiernan, J., Srikant, R., Xu, Y.: Hippocratic databases. In: Proc. of the 28th International Conference on Very Large Data Bases, pp. 143–154 (2002)
27. LeFevre, K., Agrawal, R., Ercegovac, V., Ramakrishnan, R., Xu, Y., DeWitt, D.: Limiting disclosure in Hippocratic databases. In: Proc. of the 13th Very Large Data Bases, pp. 108–119 (2004)
28. Ni, Q., Bertino, E., Lobo, J., Calo, S.B.: Privacy-aware role based access control. IEEE Security and Privacy 7(4), 35–43 (2009)
29. Tian, L.Q.: A Kind of Game-Theoretic Control Mechanism of User Behavior Trust Based on Prediction in Trustworthy Network. Chinese Journal of Computers 30(11), 1930–1938 (2007)
30. Carminati, B., Ferrari, E., Perego, A.: Rule-based access control for social networks. In: Meersman, R., Tari, Z., Herrero, P. (eds.) OTM 2006 Workshops. LNCS, vol. 4278, pp. 1734–1744. Springer, Heidelberg (2006)
31. Wang, Z., Sun, L.F., Chen, X.W., et al.: Propagation-based social-aware replication for social video contents. In: Proc. of the 20th ACM International Conference on Multimedia, Nara, Japan, pp. 29–38 (2012)
32. Ye, C.H., Ling, H.F., Zou, F.H., Liu, C.: Secure content sharing for social network using fingerprinting and encryption in the TSH transform domain. In: Proc. of the 20th ACM International Conference on Multimedia, Nara, Japan, pp. 1117–1120 (October 2012)
33. Lian, S.Q., Chen, X., Wang, J.W.: Content distribution and copyright authentication based on combined indexing and watermarking. Multimedia Tools and Applications 57(1), 49–66 (2012)
34. Chung, M.B., Ko, I.J.: Intelligent copyright protection system using a matching video retrieval algorithm. Multimedia Tools and Applications 59(1), 383–401 (2012)
35. Yeh, L.Y., Huang, Y.L., et al.: A Batch-Authenticated and Key Agreement Framework for P2P-Based Online Social Networks. IEEE Transactions on Vehicular Technology 61(4), 1907–1924 (2012)
36. Ren, Y.Z., Li, M.C., Sakurai, K.: FineTrust: A fine-grained trust model for peer-to-peer networks. Security and Communication Networks 4(1), 61–69 (2011)
37. Zhang, Z.Y., Wang, K.L.: A Trust Model for Multimedia Social Networks. Social Networks Analysis and Mining, 1–11 (2012)

Modeling of Human Saccadic Scanpaths Based on Visual Saliency

Lijuan Duan[1], Haitao Qiao[1], Chunpeng Wu[2], Zhen Yang[1], and Wei Ma[1]

[1] College of Computer Science and Technology, Beijing University of Technology,
Beijing China
{ljduan,yangzhen,mawei}@bjut.edu.cn,
qht@emails.bjut.edu.cn
[2] Fujitsu Research & Development Center Co. Ltd., Beijing, China
wuchunpeng@cn.fujitsu.com

Abstract. We propose a method to predict human saccadic scanpaths on natural images based on a bio-inspired visual attention model. The method integrates three related factors as driven forces to guide eye movements, sequentially-visual saliency, winner-takes-all and visual memory, respectively. When predicting a current fixation of saccadic scanpaths, we follow physiological visual memory characteristics to eliminate the effects of the previous selected fixation. Then, we use winner-takes-all to select the fixation on the current saliency map. Experimental results demonstrate that the proposed model outperform other methods on both static fixation locations and dynamic scanpaths.

Keywords: visual saliency, winner-takes-all, visual memory, saccadic scanpaths.

1   Introduction

Human beings are able to actively explore the environment with high resolution fovea sensors based on attention guided saccadic eye-moment, which is one of the most important mechanisms in biological vision systems. Benefitting from such unique behavior, human beings can efficiently process the information from complex environments as well as the most of primates. In this highly dynamic and cluttered world, to acquire visual information efficiently and rapidly, it is important for human beings to decide not only where we should look at, but also the sequence of fixations. In fact, both of them are essential for us to comprehend human saccadic behaviors. The computational models of visual attention and saccadic scanpaths not only help us better understand the mechanism of human cognitive behavior, but also provide us powerful tools to solve various vision related problems, such as video compression [1], scene understanding [2], object detection and recognition [3] etc. In addition, the next generation of efficient, foveated and active vision systems [4] could potentially be applied to a diverse array of problems such as automated pictorial database query,

J.-S. Pan et al. (eds.), *Genetic and Evolutionary Computing*,
Advances in Intelligent Systems and Computing 238,
DOI: 10.1007/978-3-319-01796-9_28, © Springer International Publishing Switzerland 2014

image understanding, image quality assessment [5], automated object detection, autonomous vehicle navigation, and real-time foveated video compression [6,7]. Also, the ability to understand and reproduce an expert radiologist's eye movements could be used in semi-automated detection of lesions in digital mammograms [8] - a problem of life-saving significance. Many other significant applications can be envisioned.

In the literature, it is well known that eye-movement is guided by both bottom-up (stimulus-driven) and top-down (task-driven) factors [2, 10]. The bottom-up stimulus-driven research mainly focuses on obtaining saccadic scanpaths based on visual saliency, in which a saliency map is pre-computed using low-level image features to guide task independent gaze allocation. These methods have been proven to be very effective in predicting eye fixations captured from human subjects while viewing natural images and video sequences. Itti et al. [2] proposes a computational attention model based on Koch and Ullman's attention selection mechanism[11], in which visual saliency is measured by spatial center-surround divergence across a few of feature channels in different scales. In the model of [12], two principles named winner-takes-all (WTA) and inhibition-of-return (IOR) are adopted to select fixations based on itti's saliency maps. This technique is widely used for scanning visual scenes and generating artificial saccades. For top-down research, there are also extensive studies of human saccadic behaviors during different real-world tasks, such as making a sandwich, fixing a cup of tea or learning and matching a shape. Most studies indicate that eye movements are probably made to collect task-relevant information.

In this paper, we propose a computational model to simulate human saccadic scanpaths on natural images without a particular task. The proposed model firstly computes static visual saliency maps which describe the importance of each image location. Then we adopt winner-takes-all to select a current fixation, meanwhile the obtained fixation is saved in the memory. Before the next calculation, we follow the physiological visual memory characteristics to eliminate the effects of the previous selected fixation saved in the memory.

The paper is organized as follows. We present the framework of saliency guided simulating human saccadic scanpaths method in Section 3. In Section 4, we compare our saccadic scanpaths with previous methods and the scanpaths recorded by eye tracking data. Conclusions are given in section 5.

2   Framework of Human Saccadic Scanpaths Based on Visual Saliency

The proposed framework is shown in Fig. 1. The model integrates three related factors as driven forces to guide eye movements sequentially: saliency map, visual memory and winner-takes-all. In the following, we will introduce every factor in details. To be consistent with the setting of the eye movement experiments, our model places the initial fixation at the image center and then generates a series of fixations. Firstly, we calculate the saliency map of input image.

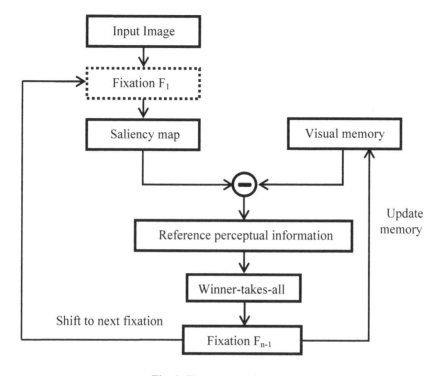

Fig. 1. The proposed framework

2.1 Review of Our Saliency Measure

In [13], we proposed a visual saliency detection method by spatially weighted dissimilarity. There are four main steps in visual saliency detection: splitting image into patches, reducing dimensionality, evaluating global dissimilarity and weighting dissimilarity by distances to centers [13]. In this paper, we use this model to detect the saliency map of the input image. But we modify the model to be lack of the central bias. The central bias is important in calculating the saliency. However, it will be useless in the simulating of the saccadic scanpaths. Therefore, we only use the following three steps mentioned in [13]. Firstly, non-overlapping patches are drawn from an image, and all of the color channels are stacked to represent each image patch as a feature vector of pixel values. All vectors are then mapped into a reduced dimensional space. The saliency of image patch p_i is calculated as

$$\text{saliency}(p_i) = \text{GD}(p_i) \tag{1}$$

$\text{GD}(p_i)$ is global dissimilarity. $\text{GD}(p_i)$ is computed as

$$GD\left(p_i\right) = \sum_{j=1}^{L}\{\omega(p_i, p_j) \cdot Dissimilarity\left(p_i, p_j\right)\} \tag{2}$$

In Eq. 2, L is total number of patches, $\omega(p_i, p_j)$ is the inverse of spatial distance, and Dissimilarity(p_i, p_j) is the dissimilarity of feature response between patches. $\omega(p_i, p_j)$ and Dissimilarity(p_i, p_j) are computed as

$$\omega(p_i, p_j) = \frac{1}{1+\text{Dist}(p_i, p_j)} \tag{3}$$

$$Dissimilarity(p_i, p_j) = ||f_i - f_j|| \tag{4}$$

In Eq. 3, Dist(p_i, p_j) is the spatial distance between patch p_i and patch p_j in the image. In Eq. 4, feature vectors f_i and f_j correspond to patch p_i and patch p_j, respectively. Finally, the saliency map is normalized and resized to the scale of the original image. Then, it is smoothed with a Gaussian filter ($\sigma = 3$).

2.2   Select the Fixation

After calculating the saliency map, then, we subtract visual memory from the saliency map to get the reference perceptual information as a reference map in human brain. Based on the reference perceptual information, we adopt a known principle named winner-takes-all (WTA) to select and locate the simulated fixation point. Meanwhile, we store the current fixation and update the visual memory.

Fig. 2. Eye-movements generated by our model. For each image, the scanpaths with three saccades and the corresponding focused regions was shown.

Once an image location is visited by the fovea, information at that fixation is acquired. Visual memory integrates the information across previous eye movements, meanwhile, it loses the stored information at a certain rate. This forgetting property will steer eyes moving back to previously visited salient spots, in other words, the information at the previous fixations has been forgotten. In our model, we multiply visual memory with a constant forgetting factor £ ($0 \leq$ £ ≤ 1) to simulate its forgetting property. If £=1, no forgetting effect; if £=0, it is memoryless.

Fig. 2 shows the visualized gaze selection process on natural images generated by the proposed model. We assign the fixation in these natural images. You can manually set the number of fixation. However, the more fixations, the more factor impacting the saccadic scanpaths, so the effect of the model will be weaken.

3   Experimental Results

To test the performance of the proposed model, we collect human eye movement data on a natural image dataset firstly. Then we compare the saccades scanpaths of fixation generated by our model with two other approaches [2, 13] against the eye movement data.

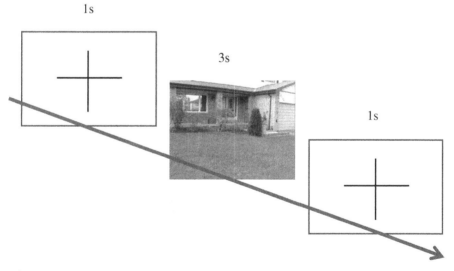

Fig. 3. The Experiment procedure

3.1   Dataset and Eye Movement Data Collection

We randomly collected a dataset of 20 color images from the Internet including natural scenes, street and buildings, and indoor images, etc. We collected eye movement data from 24 student volunteers with this dataset using a high-speed SMI eye-tracker with a 500 Hz sampling rate. Experiment was shown in Fig. 3, subjects were positioned 0.53m away from a 21-inch CRT monitor. Following calibration, the color images were presented in a random order, each was displayed for 3 seconds followed by a blank screen for 1 second. A cross was placed at the center of the blank screen so as to engage the first fixation at the center of the images. The subjects were given no particular instruction except for asking them to observe the images.

3.2   Evaluation of Fixation Order

There is a lack of literature on computational models of the dynamic scanpaths of visual attention. Itti et al. in [2] propose a scanpaths generation method from static saliency maps based on winner-takes-all (WTA) and inhibition-of-return (IOR) regulations. To our knowledge, this is the most referred method in literature. Hence, we compare our model with Itti et al.'s approach. Then we compare the generated scanpaths based on our saliency map and the one proposed in [13].

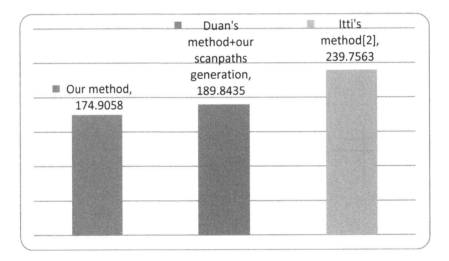

Fig. 4. Comparison results

We place the initial fixation at each image center and then generate a series of fixations. We generate three fixation sequences of a fixed length using the following three methods, Itti et al.'s scanpaths generation method [2], and the method in the paper obtained from the saliency map computed by the model in [13], and the model given in this paper, respectively. We compare the three scanpaths against human scanpaths using Hausdorff distance (H-Distance). Hausdorff distance computes the maximal value of all the minimal distances between two sets of scanpaths, which is defined as

$$H(A, B) = \max(h(A, B), h(B, A)) \tag{5}$$

$$h(A, B) = \max_{a \in B} \min_{b \in A} \|a - b\| \tag{6}$$

$$h(B, A) = \max_{b \in B} \min_{a \in A} \|b - a\| \tag{7}$$

The smaller such a distance is, the closer to human scanpaths the computed ones are. Fig.4 is the comparison results. From the comparison results, we can see that our model performs better in simulating the dynamics of saccadic scanpaths. Compared with [13], the proposed saliency map computation involves no central bias. The central bias is important in calculating the saliency in [13]. When calculating the current fixation, the area of the last fixation will be inhibited. Therefore, the central bias will be useless. This is the reason that the scanpaths obtained based on our saliency model outperform the paths using the model given in [13].

4   Conclusions and Discussion

In the paper, we proposed a computational model to simulate human saccadic scanpaths on natural images without a specific task based on human visual saliency. The proposed model uses dissimilarity and spatial distances to get the saliency map of a nature image. Then, we use winner-takes-all to select the fixation and visual memory to eliminate the effects of the selected. On a natural image dataset, we compare the saccadic scanpaths generated by the proposed model and several other visual saliency-based models against human eye movement data. Experimental results demonstrate that the proposed model achieves the best prediction accuracy on both static fixation locations and dynamic scanpaths.

Acknowledgments. This research is partially sponsored by National Basic Research Program of China (No.2009CB320900), Natural Science Foundation of China (61175115, 61003105, 61070116, 61272320, 61070149 and 61001108), and the Importation and Development of High-Caliber Talents Project of Beijing Municipal Institutions (CIT&TCD201304035).

References

1. Itti, L.: Automatic foveation for video compression using a neurobiological model of visual attention. TIP 13(10), 1304–1318 (2004)
2. Itti, L., Koch, C., Niebur, E.: A model of saliency-based visual attention for rapid scene analysis. TPAMI 20(11), 1254–1259 (1998)
3. Gao, D., Han, S., Vasconcelos, N.: Discriminant saliency, the detection of suspicious coincidences, and applications to visual recognition. TPAMI 31(6), 989–1005 (2009)
4. Klarquist, W., Bovik, A.: Fovea: a foveated vergent active stereo vision system for dynamicthree-dimensional scene recovery. IEEE Transactions on Robotics and Automation 14(5), 755–770 (1998)
5. Osberger, W., Bergmann, N., Maeder, A.: An automatic image quality assessment technique incorporating higher level perceptual factors. In: Proceedings of the 1998 International Conference on Image Processing, ICIP 1998, vol. 3, pp. 414–418 (1998)

6. Lee, S., Pattichis, M., Bovik, A.: Foveated video compression with optimal rate control. IEEE Transactions on Image Processing 10(7), 977–992 (2001)

7. Wang, Z., Lu, L., Bovik, A.: Foveation scalable video coding with automatic fixation selection. IEEE Transactions on Image Processing 12(2), 243–254 (2003)

8. Yang, G.-Z., Dempere-Marco, L., Hu, X.-P., Rowe, A.: Visual search: psychophysical models and practical applications. Image and Vision Computing 20(4), 273–287 (2002)

9. Privitera, C., Stark, L.: Human-vision-based selection of image processing algorithms for planetary exploration. IEEE Transactions on Image Processing 12(8), 917–923 (2003)

10. Tsotsos, J.K., Culhane, S.M., Kei Wai, W.Y., Lai, Y., Davis, N., Nuflo, F.: Modeling visual attention via selective tuning. Artificial Intelligence 78(1), 507–545 (1995)

11. Koch, C., Ullman, S.: Shifts in selective visual attention: towards the underlying neural circuitry. Human Neurobiology 4(4), 219–227 (1985)

12. Itti, L., Koch, C.: Computational modelling of visual attention. Nature Reviews Neuroscience 2(3), 194–203 (2001)

13. Duan, L., Wu, C., Miao, J., Qing, L., Fu, Y.: Visual saliency detection by spatially weighted dissimilarity. In: IEEE Conference on Computer Vision and Pattern Recognition, pp. 473–480 (2011)

A Social Network Information Propagation Model Considering Different Types of Social Relationships

Changwei Zhao, Zhiyong Zhang, Hanman Li, and Shiyang Zhao

Henan University of Science & Technology, Luoyang, China
zhao_chw@163.com, z.zhang@ieee.org

Abstract. In social networks, information are shared or propagated among user nodes through different links of social relationships. Considering the fact that different types of social relationships have different information propagation preference, we present a new social network information propagation model and set up dynamic equations for it. In our model, user nodes could share or propagate information according to their own preferences, and select different types of social relationships according to information preferences. The model reflects the facts that users are active and information possess propagation preferences. Simulation results proved the validity of the model.

Keywords: social network service, propagation model, dynamic equation, digital rights management.

1 Introduction

At present, a social networking service (SNS) has gained significant popularity and is among the most popular sites on the Web [1]. It has been an important platform to build social networks or social relations among people who, for example, share or propagate interests, activities, backgrounds, or real-life connections [2]. Different from the traditional web services, which are largely organized around content, SNS is user-centered, and information is disseminated via social relationships among friends. The research of information dissemination mechanism in SNS has important applications in many fields, such as opinion spread, disease control, digital rights management [3,4], and so on. In social networks, users, social relationships and information are three essential items, and users are described as nodes, social relationships of friends are called edges or links in networks graphs. Users are publishers or disseminators about information, and relationships are transmission path of information. Information is propagated through relationships among friends. If the number of edges between any two nodes is greater than two, we call the network graph as a multigraph. Comparing with the diffusion model of disease [5, 6], computer viruses, opinions, and rumor [7], the propagation model of SNS has its own characteristics. First, users might recognize activity the feature of information which will be shared or propagated, and select an appropriate path to disseminate it according its feature rather than to share indiscriminately it on all relationships. Moreover, there are many types of relationships among users, such as friends, colleagues, acquaintances, classmates, etc. And a user

J.-S. Pan et al. (eds.), *Genetic and Evolutionary Computing*,
Advances in Intelligent Systems and Computing 238,
DOI: 10.1007/978-3-319-01796-9_29, © Springer International Publishing Switzerland 2014

maybe belongs to different types of it. Also, the preference of transmission path is a basic character of information. Considering different relationship types of SNS and information dissemination preference, we present a new information dissemination model of social network, and the dynamic behavior of it is analyzed.

2   Related Works

The research of social networks pertains to the fields of complex networks. A complex network is a graph with non-trivial topological features. The study of complex networks is a young and active area of scientific research inspired largely by the empirical study of real-world networks such as computer networks and social networks. Erdős–Rényi networks, scale-free networks and small-world networks are three kinds of it. ER is a full random graph. Scale-free networks [8] and small-world networks are characterized by specific structural features—power-law degree distributions for the former and short path lengths and high clustering for the latter. Many real-world networks connection meets the feature of power-law degree distributions. In [1, 9], degree distribution, clustering coefficient, clustering coefficient and vertex degree correlation coefficients of social network were analyzed. In modeling of SNS, referring dynamics of infectious disease and using complex networks theories, [10] presented an online social network information dissemination theoretical model. However, in this model, neither preferences of information dissemination nor types of social relationships are considered. All information will disseminate on all type links and have same ability of propagating. Trust is basis of transaction among users in social networks, and is basis of information dissemination too. It is a very common situation, which multi-type relationships connect two user nodes. So far as we know, [11] first noticed this phenomenon, and a hybrid approach is presented for calculating edge trust weights. However, relationships between information dissemination preference and types of links are not concerned. Taking into account the fact that user node is an active node and the user will not share or propagate information randomly, we proposed a new model, in which user activity select suitable edges for propagating or sharing information according information features, there, we called information features as information preference. The model can accurately reflect information dissemination characteristics in SNS and is more consistent with real situation.

3   Model

3.1   Topology of Multi-type Relationships SNS

In a multigraph of social networks (as shown in Fig.1 left), according types of links, it can be transformed into several subgraphs of unique link. And corresponding nodes can share information in different type graphs. As shown in Fig.1, both subgraphs are disconnectivity, but information may spread on all network through users sharing information on different types of link.

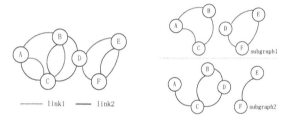

Fig. 1. Mulitgraph vs subgraph

3.2 Information Dissemination Mechanism

In SNS, users are publishers or disseminators of information, and information are spread via different types of relationships. Manner of user spreading information is active. Users select suitable path according information preference. User nodes can be divided into three classes according to function and state of it: interest node (I), disinterest node (D) and propagation node (P). For new information, propagation node is the node which receive the information from neighborhood nodes and possess dissemination capability; Interest node is the node which is a new node and has interest in it, and an interest node may become a propagation node according to a certain probability; if an interest node has not propagation interests, it will become a disinterest node. We define the following dissemination rules:

1. Information I has different dissemination preference I_p, $I_p=[p_1,p_2,...p_n]$, $p_m \in [0,1]$,m=1,2,...,n.   m is type of social relationships. p_m is dissemination preference of I about m and independent each other.
2. For different m and I, propagation node P select links whose type is m with probability p_m and disseminate information I with probability p_t, where p_t is dissemination probability. Once propagation nodes contact with interesting nodes, the latter will convert into propagation nodes and the former will convert into disinterest node with probability 1. That is, user nodes will not repeat sharing or propagating same information.
3. Propagation nodes do not always stay in the state of dissemination. It will become disinterest nodes with probability p_d, where p_d is disinterest probability. In other words, user nodes have no interest in disseminating information I, and information I cannot be propagated for ever.

3.3 Dynamic Equations of Model

Assume the state of user node N is interesting at time t. p_{ii} is probability of remaining interesting state in time slice$[$ t, t + Δt$]$, for simply, Δt is defined as one cycle time of changing or remaining state. p_{ip} is changing probability from interesting state to propagation state, and we have: $p_{ip}=1- p_{ii}$.

For information I with preference I_p, We assume user node N has k links, and has k_m links which permit to types m, then $\sum k_m = k$. g_m is the number of propagation nodes, which links to node N directly via the links of type m at time t. then:

$$p_{ii} = \prod_{m=1}^{n}[(1 - p_m)(1 - p_t)]^{g_m} \tag{1}$$

Assume type of social relationships m is independent and obey uniform distribution, and total type of social relationships is n. Then, for nodes whose degree is k, the probability is k/n those one links permits to type m. k_m, the number of links which permit to type m, obey binomial distribution:

$$P(k_m) = \binom{k}{k_m}\left(\frac{k}{n}\right)^{k_m}\left(1 - \frac{k}{n}\right)^{k-k_m} \tag{2}$$

Assume g_m obey binomial distribution too, then:

$$P(g_m) = \binom{k_m}{g_m}(\omega_{km})^{g_m}(1 - \omega_{km})^{k_m-g_m} \tag{3}$$

Where, ω_{km} is connected probability from propagation node to interesting node in networks of type m. where:

$$\omega_{km} \approx \sum_{k'} p(k' \mid k)\binom{k}{k_m}\left(\frac{k}{n}\right)^{k_m}\left(1 - \frac{k}{n}\right)^{k-k_m}\rho^p(k',t) \tag{4}$$

$p(k'|k)$ is the degree of correlation function, and represents conditional probability of adjacent nodes whose degrees are k and k' differently. In social networks, $p(k'|k) = (k'p(k'))/\bar{k}$ [8]. $\rho^p(k',t)$ is propagation nodes density, whose degree is k' at time t.

From formula (3) and (4), we can get average maintenance probability of interest node, whose degree is k, at time slice [t, t + Δt] . For type of m:

$$\bar{p}_{ii}(k_m,t) = \sum_{k_m=0}^{k}[P(k_m)\sum_{g_m=0}^{k_m}\binom{k_m}{g_m}(1 - \Delta t \cdot p_t)^{g_m}(\omega_{km})^{g_m}(1 - \omega_{km})^{k_m-g_m}] \tag{5}$$

For all networks, average maintenance probability is :

$$\bar{p}_{ii}(k,t) = 1 - \prod_{m=1}^{n}(1 - \bar{p}_{ii}(k_m,t)) \tag{6}$$

Assume N(k, t) is the total number of nodes whose degree equal k at time t, and I(k, t), D(k, t), P(k, t) is the number of interesting, disinterest and propagation nodes differently. Then we have:

$$N(k,t) = I(k,t) + D(k,t) + P(k,t) \tag{7}$$

In Δt at time t, simply we set $\Delta t=1$, then the variation of interesting node is:

$$\Delta I(k,t) = \bar{p}_{ip}(k,t) \cdot I(k,t) = \prod_{m=1}^{n}(1 - \bar{p}_{ii}(k_m,t)) \cdot I(k,t) \qquad (8)$$

Using ruler (3), we can conclude the variation of propagation node is:

$$\Delta R(k,t) = (1 - p_t - p_d) \cdot R(k,t) \qquad (9)$$

From (7), (8), (9), we can get:

$$\Delta Q(k,t) = -\Delta I(k,t) - \Delta R(k,t) \qquad (10)$$

(8), (9) and (10) compose dynamical equations of social network information propagation model.

4   Experiments

4.1   Data

In experiment, we first generate data of social relationships according to distribution and connection characteristics of actual social network data. The data includes two different types of link, which have same distribution, power-law distribution. The networks parameters as follows: The total number of nodes is 9971, the average degree is 24.92, the maximum degree is 713, the clustering coefficient is 0.019, and the degree assortativity is 0.094.

4.2   Experiment

In order to verify the model's validity, we conducted several experiments and analyzed the results. In our experiment, we set initial propagation node number equal 1, the others are interesting nodes.

Networks Stable State vs. Information and User Preference
When networks are stable, number of interesting node reflects the dissemination range of information. The results of number of interesting node vs different preferences are shown in Fig. 2.

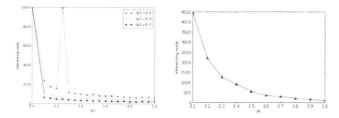

Fig. 2. Number of interesting node vs different preference

As the Fig.2 show: networks stable state is not only related to the network topology, but also the types of relationships, the information transmission preference and the users transmission preference. From above, we can draw that not all information can disseminate on whole net, the dissemination range is related to the information propagation preference. When the network is not connected with single relationship, the information can effectively increase its dissemination range through various relations.

Transmission Duration vs. Transmission Preference

Transmission duration refers to the time slice from the first time of information propagated to the system stability. Due to the randomness of transferring information of users, results which information are propagated less than twice are omitted. Parameter Setting: p_t=0.7, p_d=0.3. Under different transmission preference, transmission duration results are as Fig.3, and the number of failures is as Fig.4:

From Fig.3 and 4, we can draw that there is a relation between the transmission preference and the information transmission duration. The preference value increases, the duration decreases. That is, the information reaches its stability in a short time. At the same time, the dissemination cannot fail easily and has a better possibility to transmit on whole network.

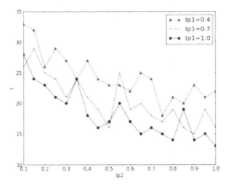

Fig. 3. Transmission duration VS Ip

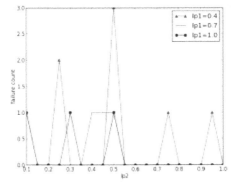

Fig. 4. Transmission failure VS Ip

Networks Stable State vs. Initial Nodes Degree

The information transmission preference I_p=[0.6,0.6], user transmission preference p_t=0.7. Initial nodes degree is 101, 324, and 558 differently. When networks are stable, number of disinterest indicates as Fig.5:

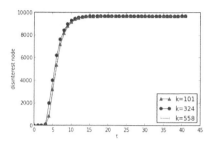

Fig. 5. Networks stable state vs initial nodes degree

As the Fig.5 shows: the initial node degree has little influence on the internet stability. This is determined by the power-law distribution and the high connectivity of social network. From above, we can draw that the network stable state is related to the transmission preference of the user and information instead of the initial node.

5   Conclusion

In this paper, a new social network information propagation model is represent. The model can reflect the facts that users and information propagated have different preference in different types of relationships, which link all user nodes. It is consistent with real social network in information dissemination, and it is a suitable model of social network information transmission. Although in this paper, information transmission preference is defined, but no real information be mapped to its spread preferences. Next step, we will study how to map text information, multimedia information to its preference, and provide forecast for real information propagation.

Acknowledgment. The work was sponsored by National Natural Science Foundation of China Grant No.61003234, Plan for Scientific Innovation Talent of Henan Province Grant No.134100510006, Program for Science & Technology Innovation Talents in Universities of Henan Province Grant No.2011HASTIT015, and Key Program for Basic Research of The Education Department of Henan Province Grant No.13A520240.

References

1. Mislove, A., Marcon, M., Gummadi, K.P., et al.: Measurement and analysis of online social networks. In: Proceedings of the 7th ACM SIGCOMM Conference on Internet Measurement, pp. 29–42. ACM (2007)
2. Social networking service, http://en.wikipedia.org/wiki/Social_networking_service

3. Zhang, Z.Y.: Digital Rights Management Ecosystem and its Usage Controls: A Survey. International Journal of Digital Content Technology & Its Applications 5(3), 255–272 (2011)
4. Zhang, Z.Y.: Security. Trust and Risk in Digital Rights Management Ecosystem. Science Press, China (2012)
5. Ni, S., Weng, W., Zhang, H.: Modeling the effects of social impact on epidemic spreading in complex networks. Physica A: Statistical Mechanics and its Applications 390(23), 4528–4534 (2011)
6. Guo, Q., Li, L., Chen, Y., et al.: Modeling dynamics of disaster spreading in community networks. Nonlinear Dynamics 64(1-2), 157–165 (2011)
7. Zhang, Y., Zhou, S., Zhang, Z., et al.: Rumor evolution in social networks. Physical Review E 87(3), 032133 (2013)
8. Barabási, A.L., Bonabeau, E.: Scale-free networks. Scientific American 288(5), 50–59 (2003)
9. Kumar, R., Novak, J., Tomkins, A.: Structure and evolution of online social networks. In: Link Mining: Models, Algorithms, and Applications, pp. 337–357. Springer, New York (2010)
10. Yan-Chao, Z., Yun, L., Hai-Feng, Z., et al.: The research of information dissemination model on online social network. Acta Phys. Sin. 60(5), 50501–50501 (2011)
11. Al-Oufi, S., Kim, H.N., Saddik, A.E.: A group trust metric for identifying people of trust in online social networks. Expert Systems with Applications 39(18), 13173–13181 (2012)

A Dynamic Intrusion Detection Mechanism Based on Smart Agents in Distributed Cognitive Radio Networks

Ma Lichuan, Min Ying, and Pei Qingqi[*]

State Key Laboratory of Integrated Service Network, Xidian University, Xi'an 710071, China
qqpei@mail.xidian.edu.cn

Abstract. To overcome the lack of available spectrum in wireless communications, the Cognitive Radio Networks arise. But this new technology also brings new threats to the whole network, especially the lion attack. Concerned with the weakness of the existing intrusion detection systems in Cognitive Radio Networks, we propose a dynamic intrusion detection mechanism on basis of smart agents with the utility of Markov chain model. And simulation results verify the efficiency of our mechanism.

Keywords: Cognitive Radio Networks, Intrusion Detection Mechanism, Markov Chain Model.

1   Introduction

Recently, there has been tremendous interest in the field of Cognitive Radio Networks (CRNs) which is an enabling technology that allows unlicensed (secondary) users to operate in the licensed spectrum bands[1]. However, the new features of CRNs have brought new threats, such as the primary user emulation (PUE) attacks[2], objective function attacks (OFA)[3], lion attacks[4] and so on. Among them, the Lion attack is a cross-layer attack aimed at disrupting TCP connections by performing a PUE attack to force frequent handoffs of the CRNs and will lead to a permanent Denial of Service.

2   Intrusion Detection Systems in CRNs and Related Works

Like other networks, cognitive wireless network security protection mechanisms are usually divided into two lines of defense. The first line of defense, such as encryption, authentication, access control, attracts more attention in cognitive wireless networks recent years. Because of the characteristics of cognitive radio networks and the attackers' intelligence, many attacks can pass through the first line of defense easily. The second line of defense focuses on detecting the attacks that pass through the first line of defense.

Nowadays, the development of intrusion detection mechanisms used in CRNs is rapid. In 2011, Olga León etc. proposed a cooperation cognitive wireless network intrusion detection system model. But this work only provided a guideline for future

---

[*] Corresponding author.

J.-S. Pan et al. (eds.), *Genetic and Evolutionary Computing*,
Advances in Intelligent Systems and Computing 238,
DOI: 10.1007/978-3-319-01796-9_30, © Springer International Publishing Switzerland 2014

CRNs intrusion detection mechanisms[5]. In 2012, they also proposed co-location detection program for PUE attack but this program relies on TDoA (Time Difference of Arrival) estimation techniques and Taylor series and this program is not able to detect joint attack effectively [6]. In the same year, Joffre Gavinho Filho et. proposed a intrusion detection mechanism in CRNs with the application of intelligence algorithms [7]. The simulation result is notable but this mechanism concentrated on the detection of PUE attack only.

3   A Dynamic Intrusion Detection Mechanism Based on Smart Agents in Distributed CRNs

3.1   CRNs Infrastructure

In this paper, just as showed in Fig.1, the CRN equipped with smart agents has a hierarchical architecture where the second users (SUs) communicate with each other in a distributed way but they are managed by a unified center and the primary networks coexist with CRNs in the same geographical range. The SUs are divided into clusters and the head of each cluster has a distance of one hop with its members.

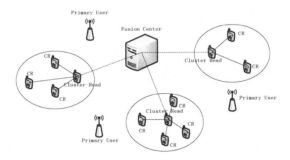

Fig. 1. CRN infrastructure

3.2   Intrusion Detection Mechanism Based on Smart Agents

In order to achieve the fast but efficient detection of lion attack, we present a structure of smart agent to be installed to SUs (as depicted in Fig.2)[8]. This structure contains seven modules:

a. Cognitive Module: Periodically sense the surrounding spectrum utilization and obtain the cluster network information from cluster head nodes.
b. Data Collection Module: Collect the network information and aggregate the real-time data and remove false ones.
c. Local Detection Engine Module: Determine whether the information acquired from the Data Collection Module suffers from the lion attack. Here, we propose an intrusion detection mechanism based on Markov chain model.

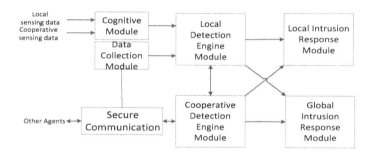

Fig. 2. Architecture of Agents

d. Local Response Module: Decide the response strategy based on the detecting result and broadcast alerts within a cluster and then inform these alerts to the network layer to take according measures.

e. Cooperative Detection Engine Module: Trigger cooperative detection when the Local Detection Engine Module cannot determine whether the current network is under an attack.

f. Secure Communication Module: To guarantee the security of communication.

g. Global Intrusion Response Module: Make the global response according to the results of collaborative intrusion detection and broadcast alerts through the entire network.

4   Intrusion Detection Mechanism

Inspired by the work of [9], we propose an intrusion detection mechanism based on Markov chain model.

4.1   Construction of Markov Chain Model

We first give the assumptions of our model: there are n channels in the environment; the time for spectrum sensing process is t_s; the time needed for spectrum handoff is t_h; the time needed for SUs to accomplish one communication successfully is t_{once}; the lion attackers can drive the SUs from the current channel in the duration of data transmission. Here, we set a countering duration T as $T = n(t_s + t_h + t_{once})$ and during the duration of T there are at most n spectrum handoffs for each SU. According to this, we can achieve the $n+1$ states of the Markov chain model. Let N stands for the amount of handoffs occurs in the duration of T and $N = 0, 1, 2, \cdots, n$. By the following algorithm, we can construct the Markov chain model.

The algorithm of constructing a Markov chain is:

```
Construction_of_Markov_chain()
Step 1: Initialize a (n+1)×(n+1) matrix N to zeros and a
parameter counter to 1;
Step 2: Record the amount of a SU's handoffs in the duration
of T as i;
Step 3: Record the amount of this SU's handoffs in the
following duration of T as j;
Step 4: Refresh the element of matrix N as N(i,j)=N(i,j)+1;
Step 5: counter=counter+1;
Step 6: If counter<w-1, switch to Step 2; else break the
procedure.
```

By this algorithm, we can compute the transition probability from state i to state j as:

$$P(i, j) = \frac{N(i, j)}{\sum_{k=0}^{n} N(i,k)}$$

In the algorithm, w stands for the window size which decides the duration for the construction of Markov chain model. To achieve the goal of dynamically adapting to the infeasible environment, we propose a refreshing process for training as:

$$(W_1, W_2, \cdots, W_i, \cdots, W_w) \rightarrow (W_{k+1}, W_{k+2}, \cdots, W_w, \cdots, W_{w+k})$$

Here, $W_i (i = 1, 2, \cdots, w)$ stands for the amount of handoffs in a duration of T under conditions without malicious entities and k stands for the slipping step size of the training window. When the refreshing process is completed, the Markov chain model would be reconstructed.

4.2   Construction of the Classifier

Before making a decision, we need to construct the classifier first and the duration is $T_D = X \cdot T$. Here, X is a parameter that depends on the widow size. And the procedure of constructing the classifier is:

a. In every duration of T, we record the amount of handoffs as i and the amount of handoffs in the next duration of T as j.

b. With reference to the Markov chain that has been already constructed, we can find the transition probability $P(i, j)$ under normal circumstance. Meanwhile, we set a threshold ε to distinguish the abnormal states.

c. Here, we set a decision parameter y which have a direct influence on making final decision and a penalty constant z which responsible for penalizing the deed of being under a lion attack with a large probability (such as $1-\varepsilon$). The computation of Y is:

i) Initialize Y to zero at the beginning of T_D.

ii) At the end of time slot of T, if $P(i, j) < \varepsilon$, $Y = Y + Z$; else, $Y = Y + 1$.

d. After the duration of T_D completed, we compute $\beta = \dfrac{Y}{X}$. Here, β means how well the actions of SUs during T_D match the constructed Markov chain. A lower β indicates the lower probability of the SU to be under a lion attack.

e. We set a decision threshold λ. If $\beta > \lambda$, the agent triggers an alert and vice versa.

4.3   Tuning the Parameters

a. Window-size w

The lager w means the more precise our Markov chain model is but also means lager memory requirement and more complicated computation. How to choose a proper w is describe in the Simulation Study part.

b. Penalty Constant Z

Setting a penalty constant Z is to distinguish the states under an attack from the normal ones. The process to determine Z can be:

i) Set a testing duration of $M \cdot (a+b) \cdot T_D$. Let $(a+b) \cdot T_D$ stand for a testing period. During the m th testing period, normal conditions last from mT_D to maT_D and conditions under a lion attack last from maT_D to $m(a+b)T_D$.

ii) Compute $D_N(m) = \dfrac{1}{a} \sum\limits_{i=m(a+b)+1}^{m(a+b)+a} \beta(i)$ and $D_A(m) = \dfrac{1}{b} \sum\limits_{i=m(a+b)+a+1}^{(m+1)(a+b)} \beta(i)$.

$D_N(m)$ indicates the discrepancy between normal conditions and the Markov chain model. $D_A(m)$ indicates the discrepancy between conditions under a lion attack and the Markov chain model.

iii) When the testing process is over, compute

$$D_N = \frac{1}{M} \sum_{i=1}^{M} D_N(m)$$

and

$$D_A = Min\{D_A(i), i = 1, 2, \cdots, m\}$$

Then we go on adjusting the value of Z until $|D_N - D_A|$ is above a predetermined threshold.

c. Decision Threshold λ

Determination of λ depends on the secure requirements of the system. If λ is set too large, it can bring high detection ratio but increase the false positive ratio in the same time. If λ is set too small, it can decrease the false positive ratio but decrease the detection ratio.

5   Simulation Study

5.1   Simulation Settings

In the simulation, 3 PUs and 1 SU with a smart agent are included with 9 channels to which they can access. Each PU has 3 licensed channels. PUs always perform regularly and we can assume that PUs have a probability of 0.4 to appear on the channels.

We choose Detection Ratio and False Positive Ratio To measure the performance of our mechanism:

Detection Ratio (DR): It is reported for intrusive behavior and is computed form dividing the total number of correct detections by the total number of victims in the anomalous data.

False Positive Ratio (FPR): The percentage of decision in which normal data are flagged as anomalous.

5.2   Results Analysis

We first set: the decision threshold λ as $\lambda = 2.0$; the penalty constant Z as 1, 2, 3; the window size as 200, 400, 600, 800, 1000. Then we set the deciding time as 100. As shown in Fig.3, it depicts the change trends of Detection Radio and False Positive Radio under different window sizes and penalty constants. Under each penalty constant, with the increase of the window size, Detection Ratio increases and False Positive Ratio decreased respectively. This means that with lager window size when constructing the Markov chain model, we can achieve a more specific model. But greater window size means larger memory requirement and much more complicated computation. Under the condition of our simulation, we can see that when the window size is greater than 600, the increase of Detection Ratio and decrease of False Positive Ratio is so slow that we can neglect. So, we can choose a suitable window size according to requirements of the real system.

Here, we show the change tendency of Detection Ratio and False Positive Ratio with different decision thresholds λ. In our simulation, we set λ as 1.5, 2.0, 2.5, 3.0. Just as described in Fig.4, we can conclude that higher λ indicates higher Detection Ratio and higher False Positive Ratio and vice versa. However, we want higher Detection Ratio and lower False Positive Ratio in practice. This demand requires us that the choice of λ should depend on the empirical knowledge of actual situations and the error-tolerance rate of the real networks.

a.

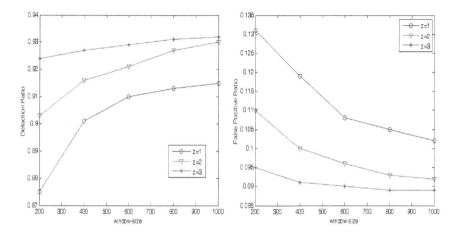

Fig. 3. DR and FPR with different window-sizes and penalty constants

b.

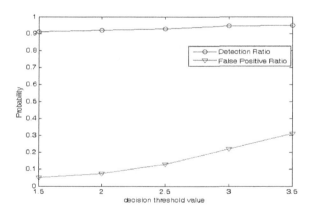

Fig. 4. DR and FPR with different decision thresholds

6   Conclusions and Future Works

In this paper, we propose an intrusion detection mechanism on base of smart agents with the utility of Markov chain model and simulation results verify the efficiency of our mechanism. However, we only focus on the lion attack and this obviously limits the popularity of our mechanism.

Our future work will concentrate on designing a mechanism which is much more compatible with much more complicated CRNs environment and can work efficiently to detecting as many kinds of intrusions as possible.

Acknowledgement. This work is supported by the National Natural Science Foundation of China under Grant No. 61172068, the Program for New Century Excellent Talents in University (Grant No.NCET-11-0691), and the Fundamental Research Funds for the Central Universities (Grant No.K50511010003).

References

1. Mitola, J.: Cognitive radio: An integrated agent architecture for software defined radio. Doctor of Technology Thesis, Royal Inst. Technol. (KTH), Stockholm, Sweden (2000)
2. Chen, R., Park, J.M.: Ensuring trustworthy spectrum sensing in cognitive radio networks. In: First IEEE Workshop on Networking Technologies for Software Defined Radio Networks (SDR), pp. 110–119 (2006)
3. Clancy, T., Goergen, N.: Security in cognitive radio networks: Threats and Mitigation. In: Third International Conference on Cognitive Radio Oriented Wireless Networks and Communications, pp. 1–8 (2008)
4. León, O., Hernandez-Serrano, J., Soriano, M.: A new cross-layer attack to TCP in cognitive radio networks. In: Proceedings of IEEE 2009 Second International Workshop on Cross Layer Design, pp. 1–5 (2009)
5. León, O., Román, R., Hernández-Serrano, J.: Towards a Cooperative Intrusion Detection System for Cognitive Radio Networks. In: Casares-Giner, V., Manzoni, P., Pont, A. (eds.) NETWORKING Workshops 2011. LNCS, vol. 6827, pp. 231–242. Springer, Heidelberg (2011)
6. León, O., Hernández-Serrano, J.: Cooperative detection of primary user emulation attacks in CRNs. Computer Networks 56, 3374–3384 (2012)
7. Gavinho Filho, J., Carmo, L.F.R.C., Machado, R.: IDS-COG- Intrusion Detection System for Cognitive Radio Network. International Journal of Computer Science and Network Security 12(3) (2012)
8. Bo, S., Kui, W.: Towards adaptive intrusion detection in mobile ad hoc networks. In: Global Telecommunications Conference, vol. 6, pp. 3551–3555. IEEE (2004)
9. Sun, W.K., Pooch, U.W.: Routing anomaly detection in mobile ad hoc networks. In: The 12th International Conference on Computer Communications and Networks, pp. 25–31. IEEE (2003)

Bio-inspired Visual Attention Model and Saliency Guided Object Segmentation

Lijuan Duan[1], Jili Gu[1], Zhen Yang[1], Jun Miao[2,*], Wei Ma[1], and Chunpeng Wu[3]

[1] College of Computer Science and Technology, Beijing University of Technology, Beijing, China
[2] Key Laboratory of Intelligent Information Processing, Institute of Computing Technology, Chinese Academy of Sciences, Beijing, China
jmiao@ict.ac.cn
[3] Fujitsu Research & Development Center Co. Ltd., Beijing, China

Abstract. In this paper, we present a saliency guided image object segment method. We suppose that saliency maps can indicate informative regions, and filter out background in images. To produce perceptual satisfactory salient objects, we use our bio-inspired saliency measure which integrating three factors: dissimilarity, spatial distance and central bias to compute saliency map. Then the saliency map is used as the importance map in the salient object segment method. Experimental results demonstrate that our method outperforms previous saliency detection method, *yielding higher precision (0.7669) and better recall rates (0.825), F-Measure (0.7545), when evaluated using one of the largest publicly available data sets.*

Keywords: visual attention, dissimilarity, spatial distance, central bias, salient object detection.

1   Introduction

Visual attention has been studied by researchers in physiology, psychology, neural systems, and computer vision for a long time. It is useful for many computer vision tasks such as content-based image retrieval, segmentation, and object detection. Computationally modeling such mechanism has been widely studied in order to identify which part of image is more useful when analyzing an image in recent years [1], [2], [3]. Applications of the models such as image classification [4], image segmentation [5] and object detection [6] have become a popular research topic. We study visual saliency by detecting a salient object in an input image. The automatic detection of visually salient regions in images is useful for image segmentation, adaptive region-of-interest based image compression, object recognition, and content aware image resizing.

   Salient object detection is defined as an image segmentation problem, where the salient object is separated from the image background [7]. It is supported by

---

* Corresponding author.

J.-S. Pan et al. (eds.), *Genetic and Evolutionary Computing,*
Advances in Intelligent Systems and Computing 238,
DOI: 10.1007/978-3-319-01796-9_31, © Springer International Publishing Switzerland 2014

research on human vision system that the human brain and visual system pay more attention to some parts of an image. The recognition and localization of searching targets in complex visual scenes is still a challenge problem for computer vision systems. However, this task is performed by humans in a more intuitive and efficient manner by selecting only a few regions to focus on. Observers never form a complete and detailed representation of their surroundings [8]. For example, in Figure 1, the tomato, dog, and woman attract the most visual attention in each respective image. Therefore, salient object detection is formulated as a binary labeling problem that separates a salient object from the background. Like face detection, we learn to detect a familiar object; unlike face detection, we detect a familiar yet unknown object in an image [7]. There are many traditional methods of saliency detection which are used in unsupervised object segmentation. To produce perceptual satisfactory salient object segmented images, we use the saliency map as the importance map in the salient object detection method. Han et al. [9] use low-level features of color, texture, and edges in a Markov random field framework to grow salient object regions from seed values in the saliency maps. Salient regions are selected by Ko and Nam [10] using a support vector machine trained on image segment features to select the salient regions of interest using Itti's maps, which are then clustered to extract the salient objects. Ma and Zhang [11] utilize fuzzy region growing on saliency maps to confine salient regions within a rectangular region. Achanta et al. [12] average saliency values within image segments produced by mean-shift segmentation, and then find salient objects by identifying image segments that have average saliency above a threshold that is set to be twice the mean saliency value of the entire image. More recently, Cheng Mingming et al. [13] propose a region-based contrast saliency detection method (RC) to show that segmentation-based method is better than their pixel-based method (HC) and then iteratively use GrabCut to refine the segmentation result initially obtained by thresholding the saliency map.

The paper is organized as follows: Review of our saliency detection is in Section 2. In Section 3, we introduce a salient object segment method. In Section 4, we compare the performance of salient object methods based on different saliency measures. The conclusions are given in Section 5.

2   Review of Our Saliency Measure

In this paper, we use our biologically inspired saliency measure proposed in [14] to detect the salient object. The saliency measure integrates three factors: dissimilarity, spatial distance and central bias, and these three factors are supported by research on human vision system. The dissimilarity is evaluated by a center-surround operator simulating the visual receptive field [1], and this structure is a general computational principle in the retina and primary visual cortex [15]. The spatial distance is supported by the research [16] on foveation of human vision system. Human samples the visual field by a variable resolution, and the resolution is highest at center (fovea) and drops rapidly toward the periphery [17]. In addition, according to previous studies on the distribution of human fixations [18], human tend to look at the center of images. This fact is also known as central bias which reflects that photographer tent to center objects of interest [19]. There are five main steps in our saliency detection method: changing an image to YCbCr color space, splitting image into patches, reducing dimensionality, evaluating the spatially-weighted dissimilarity. First Non-overlapping patches drawn from an image are represented as vectors of pixels, and all patches are mapped into a reduced dimensional space. The saliency of image patch p_i is calculated as

$$Saliency(p_i) = w_1(p_i).GD(p_i) \tag{1}$$

Where $w_1(p_i)$ represents central bias and $GD(p_i)$ is the global dissimilarity. $w_1(p_i)$ and $GD(p_i)$ are computed as

$$w_1(p_i) = 1 - \frac{DistToCenter(p_i)}{D} \tag{2}$$

$$GD(p_i) = \sum_{j=1}^{L}\{ w_2(p_i, p_j).Dissimilarity(p_i, p_j) \} \tag{3}$$

In Eq. 2, $DistToCenter(p_i)$ is the spatial distance between patch p_i and center of the original image, and $D = max_j\{(DistToCenter(p_j)\}$ is a normalization factor. In Eq. 3, L is total number of patches, $w_2(p_i, p_j)$ is inverse of spatial distance, and $Dissimilarity(p_i, p_j)$ is dissimilarity of feature response between patches. $w_2(p_i, p_j)$ and $Dissimilarity(p_i, p_j)$ are computed as

$$w_2(p_i, p_j) = \frac{1}{1 + Dist(p_i, p_j)} \tag{4}$$

$$Dissimilarity(p_i, p_j) = || f_i - f_j || \tag{5}$$

In Eq. 4, $Dist(p_i, p_j)$ is the spatial distance between patch p_i and patch p_j in the image. In Eq. 5, feature vectors f_i and f_j correspond to patch p_i and patch p_j respectively. Finally, the saliency map is normalized and resized to the scale of the original image, and then is smoothed with a Gaussian filter ($\sigma = 3$).

3   Saliency Guided Object Segmentation

The true usefulness of a saliency map is determined by the application. In this paper we consider the use of saliency maps in salient object segmentation. To produce perceptual satisfactory salient objects, we use our bio-inspired saliency measure which integrating three factors: dissimilarity, spatial distance and central bias to compute saliency map. Then the saliency map is used as the importance map in the salient object segment method. To segment a salient object, we need to binarize the saliency map such that ones (white pixels) correspond to salient object pixels while zeros (black pixels) correspond to the background [1]. Fixed parameters of 7, 10, and 20 for sigmaS, sigmaR, and minRegion area used respectively, for all the images (see [20]). In the experiment we use the image dependent adaptive threshold proposed by

[12], which average saliency values within image segments produced by mean-shift segmentation, and then find salient objects by identifying image segments that have average saliency above a threshold that is set to be twice the mean saliency value of the entire image:

$$T_a = \frac{2}{W*H} \sum_{x=0}^{W-1} \sum_{y=0}^{H-1} S(x,y) \qquad (6)$$

where W and H are the width and height of the saliency map in pixels, respectively, and $S(x,y)$ is the saliency value of the pixel at position (x, y). Using this approach, we obtain binarized maps of salient object from each of the saliency algorithms.

4   Experimental Validation

Experiment is conducted on the publicly available database provided by Achanta et al. [12] to evaluate our performance. The ground truths of this database are binary images in which salient objects are accurately marked by human. Performance of salient object detection based on different saliency detection methods including ours is compared. The same parameters of our method will be used across all images. According to our previous parameter settings [14], the color space is YCbCr, the size of image patch is 14x14 and the dimensions to which each feature vector reduced is 11. Our saliency detection method with above parameters outperforms some state-of-the-art [1], [21], [22], [23] on predicting human fixations, please see [14] for details.

4.1   Qualitative Comparison between Saliency Measures

We provide an exhaustive comparison of our approach to other six state-of-art methods on a database of 1000 images [12] with binary ground truth [2]. Saliency maps of previous works are provided by [14]. Comparison of salient object detection results between our method and other six saliency detection method in more images are shown in Fig. 1. The first row are original images, the second row are ground truth images according with the first row. From the third row, the five images on each row are segmented results by Itti et al. [1], Ma and Zhang [11], Harel et al. [22], Hou et al. [21], R. Achanta [23], R. Achanta [12]. Experiment show that our ultimate saliency maps are superior to the other saliency maps produced with a segmented result which prove the effectiveness of our saliency map evaluation method. The saliency method in [14] highlights the woman and dragon in image with well-defined border and suppresses background regions efficiently. In all experiments, our approach consistently produces results closest to ground truth. However, the image in fifth column, its ground truth image of other subjects may be same with our detected one. The key objective of attention detection should be to locate position of a salient object as accurately as possible, i.e. with high precision, recall, and F-Measure. Because background regions are successfully suppressed in our saliency map, the binary mask generated from our saliency map is more accurate than that from other methods (see also Fig. 2).

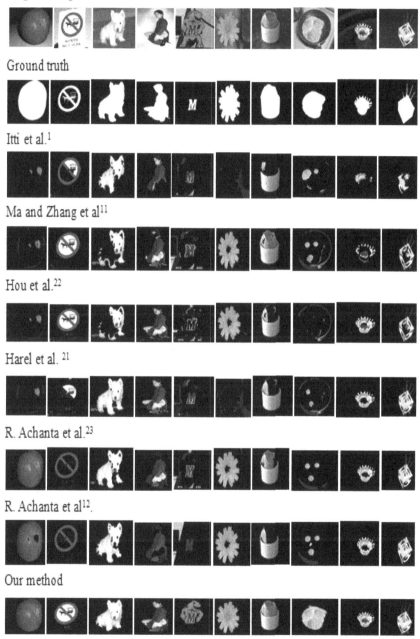

Fig. 1. Comparison between salient object detection based on seven different saliency measures as follows: Itti et al. [1], Ma and Zhang [11], Harel et al. [22], Hou et al. [21], R. Achanta [23], R. Achanta [12] and our method. The first row are original images, the second row are ground truth images according with the first row.

4.2   Quantitative Comparison between Saliency Measures

We evaluate the performance of our algorithm measuring its precision and recall rate. Precision corresponds to the percentage of salient pixels correctly assigned, while recall corresponds to the fraction of detected salient pixels in relation to the ground truth number of salient pixels. High recall can be achieved at the expense of reducing the precision and vice-versa so it is important to evaluate both measures together. With a ground-truth saliency map G, for any detected salient region mask A, we use following measurements:

$$\text{Precision} = \frac{\sum_x g_x a_x}{\sum_x a_x} \tag{7}$$

$$\text{Recall} = \frac{\sum_x g_x a_x}{\sum_x g_x} \tag{8}$$

F-Measure is the weighted harmonic mean of precision and recall, with a non-negative β :

$$F_\beta = \frac{(1+\beta^2)\text{Precision}*\text{Recall}}{\beta^2*\text{Precision}+\text{Recall}} \tag{9}$$

F_β (Eq. 9) are obtained over the same ground-truth database by Achanta et al. [12]. $\beta^2 = 0.3$ is used in our work to weigh precision more than recall. The comparison is shown in Table1 and Fig. 2 which are according with [12]. Itti's saliency detection method has a high precision (0.7919) but very poor recall (0.4643). Among all the methods, our method shows the highest recall value, third precision and second F-Measure. Compared with Achanta [12], our method has a higher recall but a lower precision. However, like all the other saliency detection methods, it can fail when the object of interest is not distinct from the background.

Table 1. Comparison between salient object detection   precision, recall and F- Measure based on seven different saliency measures as follows: Itti et al. [1], Ma and Zhang [11], Harel et al. [22], Hou et al. [21], R. Achanta [23], R. Achanta [12].

	Precision	Recall	F-Measure
Itti [1]	0.7919	0.4643	0.6336
Ma and Zhang [11]	0.675	0.6613	0.6459
Harel [22]	0.7321	0.7519	0.7104
Hou [21]	0.6581	0.5573	0.5998
R.Achanta [23]	0.7543	0.6983	0.7152
R.Achanta [12]	0.8363	0.7936	0.8048
Our Method	0.7669	0.825	0.7545

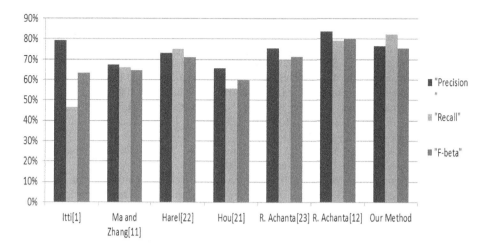

Fig. 2. Comparison between salient object detection precision, recall and F-Measure based on seven different saliency measures as follows: Itti et al. [1], Ma and Zhang [11], Harel et al. [22], Hou et al. [21], R. Achanta [23], R. Achanta [12].

5   Conclusion and Discussion

We use the saliency map as the importance map in the salient object method. Experimental results show that our model could generate high quality saliency maps that highlight the whole salient object with well-defined boundary, meanwhile successfully suppress the background regions. The resulting saliency maps of our method on Achanta's dataset of 1000 images are better suited to salient object segmentation, demonstrating highest recall value, third precision and second F-Measure values. Salient object detection has wider applications. For example, a more semantic, object-based image similarity can be defined with salient object detection for content-based image retrieval.

Acknowledgment. This research is partially sponsored by National Basic Research Program of China (No.2009CB320900), Natural Science Foundation of China (61175115, 61003105, 61070116, 61272320, 61070149 and 61001108), and the Importation and Development of High-Caliber Talents Project of Beijing Municipal Institutions (CIT&TCD201304035).

References

1. Itti, L., Koch, C., Niebur, E.: A Model of Saliency-Based Visual Attention for Rapid Scene Analysis. IEEE TPAMI 20, 1254–1259 (1998)
2. Gao, D., Vasconcelos, N.: Bottom-Up Saliency is a Discriminant Process. In: IEEE ICCV, pp. 1–6 (2007)
3. Murray, N., Vanrell, M., Otazu, X., Parraga, C.A.: Saliency Estimation Using A Non-Parametric Low-Level Vision Model. In: IEEE CVPR, pp. 433–440 (2011)

4. Kanan, C., Cottrell, G.: Robust Classification of Objects, Faces, and Flowers Using Natural Image Statistics. In: IEEE CVPR, pp. 2472–2479 (2010)

5. Yu, H., Li, J., Tian, Y., Huang, H.: Automatic Interesting Object Extraction from Images Using Complementary Saliency Maps. ACM Multimedia, 891–894 (2010)

6. Navalpakkam, V., Itti, L.: An Intergrated Model of Top-Down and Bottom-Up Attention for Optimizing Detection Speed. In: IEEE CVPR, pp. 2049–2056 (2006)

7. Liu, T., Yuan, Z., Sun, J., Wang, J., Zheng, N., Tang, X., Shum, H.-Y.: Source. Learning to detect a salient object. IEEE Transactions on Pattern Analysis and Machine Intelligence 33(2), 353–367 (2011)

8. Rensink, R., O'Regan, K., Clark, J.: To see or not to see: The need for attention to perceive changes in scenes. Psychological Sciences (1997)

9. Han, J., Ngan, K., Li, M., Zhang, H.: Unsupervised extraction of visual attention objects in color images. IEEE Transactions on Circuits and Systems for Video Technology 16(1), 141–145 (2006)

10. Ko, B.C., Nam, J.-Y.: Object-of-interest image segmentation based on human attention and semantic region clustering. Journal of Optical Society of America A 23(10), 2462–2470 (2006)

11. Ma, Y.-F., Zhang, H.-J.: Contrast-based image attention analysis by using fuzzy growing. In: ACM International Conference

12. Achanta, R., Hemami, S., Estrada, F., Süsstrunk, S.: Frequency-tuned salient region detection. In: CVPR, pp. 1597–1604 (2009); 409, 410, 412, 413, 414, 415 on Multimedia (2003)

13. Cheng, M.M., Zhang, G.X., Mitra, N.J., Huang, X.L., Hu, S.M.: Global Contrast based salient region detection. In: Proceedings of CVPR, Providence, RI, pp. 409–416 (2011)

14. Duan, L., Wu, C., Miao, J., Qing, L., Fu, Y.: Visual Saliency Detection by Spatially Weighted Dissimilarity. In: IEEE CVPR, pp. 473–480 (2011)

15. Levelthal, A.G.: The Neural Basis of Visual Function: Vision and Visual Dysfunction. CRC Press, Fla. (1991)

16. Rajashekar, U., van der Linde, I., Bovik, A.C., Cormack, L.K.: Foveated Analysis of Image Features at Fixations. Vision Research 47, 3160–3172 (2007)

17. Wandell, B.A.: Foundations of vision. Sinauer Associates (1995)

18. Tatler, B.W.: The Central Fixation Bias in Scene Viewing: Selecting an Optimal Viewing Position Independently of Motor Biased and Image Feature Distributions. J. Vision 7(4), 1–17 (2007)

19. Zhao, Q., Koch, C.: Learning A Saliency Map Using Fixated Locations in Natural Scenes. J. Vision 11(9), 1–15 (2011)

20. Christoudias, C., Georgescu, B., Meer, P.: Synergism in low level vision. In: IEEE Conference on Pattern Recognition (2002); Bruce, N.D.B., Tsotsos, J.K.: Saliency Based on Information Maximization. In: NIPS, pp. 155–162 (2005)

21. Hou, X., Zhang, L.: Dynamic Visual Attention: Searching for Coding Length Increments. In: NIPS, pp. 681–688 (2008)

22. Harel, J., Koch, C., Perona, P.: Graph-Based Visual Saliency. In: NIPS, pp. 545–552 (2006)

23. Achanta, R., Estrada, F.J., Wils, P., Süsstrunk, S.: Salient region detection and segmentation. In: Gasteratos, A., Vincze, M., Tsotsos, J.K. (eds.) ICVS 2008. LNCS, vol. 5008, pp. 66–75. Springer, Heidelberg (2008)

Tumor Cell Image Recognition Based on PCA and Two-Level SOFM

Lan Gan[*], Chunmei He, Lijuan Xie, and Wenya Lv

East China Jiaotong University, School of Information Engineering, Nanchang 330013, China
gl7046798@yahoo.com.cn

Abstract. In this paper, a method based on PCA and two-level SOFM neural network is proposed for tumor recognition. The method combines PCA with a two-level SOFM neural network in which PCA is used to reduce the dimensionality of the input tumor image sample and the two-level SOFM neural network is to extract characters and classifying. This method compromises linear dimensionality reduction, character extraction and classification. The training learning of the tumor image samples in the clinical pathological diagnosis can get the parameter of the two-level SOFM neural network. The experiment shows that the proposed method has better classifying accuracy and the classifying time is letter than the other methods such as PCA, LLE, PCA+LDA, SVM and two-level SOFM.

Keywords: tumor recognition, feature extraction, PCA, SOFM.

1   Introduction

Tumor cell images have the typical high dimension of small sample characters. As a kind of nature image, tumor cell images have quite difference in the cell structure, shape, sparse degree and spread geometry because of the irregular shape of tissues and organs and the differences between cells. On the other side, the tumor cell images contain much redundant information [1] because the image's Higher-order statistical characteristics follow Gaussian distribution. Therefore it is difficult to solve the tumor image recognition only using linear method and many scholars apply the nonlinear pattern recognition methods to tumor image recognition. The neural network is the widest applied non-linear method in pattern recognition.

Kohonen presented a unsupervised self-learning neural network: Self-organizing Feature Mapping (SOFM)[2]. This neural network reflect the memory method of the brain neural cells and the exciting rules when the nerve cells are stimulated. SOFM has many superior properties such as good stability, approximation, topological sorting, and density matching and so on [3,4]. SOFM can been use in image character extraction and pattern recognition [5]. In this paper, a new two-level SOFM is proposed to tumor recognition. But if the original image samples are directly used to train the SOFM, the training process takes much time. We should reduce the

---

[*] Corresponding author.

J.-S. Pan et al. (eds.), *Genetic and Evolutionary Computing,*
Advances in Intelligent Systems and Computing 238,
DOI: 10.1007/978-3-319-01796-9_32, © Springer International Publishing Switzerland 2014

dimension of the original image sample to accelerate the training process. PCA is an efficient linear image process technology to reduce dimension and eliminate redundancy. It can not only reduce the dimension of the images, but also make the image distortion to a minimum [6]. Therefore in this paper, we proposed a new method combining PCA with a two-level SOFM to classify gastric epithelial tumor cells where PCA are applied to reduce dimension of the original acquired tumor cell images and the first level of the SOFM is used for character extraction and the second level of the SOFM is used for classifying and recognition. The simulations show that this method in efficient in tumor recognition.

The rest of this paper is organized as follows. Some basic theories about PCA and SOFM are introduced in section 2. Section 3 presents our proposed methods combining PCA and two-level SOFM for tumor recognition. Section 4 contains our experiment results. Finally, some conclusions are given in section 5.

2   Previous Work

2.1   PCA

Principal Component Analysis (PCA) is a very classical character extraction method [7-8] in statistical pattern recognition theory. The brief mathematic theory about PCA will be given in this section.

We suppose x is a stochastic variable with m dimension and $\xi_1, \xi_2, \cdots, \xi_d$ are d unit vectors with m dimension which the units are Orthogonal to each other. PCA can acquire $\xi_1, \xi_2, \cdots, \xi_d$ to minimize the error of mean square as follows:

$$\min E\left(\left\| x - \sum_{i=1}^{d} y_i \xi_i \right\|^2 \right). \tag{1}$$

where $y_i = \xi_i^T x$ is denoted as the ith principal component of sample x, $\xi_1, \xi_2, \cdots, \xi_d$ is the d largest eigenvalues of the covariance matrix S of x which satisfy the following equation:

$$S\xi_i = \lambda_i \xi_i \quad (i = 1, 2, \cdots, d) \tag{2}$$

where $\lambda_1 \geq \lambda_2 \geq \cdots \geq \lambda_d > 0$.

2.2   SOFM

2.2.1   The Neural Network Model of SOFM

SOFM neural network is a kind of competitive neural network [9] which contains input layer and competitive layer. In SOFM, the input layer is the sample space receiving the outside stimulations and the competitive layer (also is known as output layer) is one or two-dimensional planar array made up of many nerves.

The common SOFM neural networks are one-dimensional or two-dimensional planar array. The topology of the SOFM with two-dimensional planar array is shown in Fig. 1.

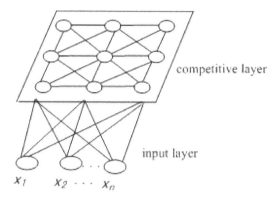

Fig. 1. Topology of SOFM neural network

2.2.2   The Learning Algorithm of SOFM

In SOFM, for an input sample pattern, the sample is assigned to the class of neuron which acquires the maximum response. If the input sample pattern don't belong to any class of neuron, it belongs to the closest class according to the nearest neighbor rule. The learning algorithm of SOFM is concluded as follows [10].

(1)Initialization

Let N be the input neurons number of the SOFM and O be the output neuron number. Let $0 < \beta < 1$ be a random number. We give a initial value for the above-motioned parameters and normalize them.

(2)Acquire a new input pattern $\mathbf{X}_k = (\mathbf{X}_{1k}, \mathbf{X}_{2k}, ..., \mathbf{X}_{Nk}), k = 1,2,..., N$.

(3)Compute the distance d_{jk} between the kth input pattern $\mathbf{X}_k$ and the jth neuron W_j according to the following equation.

$$d_{jk} =\| X_k - W_j \|= \sqrt{\sum_{i=1}^{N}[X_{ik}(t) - W_{ij}(t)]^2} , j = 1,2,....,O \qquad (3)$$

(4)Determine the winning neuron P. The neuron P is the one which has the minimum distance with $\mathbf{X}_k$. Let W_P be the weight vector of the winning neuron. W_P should satisfy the following equation.

$$\| \mathbf{X}_k - \mathbf{W}_P \|= Min\{d_{jk}\} \qquad (4)$$

(5)Define the winning neighborhood $N_j(t)$, where t stands for the time and j is the jth neuron. The initial value of neighborhood $N_j(t)$ (t=0) is large and it will gradually become smaller over time in the training. Line, square and hexagon are the common shape of the neighborhood in SOFM neural network. The methods determining the neighborhood have Euclidean method and Manhattan method.

(6)Adjust the weight. The weights $N_j(t)$ in the neighborhood are adjusted as follows.

$$W_{ij}(t+1) = W_{ij} + \eta(t)[X_i(t) - W_{ij}(t)] \tag{5}$$

where $\eta(t)$ is a monotone decreasing function of time and $0 < \eta(t) < 1$.

The weights outside the neighborhood aren't adjusted and remain the same.

$$W_{ij}(t+1) = W_{ij}(t) \tag{6}$$

(7)Termination condition. If $\eta(t)$ reduces to zero or a designed small positive decimal then the algorithm terminates, otherwise return to Step (2).

3   Our Proposed Method

SOFM neural networks reflect the characters of human brain nerve cells' memory method and the nerve cells exciting rule when the nerve cells are stimulated from outside world. SOFM neural networks are often applied to character extraction and pattern recognition. In this section, we propose a two-level SOFM for character extraction and pattern recognition. The two-level SOFM combining with PCA are constructed as the tumor recognition classifier.

3.1   Idea of Our Method

The dyeing microscopic section tumor cell images used in hospital clinical diagnosis are complex and high-dimensional. Moreover they contain little useful information for tumor classifying. Therefore the dyeing images are grayed and preprocessed to get high-quality gray images. Secondly, PCA is used to reduce the dimension of the gray tumor cell images and acquire low-dimension sample patterns. Then the first-level SOFM receives the low-dimension sample patterns as the inputs and this level SOFM further reduce the dimension of   the tumor cell image and extract characters because SOFM can response to some sample patterns selectively. At last the second-level SOFM classifies the sample character pattern from the first-level SOFM for the SOFM neural network   will response some complex character pattern, strengthen the neuron in the neighborhood and restrain the neuron outside the neighborhood. The two levels SOFM neural networks are all two-dimension planar array for we consider that the tumor cell images are very complex. The tumor cell image classifier based on PCA and two-level SOFM neural network is shown in Fig.4.

Next we present the whole process of the classifier and give the algorithm.

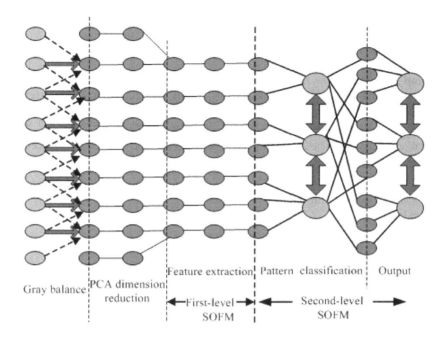

Gray balance   PCA dimension   Feature extraction   Pattern classification   Output
                reduction

←—First-level—→|←   Second-level   —→
        SOFM   SOFM

Fig. 2. Classifier based on PCA and two-level SOFM network

3.2   Algorithm of Our Method

We divide the tumor cell image samples into training sample set and testing sample set. In this paper, we classify the tumor cell into three categories: cancerous, hyperplasic and normal.

We denote the training sample set as X which has N tumor cell image samples. In the training sample set, the category X_1 contains the samples classified into cancerous and the sample number of class X_1 is denoted as N_1. Similarly, the category X_2 contains the samples classified into hyperplasic and its number is denoted as N_2; the category X_3 contains the samples classified into normal and its number is denoted as N_3. Therefore, in the training sample set, the tumor cell categories number are $C = 3$ and the sample number $N = N_1 + N_2 + N_3$.

The testing sample set is denoted as Y and the sample number of Y is denoted as M. Similarly to the training sample set, we respectively denote the categories containing the cancerous, hyperplasic and normal samples as Y_1, Y_2 and Y_3. And the sample number of Y_1, Y_2 and Y_3 are accordingly denoted as M_1, M_2 and M_3.

Let the dimension of the original high-dimensional tumor gray image be D and the dimension reduced be d. The algorithm of our method is explained as follows.

Step 1: Do PCA transformation for each tumor cell image training sample. Set the constant to make sure the accumulating contribution rate not less than 90%. After PCA transformation, the sample dimension of the character subspace reduces from D to d.

Step 2: Construct the first-level SOFM neural network and process the character samples in first-level SOFM. Input N d-dimensional training samples into the first-level SOFM neural network. Let the neuron number in the competitive layer be C. That means sample dimension becomes C after processed by the first-level SOFM neural network. We consider the computation efficiency and set $C \in (\frac{d}{4}, \frac{d}{2})$ according to the statistical experiment results. Let the tumor training sample matrix be $\mathbf{P} = [\mathbf{P}_{i1}, ..., \mathbf{P}_{iC}], i = 1, ..., N$ and normalize P.

The initial weight and initial learning rate are a random number Severally in the interval $[0,1]$ and $[0.01, 0.1]$. We set the initial learning rate $\eta = 0.01$ because we compare the different η in the experiment and find the classifying accuracy is better when $\eta = 0.01$. We suppose the character sample acquired by the first-level SOFM is T.

Step 3: Construct the second-level SOFM and classify the character samples. We set the neuron number of competitive layer in the second-level SOFM as 3 for the sample category is 3. Input the character sample T into the second-level SOFM and the construction of the second-level SOFM according to the following equation.

$$net = newsom(\max \min(T), [3, C]) \tag{7}$$

Step 4: Let the iteration number be 1000 and train the two-level SOFM neural network according the weight adjustment rule until the topology and the weight in the SOFM don't change.

Step 5: Respectively compute the center of clustering of the three categories character samples and denote as c_1, c_2, c_3.

Step 6: Choose the Euclidean distance as the criterion whether two neuron belong to one category and the Euclidean distance is computed by

$$d = \sqrt{\sum_{i=1}^{n}(x_i - y_j)^2}, n = N_1 = N_2 = N_3, \forall x_i \in X_k, \forall y_j \in X_k, k = 1,2,3. \tag{8}$$

If $d \geq c_1$, then the tumor image sample is classified into cancerous category C1; if $d \geq c_2$, then the tumor image sample is classified into hyperplasia category C2; if $d \geq c_3$, then the tumor image sample is classified into normal category C3.

Step 7: Repeat Step2 to Step6 for the any testing tumor image sample and test the testing sample.

4   Experiments

The experiment samples in this paper are the dyeing micro section gastric epithelial cell images used in the hospital clinical diagnosis. The gastric epithelial cell images are roughly classified into three categories: cancerous, hyperplasia and normal according to the medical experts' experience. The hyperplasia category is further classified into mild hyperplasia, moderate hyperplasia and severe hyperplasia according to the hyperplasia degree. The five categories of gastric epithelial tumor cell images are shown in Fig.3. But in this paper we simply classify the sample into three categories: cancerous, hyperplasia and normal.

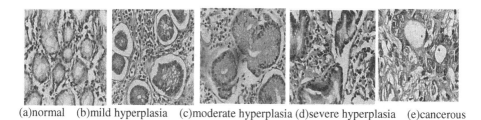

(a)normal   (b)mild hyperplasia   (c)moderate hyperplasia (d)severe hyperplasia   (e)cancerous

Fig. 3. Five categories of gastric epithelial tumor cell images

The training sample number of every category is 55 and the testing sample number of every category is 30. The original gastric epithelial tumor cell image's pixel is 320 * 240 = 76800. The dimension C of the low-dimensional sample pattern reduced by PCA satisfies the inequality $50 < C < 200$ according to our statistical experiment. We set C=55 in the experiment.

For comparisons, we classify the gastric epithelial tumor cell images using the other classifying methods such as LLE, PCA+LDA, LLE+LDA, SVM and two-level SOFM neural network in the similar condition. The comparison results are shown in Table 1. In two-level SOFM method, the learning rate and the iteration are the same to our proposed method, that is, the learning rate is 0.01 and the iteration is 1000.

Table 1. Comparisions with other classification methods

classifying method	Average training time(s)	Average testing time(s)	Average total time(s)	classifying accuracy
LLE	44.28	1.43	65.46	73.3%
PCA+LDA	33.22	1.9	50.89	80.1%
LLE+LDA	28.62	1.87	30.15	81.6%
SVM	31.26	1.03	41.78	83.36%
2SOFM	29.56	0.97	31.48	87.68%
PCA+2SOFM	24.89	0.88	26.28	89.09%

Table 1 indicates that our proposed method: PCA+2SOFM is superior to the other methods because our method improves the classifying accuracy and decrease the classifying time.

5   Conclusions

A new tumor cell image recognition method based on PCA and a two-level SOFM is proposed in this paper. The two-level SOFM can not only reduce dimension but also extract characters and classify. The experiment shows that our method can improve the classifying accuracy and decrease the samples' classifying time. The comparison with the other classifying methods demonstrates that our method is superior to the others.

Acknowledgements. This work is supported by NSFC of China (Grant No. 61163040) and the Science Foundation of Jiangxi Provincial Department of Education (Grant No. GJJ10451).

References

1. Suzuki, K., Abe, H., MacMahon, H., Doi, K.: Image-processing technique for suppressing ribs in chest radiographs by means of massive training artificial neural network (MTANN). IEEE Transactions on Medical Imaging 25(4), 406–416 (2006)
2. Fukushima, K.: Neocognitron: A Self-organizing Neural Network Model for a Mechanism of Pattern Recognition Unaffected by Shift in Position. Biological Cybernetics, 193–202 (1980)
3. Ng, W.W.Y., Dorado, A., Yeung, D.S., Pedrycz, W., lzquierdo, E.: Image classification with the use of radial basis function neural networks and the minimization of the localized generalization error. Pattern Recognition 40(1), 19–32 (2007)
4. Obimbo, C., Zhou, H., Wilson, R.: Multiple SOFMs working cooperatively in a vote-based ranking system for network intrusion detection. Procedia Computer Science 6, 219–224 (2011)
5. Ebied, H.M., Revett, K., Tolba, M.F.: Evaluation of unsupervised feature extraction neural networks for face recognition. Neural Comput. & Applic. 22, 1211–1222 (2013)
6. Feng, X., Tao, F., Gang, L.J., Xian, S.Y.: An Image Compressing Algorithm Based on PCA/SOFM Hybrid Neural Network. Journal of Image and Graphics 8(9), 1100–1104 (2003)
7. Fukunaga, K.: Introduction to Statistical Pattern Recognition. Academic Press, New York (1990)
8. Levy, A., Lindenbaum, M.: Sequential Karhunen-Loeve basis extraction and its application to images. IEEE Trans. on Image Processing 9, 1371–1374 (2000)
9. Shi, Y., Han, L., Lian, X.: Neural network design method and instance analysis. Beijing University of posts and telecommunications press, Beijing (2009)
10. Hsu, A.L., Tang, S.-L., Halgamuge, S.K.: An unsupervised hierarchical dynamic self-organizing approach to cancer class discovery and marker gene identification in microarray data. Bioinformatics 19(16), 2131–2140 (2003)

Face Recognition Based on Representation with Reject Option

Min Wang[1], Yuyao Wang[1], Jinrong Cui[1], Shu Liu[1], and Yuan Tian[2]

[1] Bio-Computing Research Center, Shenzhen Graduate School,
Harbin Institute of Technology, China
{leafcano,shinewyy}@gmail.com, {tweety1028,exorcist_liu}@163.com
[2] Industrial Engineering Department, Shenzhen Graduate School,
Tsinghua University, China
tq0agme6@163.com

Abstract. In this paper, we proposed a method for face recognition with reject option. First, by setting a threshold, we use sparse representation (SR) method to find out candidates who should be rejected, and choose their nearest neighbors in the training set based on contribution in SR. Then we extract the Locally Adaptive Regression Kernels (LARK) feature of each candidate sample and its neighbors respectively. At last, we determine whether a candidate should be rejected via calculating the matrix cosine similarity measure. A number of experiments show that combining with sparse and LARK representation can obtain good performs for rejection.

Keywords: Sparse representation, Lark feature, Reject option.

1 Introduction

In recent years, many excellent methods for face recognition have been proposed [1–5]. In most cases, these methods choose images of some subjects as test samples, while these subjects are included in the training set. However, there is a situation they do not take into consideration: some unknown people in a test set are not included in the corresponding training set. Generally, the unknown people excluded in a predefined training set are called "outlier" and they often appear in many application. Hence, the reject option is very essential in face recognition. Most face recognition methods are focused more on the correct recognition rate rather than rejection rate, not a lot of methods for rejection have been proposed [6]. Eickeler et al. [7] compare different confidence measures for rejection, and propose a new confidence measure based on ranking. This method gives the best result of reject recognition for outliers in their experiments. Su-utala et al. [8] set a threshold for rejection, and the threshold is assigned based on the conditional posterior probabilities of classifier outputs. Kim et al. [9] also set a threshold for rejection which based on a formula on the distances of nearest neighbors. Chen et al. [10] propose a new subspace called as Margin Enhance Space, and use it to model the acceptance and rejection likelihood probability. In the second section of this paper, details of our method are described. In the third section, the experiments results are shown. In the last section, we give a brief conclusion.

J.-S. Pan et al. (eds.), *Genetic and Evolutionary Computing,*
Advances in Intelligent Systems and Computing 238,
DOI: 10.1007/978-3-319-01796-9_33, © Springer International Publishing Switzerland 2014

2   Description of Our Method

Sparse representation (SR) is very popular in face recognition. Extensive researches demonstrate that SR outperforms other methods in face recognition and is nature to reject [11]. Many researchers have carried on further research on it and have made remarkable achievements [12–15]. Locally Adaptive Regression Kernels (LARK) [16] can efficiently capture the local geometric structure of an image, which makes LARK distinguish two images effectively. Hence, SR is applied in the first step of our method to pick out the "outliers". However, the "outliers" are always mixed with some "non-outliers". Due to the excellent performance in face verification of LARK, we use it to confirm the real "outliers" from those picked out in the first step. The details of our method are described as follows.

2.1   The First Step: Pick Out "Outliers" Using SR

In this step, we represent a test sample as linear combination of all train samples:

$$\mathbf{y} = a_1\mathbf{x}_1 + a_2\mathbf{x}_2 + ... + a_n\mathbf{x}_n \tag{1}$$

where $\mathbf{y}$ is a test sample, $\mathbf{x}_i \in R^m (i = 1,2,...,n)$ is the column vector of a training sample. $a_i (i = 1,2,...,n)$ is the corresponding coefficient of each training sample in representation. Eq. 1 is rewritten as $\mathbf{y} = \mathbf{AX}$. The coefficient matrix $\mathbf{A} = [a_1, a_2, ..., a_n] \in R^n$ is solved by the method proposed in [13].

$$\mathbf{A} = \left(\mathbf{X}^T\mathbf{X} + \delta\mathbf{I}\right)^{-1}\mathbf{X}^T\mathbf{y} \tag{2}$$

where $\mathbf{X} = [\mathbf{x}_1, \mathbf{x}_2, ..., \mathbf{x}_n] \in R^{m \times n}$ denotes a matrix of all train samples. δ is a very small positive constant and $\mathbf{I}$ is an identity matrix.

We use each class i to represent the test sample $\mathbf{y}$ and obtain a represented result:

$$\hat{\mathbf{y}}_i = a_i\mathbf{x}_i \tag{3}$$

Then we calculate the residual $\lambda_i(\mathbf{y})$ between $\hat{\mathbf{y}}_i$ and using $\lambda_i(\mathbf{y}) = \|\mathbf{y} - \hat{\mathbf{y}}_i\|$. If $\mathbf{y}$ belongs to the class i, $\lambda_i(\mathbf{y})$ will be the minimum. Thus, we can label the test sample as a class with the minimum residual, shown as Eq. 4.

$$Predict\_label\_1(\mathbf{y}) = i \ for \ \min_i \lambda(\mathbf{y}) \tag{4}$$

To an "outlier", no one training class can sufficiently represent it. This equals that several residuals are similar to each other. Assume that $\lambda_i(\mathbf{y})$ is the minimum residual, $\lambda_j(\mathbf{y})$ is the second minimum residual. If a test sample is an "outlier", the difference $\mu(\mathbf{y})$ between $\lambda_i(\mathbf{y})$ and $\lambda_j(\mathbf{y})$ will be very small; otherwise, it will be large. According to this point, we set a threshold θ to reject "outliers":

$$Predict\_label\_1(\mathbf{y}) = \begin{cases} i, & \mu(\mathbf{y}) > \theta \\ 0\,(rejection), & \mu(\mathbf{y}) \leq \theta \end{cases} \tag{5}$$

In this paper, we set θ as 0.2 and get a satisfactory result for both recognition and rejection, as shown in Tabel 1.

2.2 The Second Step: Verify "Outliers" Using LARK

Due to the variation in pose, illumination and expression .etc of faces, different images of a same person vary apparently. Just using a training class with the minimum residual to label a test sample, as described above, is not accurate. Thus, we chose the first nearest neighbors which have the smallest residuals in the training set, and extract their LARK features. Then matrix cosine similarity measure is applied to measure the similarity between neighbors and the test sample. Steps of calculating the local kernels are described as follows:

Step1. Calculate the local gradient covariance matrix $\mathbf{C}$ for pixel x_m:

$$\mathbf{C}_{x_m} = \sum_{x_k \in M} \begin{bmatrix} z_x^2(x_k) & z_x(x_k) z_y(x_k) \\ z_x(x_k) z_y(x_k) & z_y^2(x_k) \end{bmatrix} \tag{6}$$

where $z_x^2(x_k)$ represents the degree change of pixel x_k in the horizontal direction, and $z_y^2(x_k)$ represents the degree change of x_k in the vertical direction. Usually, M is a path of pixels centered at x_m.

Step2. Divide an image into some 13*13 local windows centered at x_m. The local kernels of the pixel x_k in a local window are calculated using Eq. 7:

$$K(C_{x_m}, x_k - x_m) = \exp\left((x_k - x_m)^T C_{x_m}(x_k - x_m)\right) \tag{7}$$

Step3. LARK feature of the whole image is represented as a set of local kernels in local windows:

$$\mathbf{K} = [K_1, K_2, ..., K_n] \tag{8}$$

where n denotes the number of local windows.

Step4. Measure the similarity between a neighbor $\mathbf{K}^{(l)}$ and the test sample $\mathbf{K}^{(y)}$ using matrix cosine similarity measure shown as Eq. 9:

$$\psi\left(\mathbf{K}^{(y)}, \mathbf{K}^{(l)}\right) = \frac{\mathbf{K}^{(y)} \cdot \mathbf{K}^{(l)}}{\|\mathbf{K}^{(y)}\| \cdot \|\mathbf{K}^{(l)}\|} \tag{9}$$

Step5. Obtain the label by using Eq. 10:

$$\mathrm{Pr}edict\_label\_2(\mathbf{y}) = l \; for \; \max\left(\psi\left(\mathbf{K}^{(y)}, \mathbf{K}^{(l)}\right)\right). \tag{10}$$

2.3 The Third Step: Get the Final Result

If the labels of a test sample obtained by the former two steps are the same, the test sample should be recognized. Otherwise, the test sample should be rejected. The final label of a test sample is represented as Eq. 11:

$$Final\_label(\mathbf{y}) = \begin{cases} l, & l = i \\ 0\,(rejection), & others \end{cases}. \tag{11}$$

3   Experimental Results and Analysis

We compare the proposed method with other face recognition methods with rejection, including PCA, R-LDA, ME Space. All experiments are tested on Extended YaleB Database [17] and ORL Database. Extended YaleB contains 38 subjects with each providing 64 images in different illumination conditions, poses, etc. ORL consists of 400 images for 40 subjects with each providing 10 images in different poses and expressions. In order to evaluate the performance of rejection, we choose 18 subjects in the Extended YaleB for test and 25 subjects for training.In the 25 subjects for training,there are 5 subjects which are also subjects for test. Each subject provides 30 images as training samples and the remaining ones as test samples, which is expressed as G30/P34. In ORL database, 25 subjects are used for training while 20 subjects are used for test. There are 5 subjects included in the test set as well as the training set. 5 images are selected randomly from each subject as training samples and the remaining ones are test samples. Table 1 lists the correct recognition rate and correct rejection rate on YaleB. It shows that our proposed method obtains higher recognition accuracy and rejection accuracy than both PCA and ME. Fig. 1 shows the ROC curve of different methods. These results are obtained on ORL. We can find that our method has the best performance on rejecting "outliers".

Table 1. Correct recognition rate and correct rejection rate of different methods in Extended YaleB with G30/P34

	PCA	R-LDA	ME[10]	SR[13]	Our method
Rejection rate	63.5	71.4	87.2	85.4	**89.6**
Recognition rate	67.9	85.2	91.6	89.7	**93.1**

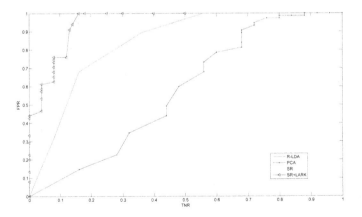

Fig. 1. ROC curves of PCA,R-LDA,SR and our method(SR+LARK)

4 Conclusion

In our method, SR is used to find out "outliers". LARK is applied to further confirm whether "outliers" found by SR are real "outliers". This is the first time to combine SR and LARK for face recognition with rejection. Experiments on public databases show that our method is simple and effective. Meanwhile, it outperforms in face recognition with rejection.

Acknowledgments. This paper is partially supported by Shenzhen Municipal Science and Technology Innovation Council (Nos. JC201005260122A, JCYJ201206131533-52732 and CXZZ20120613141657279).

References

1. Zhang, D., Song, F., Xu, Y., Liang, Z.: Advanced Pattern Recognition Technologies with Applications to Biometrics. Medical Information Science Reference (2009)
2. Xu, Y., Zhong, A., Yang, J., Zhang, D.: LPP Solution Schemes for Use with Face Recognition. Pattern Recognition 43, 4165–4176 (2010)
3. Xu, Y., Zhang, D., Yang, J.Y.: A Feature Extraction Method for Use with Bimodal Biometrics. Pattern Recognition 43, 1106–1115 (2010)
4. Wang, J., You, J., Li, Q., Xu, Y.: Orthogonal Discriminant Vector for Face Recognition Across Pose. Pattern Recognition 45, 4069–4079 (2012)
5. Xu, Y., Zhu, X., Li, Z., Liu, G., Lu, Y., Liu, H.: Using the Original and 'Symmetrical Face' Training Samples to Perform Representation Based Two-step Face Recognition. Pattern Recognition 46, 1151–1158 (2013)
6. Landgrebe, T.C.W., Tax, D.M.J., Paclík, P., et al.: The Interaction Between Classification and Reject Performance for Distance-based Reject-option Classifiers. Pattern Recognition Letters 27, 908–917 (2006)
7. Eickeler, S., Jabs, M., Rigoll, G.: Comparison of Confidence Measures for Face Recognition. In: 4th IEEE International Conference on Automatic Face and Gesture Recognition, pp. 257–262. IEEE (2000)
8. Suutala, J., Roning, J.: Methods for Person Identification on a Pressure-sensitive Floor: Experiments with Multiple Cclassifiers and Rreject Option. Information Fusion 9, 21–40 (2008)
9. Kim, C., Choi, C.H.: Image Covariance-based Subspace Method for Face Recognition. Pattern Recognition 40, 1592–1604 (2007)
10. Chen, J.C., Shi, S.Y., Lien, J.J.: Face Recognition and Unseen Subject Rejection in Margin-enhanced Space. In: International Conference on System Science and Engineering (ICSSE), pp. 631–636. IEEE (2010)
11. Wright, J., Yang, A.Y., Ganesh, A., Sastry, S.S., Ma, Y.: Robust Face Recognition via Sparse Representation. IEEE Transactions on Pattern Analysis and Machine Intelligence 31, 210–227 (2009)
12. Karimi, M.M., Soltanian-Zadeh, H.: Face Recognition: A Sparse Representation-based Classification using Independent Component Analysis. In: 6th International Symposium on Telecommunications (IST), pp. 1170–1174. IEEE (2012)
13. Xu, Y., Zhang, D., Yang, J., Yang, J.Y.: A Two-phase Test Sample Sparse Representation Method for Uuse with Face Rrecognition. IEEE Transactions on Circuits and Systems for Video Technology 21, 1255–1262 (2011)

14. Wagner, A., Wright, J., Ganesh, A., Zhou, Z., Mobahi, H., Ma, Y.: Toward a Practical Face Recognition System: Robust Alignment and Illumination by Sparse Representation. IEEE Transactions on Pattern Analysis and Machine Intelligence 34, 372–386 (2012)
15. Donoho, D.L.: For Most Large Underdetermined Systems of Linear Equations the Minimal l1-norm Ssolution is Also the Sparsest Solution. Communications on Pure and Applied Mathematics 59, 797–829 (2006)
16. Seo, H.J., Milanfar, P.: Face Verification Using the Lark Representation. IEEE Transactions on Information Forensics and Security 6, 1275–1286 (2011)
17. Information on,
    http://www.cl.cam.ac.uk/research/dtg/attrachive/facedatabase.html

The Fusion of SRC and SRRC Algorithms

Ke Yan[1] and Jian Cao[2]

[1] School of Information Technology and Communication, Qufu Normal University, China
1249640576@qq.com
[2] Bio-Computing Research Center, Shenzhen Graduate School,
Harbin Institute of Technology, China
249960486@qq.com

Abstract. As a recently proposed technique, sparse representation based classi-fication(SRC) and sparse residue representation classification SRRChave been widely used for face recognition(FR).SRC and SRRC represent the test sample as a linear combination of training samples. SRC first calculates the coefficient so-lution via -minimization, and then we calculate the reconstruction residue errors of the test sample generated from each class respectively. However, the SRRC algorithm first exploits the training samples to constructs the error-free samples of the test sample and then calculates the residue between the error-free sample and original test samples. The residue coefficients are also solved by -minimization. In order to integrate the advantages of SRC and SRRC, we use the score level fusion to combine the residues of SRC and SRRC and the test sample is classified into the class. The minimum final residual is the experiment result of the test sample. Compared with previous methods, the fusion algorithm has very competitive FR accuracy and well performance in robustness to occlusion.

Keywords: Sparse Representation, SRC, SRRC.

1 Introduction

The sparse representation has been widely used for different applications, such as pat-tern recognition and computer vision [1], signal separation [2], denoising [3], mo-tion and data segmentation [4], face recognition [5–7]. In this paper, we mainly study the issue of calculating sparse representation for classification. The first method proposed by Wright et al. [8], called Sparse Represen-tation based Classification (SRC). The sec-ond method proposed by Jinghua Wang et al. [9], called Sparse Residue Representation based Classification (SRRC). The me-thods of SRC and SRRC represent the test sam-ple as a linear combination of training samples, but the difference of two methods is calculating coefficient of linear expres-sion. The basic idea of SRC is that: (a) to calcu-late the expression coefficient by -norm at first, and then to reconstruct the test sample by sparse decomposition coeffi-cients that related to the each class; (b) to compute the residue between the test sam-ple and reconstruction test sample; (c) to classify test sam-ple into the class which has the minimum reconstruction residual error. The algorithm can achieve a good perfor-mance in severely occlusion and disguise. The accuracies of SRC are higher than the traditional classification methods (such as Nearest Neighbor (NN), Nearest Subspace (NS) [8]. The basic idea of SRRC is that: (a) to reconstruct the test sample by linear combination of the training samples, and then we calculate the residue between test sample and reconstruction sample, the expression coefficients are solved by -minimization residue; (b) to compute the residue between test sample and reconstruc-tion sample by sparse decomposition coefficients that related to the each class respectively; (c) to classify test sample into the class which has the minimum reconstruction residual error. The algorithm has a good performance to deal with occlusion. The occlusion is the residue of the test sample, so we can accurately classify the test sample. With no assumption on the shape of the occlusion, the proposed method can work well for kinds of occlusions [8]. The occlusion degrades the performance the traditional subspace methods. As a result, the accuracies of SRRC are higher than the traditional subspace methods [8]. In this paper we proposed a new algorithm scheme based on sparse representation, namely the fusion of SRC and SRRC algorithm. The accuracies of the proposed algorithm are higher than the traditional methods, especially in FR occlusion and disguise. We will demonstrate that the new validation rule outperforms the subspace methods (such as PCA). Experiments conducted on face recognition data sets ORL, YaleB and AR demonstrated that proposed algorithm can remain the comparative classification accuracy and robustness. Section 2 briefly reviews SRC and SRRC. Section 3 mainly analyzes the SRC and SRRC algorithm. Section 4 performs experiments. Section 5 concludes this paper.

2   A Brief Reviews of SRC and SRRC

SRC is the reconstruction classification approach which aim at tacking the classification problem on data with corruption (such as missing data and noising). Assuming there are N distinct object classes. Denoted by $a_i \in {}^{m*n_i}$ the data set of ith class, and each column of a_i is the sample of class i.As a result, the training data set $A = \{a_1, a_2, ..., a_N\} \in {}^{m*n}$, where $n = \sum_{i=1}^{N} n_i$. Given a test sample $y \in {}^{m*1}$, the SRC algorithm can be summarized as follow:

1. Normalize: each column of A to have unit ℓ_2-norm;
2. Solve the ℓ_2-minization problem:

$$\hat{x}_1^* = \arg\min \|x\|_1 \, s.t. Ax = y \tag{1}$$

3. Compute the residuals:

$$r_i(y) = \|y - A\sigma_i(\hat{x}_1^*)\|_2, i = 1, 2 ..., N \tag{2}$$

4. Output: identify$(y) = \arg\min_i r_i(y)$, where the result is the class label of the test sample y.

Let the coefficients matrix $\hat{x}_1^* = \{x_1, x_2, ..., x_i, ..., x_n\}$, x_i is the coding vector associated with class i. Ideally, the matrix $\hat{x}_1^*$ is nonzero entries are associated with the column of training sample A from a single class i,so we can assign the test sample y to that class. As a result, if y is from the ith class, usually $y = A_i X_i$ holds well, implying that most efficient are zeros and only x_i has significant entries [10]. The number of class which responds to the minimum residue is the result of the test sample. In SRC, the sparse representation is used for robust face recognition. When the coefficients were solved by ℓ_1-norm, the approach separates the information required for these tasks: the residuals for identification and the sparse coefficients for validation [8]. The class label of the test sample can be decided by residue measurement. It has been proved that the solution of ℓ_1-minimization is equal to the solution of ℓ_0-minimization [11], The coefficients that solved by -norm are much sparse than those given by ℓ_2-norm, and the dominant coefficients are associated with the target class [8]. The residual that solved by reconstructing sample has smaller than these which calculated by training samples, so the new rule has a higher accuracy rate. Consequently, SRC is used for robust face recognition, outlier detection and so on. The algorithm of SRRC is the reconstruction classification approach which aim at dealing with the classification problem on data with occlusion. Assuming there are n distinct object classes, denoted by $a_i \in {}^{m*n_i}$ the data set of ith class, and each column of a_i is the sample of class i. As a result, the training data set $A = \{a_1, a_2, ..., a_N\} \in {}^{m*n}$, where $n = \sum_{i=1}^{N} n_i$. Given a test sample $y \in {}^{m*1}$, the SRRC algorithm can be summarized as follow:

1. Normalize the columns of A and y to have unit ℓ_2-norm;
2. Solve the ℓ_2-minization problem:

$$\hat{x}_1^* = \min \|y - Ax\|_1 \ s.t. \ y = Ax \tag{3}$$

3. Compute the residuals:

$$r_i(y) = \|y - A\sigma_i(\hat{x}_1^*)\|_2, i = 1, 2 ..., N \tag{4}$$

4. Output: identify $(y) = \arg\min_i r_i(y)$, where the result is the class label of the test sample y.

The observation sample matrix consists of non-error sample matrix and error sample matrix. We consider the non-error sample and the observation sample are of the same size. The algorithm of SRRC represents the test sample as a linear combination of training samples, and then we calculate the residue between the test sample and the representing sample. The optimal solution X can be solved by ell_1 -minimization residue. Ideally, the matrix $\hat{x}$ is nonzero entries are associated with the column of training sample A from a single class i, so we can assign the test sample y to that class. As a result, if y is from the ith class, usually $y = A_i X_i$ holds well, implying that most efficient are zeros and only x_i has significant entries. The number of class which responds to the minimum residue is the result of the test sample. In SRRC, the sparse representation is used for robust face recognition, especially in the occlusion problem. The reconstructed sample is same to the test sample in most pixels, and the most difference of pixels is the occluded part of the test sample. As a result, the residue denotes the local deviations. Differently, the SRC emphasize the globally residue by the ell_2-norm [8]. The SRRC algorithm is a kind of supervised sparsity, but SRC is a kind of unsupervised sparsity contrarily [12]. So the new rule can deal with the robustness to occlusion.

3   The Fusion of SRC and SRRC Algorithm

In order to integrate the advantage of SRC and SRRC, we propose one more method namely the fusion of SRC and SRRC algorithm. According to the residuals of the SRC and SRRC, we calculate the summation of the residuals by class, The class label which responds to the minimum summation is the result of the test sample.

The Fusion OF SRC And SRRC Algorithm:

1. Input: a matrix of training samples $A = [A_1, A_2, ..., A_K] \in \mathbb{R}^{m*n}$ for K classes, a test sample $y \in \mathbb{R}^m$;
2. Normalize: the columns of A to have ℓ_2-norm;
3. Solve the minimization problem in residual using ℓ_1-norm:

$$\hat{x} = \min \|y - Ax\|_1 s.t. y = Ax \tag{5}$$

4. Compute the residual $g_i(y) = \|y - A\sigma_i(\hat{x})\|_2, i = 1, 2, ..., K$ is the character function that selects the coefficients associated with class i;
5. Solve the ℓ_1-norm minimization problem:

$$\hat{x}^* = \arg\min \|x\|_1 s.t. Ax = y \tag{6}$$

6. Compute the residual $r_i(y) = \|y - A\sigma_i(\hat{x}^*)\|_2, i = 1, 2, ..., K$, where $\sigma_i : \mathbb{R}^n \to \mathbb{R}^n$ is the character function that selects the coefficients associated with class i.
7. Calculate the summation of the residual g_i and r_i with different weights:

$$p_i(y) = r_i(y) + g_i(y) = \lambda_1 * \|y - A\sigma_i(\hat{x}^*)\|_2 + \lambda_2 * \|y - A\sigma_i(\hat{x})\|_2 \tag{7}$$

   where $i = 1, 2, ..., k, \lambda_1 + \lambda_2 = 1$;
8. Output: identify $y = \arg\min_i p_i(y)$, where the result is the class label of the test sample y.

The framework of SRC exploits the discriminative nature of sparse representation to perform robustness. However SRRC has a better performance than SRC in the robustness to occlusion. As a result, if we make the fusion of the two algorithms, we will integrate the advantages of the two algorithms. The proposed method has a better performance in occlusion face recognition.

The fusion of SRC and SRRC algorithm calculates the coefficient matrix by ℓ_1-norm. The coefficients are much sparse than those given by ℓ_2-minimization, and the main coefficients are associated with target class. Different from the traditional algorithms such as PCA and LDA represent the test sample by all training samples, the proposed method uses the training samples that are associated with the teat sample. Consequently, the proposed methods efficiency can be higher.

The proposed method is a kind of NS classifier. Compared with the traditional NS [13], the proposed method uses a new measurement to calculate the distance between a test sample and a class. A natural question arose is that the fusion of SRC and SRRC can inherit the advantages(such as the robustness to data corruption) of SRC and SRRC. We proposed some experiments to solve this problem.

4 Experiments

We conduct experiments on three popular databases: AR [10], ORL, YaleB. The sizes of ORL, AR and YaleB images are respectively 46×56, $40*50$ and $32*32$. We compare our method with five popular methods: PCA, LRC, CRC, SRC, SRRC.

4.1 Face Recognition without Occlusion

ORL database consists of 400 images of 40 persons (10 images for each person).We randomly select 5 images of each person for training and the rest for testing. We conduct experiments with different feature dimensions20, 30, 40 respectively. We set the parameter λ to be 1e-2 in CRC. In the proposed method, the parameters λ_1 and λ_2 are 5e-1 respectively. Fig. 1 lists the classification accuracy of different methods on ORL images.

A subset of AR database consists of 3120 images of 120 persons (about 26 images for each person).The images from the first to the seventh and the images from the sixteenth to the twentieth are non-occlusion samples, so the seven samples from the first to the seventh of each person are used for training samples, and the other five samples from the sixteenth to the twentieth of each person are used for test samples. We conduct experiments with different feature dimensions 20, 30, 40 respectively. We set the parameter λ to be 1e-2 in CRC and the parameter λ_1 and λ_2 to be 5e-2 in proposed method. Fig. 2 lists the classification accuracy of different methods on AR images without occlusion.

YaleB database consists of 165 frontal images of 11 persons (about 15 images for each person). We randomly select 8 images of each person for training and the rest for test. We set the parameter λ to be 1e-2 in CRC. In the proposed method, the values of λ_1 and λ_2 are assigned to be 0.3 and 0.7 respectively. Table 1 lists the classification accuracy of different methods on YaleB images.

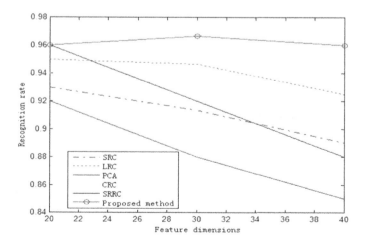

Fig. 1. The classification accuracy of different methods on ORL images

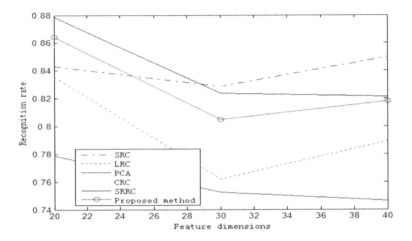

Fig. 2. The classification accuracy of different methods on AR images without occlusion

Table 1. The classification accuracy of different methods on YaleB images

methods	SRC	LRC	CRC	PCA	SRRC	Propesed method
Recognition Rate	85%	91.57%	90%	90%	97.86%	97.14%

The accuracy of the proposed method is the first highest on ORL. When we conduct experiments on AR images without occlusion, the accuracy of the proposed method is fourth in the all methods. As can be seen from Table 1, we know the proposed methods classification accuracy is same to those that solved by SRRC. The accuracy of the proposed method is higher than some traditional algorithm clearly.

4.2 Face Recognition with Occlusion

This section proves the robustness of different methods against occlusion on a subject of the AR databases. A subject of AR database consists of 3120 images of 120 persons (about 26 images per person). There are 12 images without occlusion, 6 images occluded by sunglass, 6 by scarf of each person. In this experiment, the samples from the first to the seventh of each person are used for training samples, the samples from the eighth to the twelfth of each person are used for test samples. The training samples are non-occlusion, and the test samples are occlusion. We conduct experiments with different feature dimensions 20, 30, 40 respectively. Fig. 3 lists the classification accuracy of different methods on AR images with test samples occlusion.

We exchange the value of the training samples and testing samples. The samples from the eighth to the twelfth of each person are used for training sample, the samples from the first to the seventh of each person are used for test sample. We conduct experiments with different feature dimensions 20, 30, 40 respectively. Fig. 4 lists the classification accuracy of different methods on AR images with training samples occlusion.

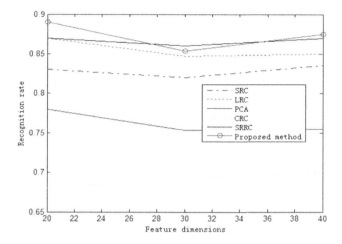

Fig. 3. The classification accuracy of different methods on AR images with training samples occlusion

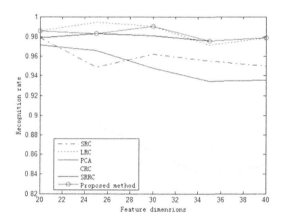

Fig. 4. The classification accuracy of different methods on AR images with training samples occlusion

We conduct another experiment to simulate the random noise. The training samples consist of the 14 non-occluded images. We pollute the training samples with random noise and take them as the test samples. We choose 15 lines and replace each pixels of them with random numbers. We randomly pollute the line 35-50 in order to imitate the sunglass. We randomly pollute the line 24-39 in order to imitate the scarf. Fig. 5 shows the 14 samples with synthetic occlusion. We conduct experiments with 10 persons. Table 2 lists the classification accuracy of different methods on AR images with synthetic occlusions.

Fig. 5. Samples of synthetic occlusion

The accuracy of the proposed method is highest on AR images with test samples occlusion. When we conduct the experiment on AR images with training samples occlusion, the accuracy of the proposed method is similar to those solved by LRC and the fusion of the SRC and SRRCs accuracy is the second highest on AR images with training samples occlusion. As can been seen from Table 2, we know the proposed method can achieve the second classification accuracy.

Table 2. The classification accuracy of different methods on AR images with synthetic occlusions

methods	SRC	CRC	PCA	LRC	SRRC	Propesed method
Recognition Rate	86.67%	91.11%	77.78%	88.89%	97.78%	97.78%

5   Conclusion

In this paper, we proposed the fusion of SRC and SRRC to improve the accuracy of face recognition. In the fusion of SRC and SRRC, we calculate the residuals that are solved by SRC and SRRC. We calculate the summation of the residuals by class, the class label which responds to the minimum summation is the result of the test sample. Experiments on face recognition data sets ORL, YaleB and AR demonstrated that the proposed algorithm can improve the comparative classification accuracy. The appearance of the occlusion degrades the performance the traditional subspace methods, the proposed method linearly express the test sample by the training sample of the different person, and classifies the sample based on the residue. The recognition rates of the proposed algorithm are more competitive than these traditional subspace algorithms. However, more investigations are to be made to further study the fusion of SRC and SRRC scheme for various pattern classification problems, and this is one of our main tasks in the future work.

Acknowledgments. This paper is partially supported by Shenzhen Municipal Science and Technology Innovation Council (Nos. JC201005260122A, JCYJ2012061315335 2732 and CXZZ20120613141657279).

References

1. Wright, J., Yang, A.Y., Ganesh, A., Sastry, S.S., Ma, Y.: Robust face recognition via sparse representation. IEEE Trans., Pattern Anal. Mach. Intell. 31(2), 210–227 (2009)
2. Elad, M., Matalon, B., Zibulevsky, M.: Image denoising with shrinkage and redundant representations. In: IEEE Conference on Computer Society, vol. 2, pp. 1924–1931 (2006)
3. Elad, M., Aharon, M.: Image denoising via learned dictionaries and sparse representation. In: IEEE Conference on Computer Society, vol. 1, pp. 895–900 (2006)
4. Rao, S.R., Tron, R., Vidal, R., et al.: Motion segmentation via robust subspace separation in the presence of outlying, incomplete, or corrupted trajectories. In: IEEE Conference on Computer Vision and Pattern Recognition, pp. 1–8 (2008)
5. Li, X., Jia, T., Zhang, H.: Expression-insensitive 3D face recognition using sparse representation. In: IEEE Conference on Computer Vision and Pattern Recognition, pp. 2575–2582 (2009)
6. Mei, X., Ling, H., Jacobs, D.W.: Sparse representation of cast shadows via ℓ_1-regularized least squares. In: IEEE 12th International Conference on Computer Vision, pp. 583–590 (2009)
7. Nagesh, P., Li, B.: A compressive sensing approach for expression-invariant face recognition. In: IEEE Conference on Computer Vision and Pattern Recognition, pp. 1518–1525 (2009)
8. Wright, J., Yang, A.Y., Ganesh, A., et al.: Robust face recognition via sparse representation. IEEE Transactions on Pattern Analysis and Machine Intelligence 31(2), 210–227 (2009)
9. Wang, J., Xu, Y., You, J.: Sparse residue for occluded face image reconstruction and Classification. In: 21st IEEE International Conference on Pattern Recognition (ICPR), pp. 1707–1710 (2012)
10. Zhang, L., Yang, M., Feng, X.: Sparse representation or collaborative representation: Which helps face recognition? In: IEEE International Conference on Computer Vision (ICCV), pp. 471–478 (2011)
11. Li, C.G., Guo, J., Zhang, H.G.: Local sparse representation based classification. In: 20th International Conference on Pattern Recognition (ICPR), pp. 649–652 (2010)
12. Candes, E.J., Romberg, J.K., Tao, T.: Stable signal recovery from incomplete and inaccurate measurements. Communications on Pure and Applied Mathematics 59(8), 1207–1223 (2006)
13. Martinez, A., Benavente, R.: The AR face database. Technique Report (1998)

Posterior Probability Based Multi-classifier Fusion in Pedestrian Detection

Jialu Zhao, Yan Chen, Xuanyi Zhuang, and Yong Xu

Shenzhen Graduate School, Harbin Institute of Technology, Shenzhen 518055, China
sxxazhaojialu@126.com, jadechenyan@gmail.com, zhxy_lina@163.com,
laterfall2@yahoo.com.cn

Abstract. This paper presents a novel method for pedestrian detection at measurement level. At feature extraction stage, we use Histogram of Oriented Gradient to describe the feature of pedestrian and non-pedestrian. To decrease the time cost, we reduce the dimension by using PCA. The base classifiers used in posterior probability based multi-classifier fusion are posterior probability based SVM, Naïve Bayesian and Minimum Distance Classifier, respectively. To estimate the accuracy of fusion result, stratified cross-validation is used. Experimental results on pedestrian databases prove the efficiency of this work.

Keywords: pedestrian detection, multi-classifier fusion, posterior probability, stratified cross-validation.

1 Introduction

In machine learning field, the ensemble learning methods show that, fusions of several single classifiers can achieve much higher accuracy than a single classifier. Some researchers consider that it is necessary to introduce classifiers fusion methods into pedestrian detection [1,2]. For a classifiers fusion method, it is assumed that all classifiers should be trained over a same training dataset [3,4], and the base classifiers should be different [5]. However, choosing an appropriate fusion method can further improve the performance.

According to the output level of classifier outputs the results, multiple classifiers fusion can be divided into the following three categories [3]: abstract level fusion, rank level fusion and measurement level fusion. At abstract level, classifiers output the class labels or an unordered set of candidate classes, which contain the least classification information. At rank level, classifiers export n-best candidate classes in order. The most likely candidate class is listed at the top, while the most unlikely one is listed at bottom. At measurement level, classifiers output not only the n-best candidate classes, but also their corresponding confidence values. For all these three levels, the measurement level fusion contains the most information. Hence fusion at measurement level is possible for most applications.

For measurement level fusion, many methods have been proposed. The fusion methods usually can be divided into fixed rules and trained rules by tunable parameters [6]. The fixed rules include sum rule, product rule, maximum rule, minimum rule, median rule and so on [7]. The trained rules include trainable classifiers (neural networks [8], SVM [9], etc.), optimized evidence fusion [10] and the weighted fusion.

In this work, we focus on the application of posterior probability based multiple classifier fusion in pedestrian detection. In order to obtain good performance, classifier fusion is carried out at measurement level in this paper. At feature extraction stage, taking the advantage of the highly discriminative power, the Histogram of Oriented Gradients (HOG) [11] descriptor is used to calculate the image feature. At learning stage, several different classifiers are trained respectively. At classification stage, samples are tested with different models, and a decision template (DT) that consists of posterior probability is given for further fusion. At fusion stage, the weighted sum rule is used to deal with the decision template, and the final posterior probability is calculated for each class. The maximum rule is used to make the final decision of the classification, and the sample will be classified to the class that has the maximum posterior probability.

The rest of the paper is organized as follows: in Section 2, we briefly introduce the feature descriptor used in this work. Section 3 shows some posterior probability based classifiers. The fusion scheme is introduced in Section 4 and the experimental results are shown in Section 5. Finally, we draw our conclusion in Section 6.

2   Feature Extraction

At feature extraction stage, Histogram of oriented gradient (HOG) [11] is used as appearance feature in this work. HOG is an efficient feature. It has been proved that HOG gives nearly perfect results on MIT database [12]. The extraction steps of HOG feature are illustrated as follows:

Step1 Calculate the horizontal and vertical gradient of each pixel.

$$\begin{cases} f_x(x,y) = I(x+1,y) - I(x-1,y) \\ f_y(x,y) = I(x,y+1) - I(x,y-1) \end{cases} \forall x,y \tag{1}$$

where $f_x(x,y)$ and $f_y(x,y)$ denote the x and y components of the image gradient, respectively. $I(x,y)$ is the gray value of a pixel at position (x,y).

Step2 Compute the magnitude $m(x,y)$ and the orientation $\theta(x,y)$ of each pixel.

$$m(x,y) = \sqrt{f_x(x,y)^2 + f_y(x,y)^2} \tag{2}$$

$$\theta(x,y) = \arctan \frac{f_y(x,y)}{f_x(x,y)} \tag{3}$$

For each cell, the computed orientations $\theta(x,y)$ are quantized into c_b orientation bins. All the magnitudes $m(x,y)$ in each bin are summed up to make a histogram.

Step3 Normalize gradient magnitude within the block.

While calculating the histograms in each cell, the block slides a cell each time. The gradient image consists of several overlapping blocks, with each block consists of $b_w \times b_h$ cells, and each cell consists of $c_w \times c_h$ pixels. Let f be a feature vector of a block, h_{ij} denote the overlapping histogram of the cell in position (i,j) from an overlapping block ($1 \leq i \leq b_w, 1 \leq j \leq b_h$). The feature vector of the overlapping block is normalized with L2-norm as follows:

$$h'_{ij} = \frac{h_{ij}}{\sqrt{\|f\|^2 + 1}} \qquad (4)$$

3 Posterior Probability Based Classifiers

For pedestrian detection, there are two classes $\Omega = \{\omega_1, \omega_2\}$. Given an unknown sample $X \in \Re^n$, assume $X = [x_1, x_2, \ldots, x_n]$ is a feature vector and $\Re^n$ is its feature space. Let $C = \{C_1, C_2, \ldots, C_l\}$ be the chosen classifiers. Each classifier C_i outputs a set of values $[c_{i1}(X), c_{i2}(X)]$. Each $c_{ij}(X) \in [0, 1]$ is a confidence value supporting the evidence that X belongs to ω_j (where $i = 1, 2, \ldots, l$ and $j = 1, 2$). In this work, we take $c_{ij}(X)$ as the posterior probability, so we have

$$c_{i1}(X) + c_{i2}(X) = 1 \qquad (5)$$

In this work, three different classifiers are used, they are a posterior probability based Support Vector Machine (SVM), a Minimum Distance Classifier (MDC) and a Naïve Bayesian (NB) classifier. For the first case, to realize the post processing of SVM, we transform the standard outputs to posterior probability by using a sigmoid function [13].

$$P(y = 1|f(x)) = \frac{1}{1 + \exp(Af(x) + B)} \qquad (6)$$

where $f(x)$ is the standard output of SVM, $P(y = 1|f(x))$ is posterior probability that the sample x belonging to class 1 when the output is $f(x)$. [13] shows the method to calculate parameters A and B.

For MDC, given a test sample X, the classifier calculates the distance d_k^2 (Manhattan Distance is chosen) between X and the mean vector U_k, and classifies X to the class with minimum distance [14]. For the Naïve Bayesian classifier, it classifies the sample to the class that has the highest posterior probability. That is to say, the Naïve Bayesian classifies the test sample to class ω_i, if and only if $P(\omega_i|X)$ is the maximum probability.

Each classifier is trained by using the same training set, and the posterior probability is estimated on the same test set. To train the classifiers, we use a 5-fold cross validation method at validation stage. In order to ensure each fold has a similar class distribution with the whole dataset, a stratified cross validation method is used. For our binary classification problem, we have two subsets: pedestrian data labeled with 1 and non-pedestrian data labeled with -1. For each subset, run 5-fold partition algorithm and partition each subset into 5 equal-sized folds. Each time, take 3 folds from pedestrian data and non-pedestrian data respectively to form the training dataset, and take all the rest folds to form the test dataset. Each classifier runs 10 times with different training and test sets. The accuracy rate of each classifier is estimated by counting the total right classified ones, divided by the total number of the test samples.

4   Fusion Method

4.1   Decision Template

We use the Decision Templates (DT) [15] as classifiers fusion scheme in this work. For a given pattern X, the outputs of all classifiers can be represented as a decision template by a $l \times 2$ matrix:

$$DT(X) = \begin{bmatrix} c_{11}(X) & \cdots & c_{i1}(X) & \cdots & c_{l1}(X) \\ c_{12}(X) & \cdots & c_{i2}(X) & \cdots & c_{l2}(X) \end{bmatrix}^T \tag{7}$$

In this template, the ith row is the outputs of the ith classifier $C_i(X)$, and the jth column is the probability that X belonging to class ω_j. Then the fusion result of the l classifiers is expressed as

$$C(X) = F(C_1(X), C_2(X), \ldots, C_l(X)) = [d_1(X), d_2(X)] \tag{8}$$

where F is the fusion rule, $d_j(X)$ is the fused value that X belonging to class ω_j.

4.2   Linear Fusion Scheme

In this work, the Weighted Majority Vote (WMV) theorem [16] is used as the voting scheme. Therefore, the weight of the ith classifier is calculated by

$$w_i = \log \frac{a_i}{1 - a_i} \tag{9}$$

where a_i is the individual accuracy of the ith classifier.

By using weighted sum rule, we calculate the fusion result for each class

$$d_j(X) = \sum_{i=1}^{l} w_i c_{ij}(X), j = 1, 2 \tag{10}$$

After gaining the fusion results, we make a decision for the input pattern, and we use the maximum rule in this work, that is, if $d_K(X) = \max\limits_{i=1}^{2} d_i(X)$, then $X \in \omega_K$.

5   Experimental Results

5.1   Dataset and Feature Extraction

In this work, three pedestrian databases are used. They are INRIA Pedestrian dataset [11], CVC-CER-01 Pedestrian Database [17] and DaimlerChrysler Pedestrian Classification Benchmark Dataset [18]. The division of each database is shown in Table 1. The pedestrian images are labeled with 1 and non-pedestrian images are labeled with -1. Before feature extraction, images in INRIA and CVC01 are scaled into 64×128 pixels, images in Daimler are scaled into 32×64.

At feature extraction stage, the block is divided into 2×2 cells, and each cell consists of 8×8 pixels. Hence the block has $2 \times 2 \times 9 = 36$-dimensional features. As the block slips a cell each time, we can obtain $(64/8 - 1) \times (128/8 - 1) = 105$ blocks for a 64×128 image and 21 blocks for a 32×64 image. Hence, we obtain $36 \times 105 = 3780$-dimensional feature vector for INRIA & CVC01 database and 756-dimensional feature vector for Daimler database.

Table 1. The division of each database used in this work

Dataset	Image size	Feature dimensional	Training set		Test set	
			Pedestrian	Non-pedestrian	Pedestrian	Non-pedestrian
INRIA	64×128	3780	2416	12180	1126	4530
CVC01	64×128	3780	1200	3705	800	2470
Daimler	32×64	756	14400	15000	9600	10000

5.2 Experiments

In order to test the proposed multiple classifiers fusion scheme, we construct a system as shown in Fig. 1

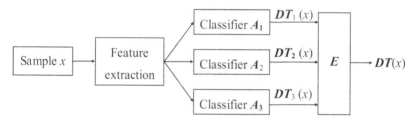

Fig. 1. Multi-classifier system. In our experiments, A_1, A_2, A_3 are posterior probability based classifiers introduced in Section 3, E is the weighted sum rule.

Table 2 and Table 3 show different result on INRIA database when using different dimension of PCA based HOG feature. We take the average of 10 runs as the final result.

Table 2. Different results when using different dimension of PCA based HOG feature on INRIA database

PCA based HOG feature	Our method	SVM	Naïve Bayesian	MDC
50-dimensional	**95.38%**	91.64%	92.54%	90.79%
100-dimensional	**95.52%**	92.40%	92.89%	90.86%
150-dimensional	**95.57%**	92.67%	92.92%	90.84%
200-dimensional	**95.43%**	92.80%	92.68%	91.69%

From Table 2 we can see that, multiple classifiers fusion can get higher performance than single classifier does. The precision rate of our fusion method is about **2.5%** higher than single posterior probability based SVM and Naïve Bayesian, about **5%** higher than single minimum distance classifier.

In order to evaluate measurement level fusion, rank level fusion and decision level fusion, we set up an experiment to compare the fusion results at different levels. As shown in Table 3, the measurement level fusion performs better than rank level fusion

Table 3. The comparisons of fusion at different levels

PCA based HOG feature	Measurement level	Rank level	Decision level
50-dimensional	**95.38%**	91.69%	93.59%
100-dimensional	**95.52%**	92.10%	93.94%
150-dimensional	**95.57%**	92.20%	93.90%
200-dimensional	**95.43%**	92.17%	93.78%

Table 4. Experimental results on other pedestrian database

Database	Method	TPR	FPR	FNR
CVC01 database	Method in [17]	92.50%	1.00%	22.81%
	Our method	**99.00%**	**0.80%**	**16.33%**
Daimler database	Method in [18]	**90.91%**	33.62%	73.61%
	Our method	87.74%	**10.49%**	**11.55%**

and decision level fusion. The precision rate of measurement level fusion is about **3.4%** higher than rank level fusion and about **2.5%** higher than decision level fusion.

Table 4 shows the comparisons with different methods on different database. The performance evaluation methods are true positive rate (TPR), false positive rate (FPR) and the false negative rate (FNR). The results indicate that our method outperforms over methods in [17] and [18]. On CVC01 database, the TPR is 99% which is higher than that of method in [17] about **6.5%**. The FPR and FNR is lower as well. On Daimler database, though the TPR is lower than that in [18], but the false detections of our method is much lower (FPR is about **23.1%** lower and FNR is about **62%** lower than method in [18]).

6 Conclusion

In this paper, we proposed a posterior probability based multi-classifier fusion method at measurement level. The HOG detector was used to describe the feature of pedestrian and non-pedestrian. The base classifiers used in this work were posterior probability based SVM, Naïve Bayesian and Minimum Distance Classifier, respectively. The comparisons of measurement level fusion and other level fusion have shown the better performance of our work. The experimental results on INRIA pedestrian database have shown a more robust performance in general than individual classifiers. The comparisons results on CVC01 and Daimler pedestrian database have shown that, our method has less misclassification rate.

Acknowledgments. This paper is partially supported by Shenzhen Municipal Science and Technology Innovation Council (Nos. JC201005260122A, JCYJ201206131533 52732 and CXZZ20120613141657279).

References

1. Gavrila, D.M., Munder, S.: Multi-Cue Pedestrian Detection and Tracking from a Moving Vehicle. International Journal of Computer Vision 73(1), 41–59 (2007)
2. Parra, I., Fernandez, D., Bergasa, M.A.: Combination of Feature Extraction Methods for SVM Pedestrian Detection. IEEE Trans. on Intelligent Transportation Systems 73(1), 292–307 (2007)
3. Xu, L., Krzyzak, A., Suen, C.Y.: Methods of Combining Multiple Classifiers and their Application to Hand Writing Recognition. IEEE Trans. on Systems, Man, and Cybernetics 22, 418–435 (1992)
4. Ng, K.C., Abramson, B.: Consensus diagnosis: A simulation study. IEEE Trans. on Systems, Man, and Cybernetics 22, 916–928 (1992)
5. Duin, R.: The Combining Classifier: to Train or not to Train? In: Proceedings of the International Conference on Pattern Recognition, vol. 2, pp. 765–770 (2002)
6. Duin, R.P.W., Tax, D.M.J.: Experiments with Classifier Combining Rules. In: Kittler, J., Roli, F. (eds.) MCS 2000. LNCS, vol. 1857, pp. 16–29. Springer, Heidelberg (2000)
7. Dass, S.C., Nandarumar, K., Jain, A.K.: A Principled Approach to Score Level Fusion in Multimodal Biometric Systems. In: Proceedings of International Conference on Audio and Video Based Biometric Person Authentication, pp. 1049–1058 (2005)
8. Lee, D.S., Srihari, S.N.: A Theory of Classifier Combination: the Neural Network Approach. In: The 3rd International Conference on Document Analysis and Recognition, pp. 42–45 (1995)
9. Burges, C.J.C.: A Tutorial on Support Vector Machines for Pattern Recognition. Knowledge Discovery Data Mining 2(2), 1–43 (1998)
10. Al-Ani, A., Deriche, M.: A New Technique for Combining Multiple Classifiers Using the Demspter Shafer Theory of Evidence 17, 333–361 (2002)
11. Dalal, N., Triggs, B.: Histograms of Oriented Gradients for Human Detection. In: IEEE Computer Society Conference on Computer Vision and Pattern Recognition, pp. 886–893 (2005)
12. Wang, Z.R., Jia, U.L., Huang, H., Tang, S.M.: Pedestrian Detection Using Boosted HOG Features. In: Proceedings of the 11th International IEEE Conference on Intelligent Transportation Systems, pp. 1155–1160 (2008)
13. John, C.P.: Probabilistic Outputs for Support Vector Machines and Comparisons to Regularized Likelihood Methods. Advances in Large Margin Classifiers, 61–73 (1999)
14. Jain, A.K., Duin, R., Mao, J.: Statistical Pattern Recognition: A Review. IEEE Trans. on Pattern Analysis and Machine Intelligence 22(1), 4–37 (2000)
15. Kuncheva, L.I., Bezdek, J.C., Duin, R.P.: Decision Templates for Multiple Classifier Fusion: An Experimental Comparison. Pattern Recognition 34(2), 299–314 (2001)
16. Kuncheva, L.: Combining Pattern Classifiers: methods and algorithms, p. 124. Wiley (2004)
17. Gerónimo, O., Sappa, A.D., López, A., Ponsa, D.: Adaptive Image Sampling and Windows Classification for on-board Pedestrian Detection. In: Proceedings of the International Conference on Computer Vision Systems, Bielefeld, Germany (2007)
18. Munder, S., Gavrila, D.M.: An Experimental Study on Pedestrian Classification. IEEE Trans. on Pattern Analysis and Machine Intelligence (2006)

Quantification-Based Ant Colony System for TSP

Ming Zhao, Jeng-Shyang Pan, Chun-Wei Lin, and Lijun Yan

School of Computer Science and Technology,
Harbin Institute of Technology Shenzhen Graduate School
mingzhao@yangtzeu.edu.cn

Abstract. Ant Colony Optimization (ACO) is one of the swarm intelligent methods for solving computational problems, especially in finding the optimal paths through graphs. In the past, floating point is widely used to represent the pheromone in ACO, thus requiring large amounts of memory to find the optimal solutions. In this paper, the quantification-based ACS (QACS) is thus proposed to reduce the space complexity. New updating rules of pheromone with no decay parameters are also designed in the proposed QACS for simplifying the updating processing of pheromone. Based on the experimental results of proposed QACS, the convergence rate can be improved with less memory space for solving Traveling Salesman Problem (TSP).

Keywords: Ant colony Optimization, pheromone updating rule, Qualification-based, TSP.

1 Introduction

Ant colony optimization (ACO) is a popular bio-inspired optimization algorithm with swarm intelligence for deriving the optimal solutions in different applications, such as Traveling Salesman Problem (TSP) [1, 3, 10], vehicle routing problem [2, 15], job-shop scheduling [5, 13], system identification [4, 9], image processing [9, 16], among others [7, 18]. Among them, ACO is used to simulate the searching behavior of real ants, thus depositing chemical pheromone trail on the ground to communicate with each other for finding the shortest paths between their nest and food. Recently, many researches have been developed with ACO-based algorithms for learning and optimization. In the past, Dorigo et al. proposed an elitist ant system (EAS) [10] to reinforce the pheromone of edges the possible optimal path, thus speeding up the convergence rate. Bullnheimer et al. then further proposed the rank-based ant system (AS-RANK) [3], assigning the rank order to each ant by tour length. In AS-RANK approach, the top (k-1) ants can deposit extra pheromone on the visited tour, which is used to decide the rank order of the ants, thus speeding up the convergence rate. Stützle et al. proposed MAX-MIN ant system [17, 18] to achieve a strong exploitation ability from search history, thus accumulating the pheromone of the best solutions while the updating process. The strengths of pheromone trails are also limited to avoid the premature convergence. Other ACO-based algorithms are still in the progress and have been extended to other issues [7, 8, 19].

J.-S. Pan et al. (eds.), *Genetic and Evolutionary Computing*,
Advances in Intelligent Systems and Computing 238,
DOI: 10.1007/978-3-319-01796-9_36, © Springer International Publishing Switzerland 2014

The above ACO-based algorithms, use, however, floating points to represent the pheromone in the memory, thus requiring highly space complexity. In this paper, a novel qualification-based ant colony system (QACS) is proposed. It uses integers to represent the pheromone, adopts new updating rules of pheromone and designs a new cycling strategy to solve the overflow problem. Experimental results demonstrate that the proposed QACS has a better performance in execution time and less memory, comparing to the traditional ACO-based algorithms. The remainder of this paper is organized as follows. The detail of the proposed Qualification-based ant colony system (QACS) is stated in Section 2. Numerical results are shown in Section 3. Conclusion and future works are given in Section 4.

2   The Proposed Qualification-Based Ant Colony System (QACS)

The QACS follows the same state transition rules, which is the same to the Ant Colony System (ACS) [9]. The details of the proposed QACS are given below.

2.1   TSP Problem

The size of m also have great effects on the optimal solution, this could be discussed section 3. The first Ant Colony Optimization algorithm was proposed for solving Traveling salesman problem (Tsp), it could be presented as follows:

Let $c = \{c_1, c_2, \ldots c_n\}$ be a set of cities, $A = \{(r,s) : r, s \in c\}$ be the edge set, and $d_{ij} (1 \le i, j \le n, i \ne j)$ is the Euclidean distance between City i and city j, and $d_{ij} \ne d_{ji}$,The Tsp is a problem of finding a minimal distance of the closed tour that visits each city once.

2.2   State Transition Rule

In the proposed QACS, to each ant k, the path vector R^k, which keeps the Sequence Numbers of all visited cities based on visiting Sequence, the ant k positioned on city i chooses the city j to visit by applying the rule given by formula (1).

$$j = \begin{cases} \arg\max_j \left\{ \tau(i,j)^\alpha \left[\eta(i,j) \right]^\beta \right\} (j \in J_k(i)), q \le q_0 \\ S \qquad\qquad\qquad\qquad\qquad\qquad , q > q_0 \end{cases} \qquad (1)$$

Where $J_k(i)$ is the set of cities that remain to be visited by ant k positioned on city i, and $\tau(i,j)$ is the pheromone of the edge (i,j), $\eta(i,j)$ is heuristic information, it could normally be calculated by using $\eta(i,j) = \frac{1}{d_{ij}}$. α and β are two parameters which are used to control and adjust the weight of the heuristic information and the dosages of pheromone [7]. q_0 is a parameter in the interval[0, 1] and q is a random number.

When $q \leq q_0$, the ant k will select the max j as the next city to visit; When $q > q_0$, biased exploration strategy will be used to decide which is the next city j. The bias exploration is presented as formula (2).

$$P_k(i,j) = \begin{cases} \dfrac{[\tau(i,j)]^\alpha \cdot [\eta(i,j)]^\beta}{\displaystyle\sum_{u \in J_k(i)} [\tau(i,u)]^\alpha \cdot [\eta(j,u)]^\beta}, & if \quad s \in J_k(i) \\ \\ 0 \quad , & otherwise \end{cases} \tag{2}$$

When $q \leq q_0$, ant k choose the next city which has the largest product of $\tau(i,j)^\alpha [\eta(i,j)]^\beta$, and we called this "exploitation". When $q > q_0$, next city j would be selected according to formula (2), and it is called as "biased exploration".

q_0 is a very important control parameter in the state transition rule, the probability which ant k choose the optimal direction is q_0. When $q \geq q_0$, ant k also has a chance to explore new path. It can adjust q_0 to keep the balance of "exploitation" and "exploration".

Formula (1) and formula (2) together is called pseudo-random proportional rule.

2.3   The Representation of Pheromone

In qualification-based ant colony system, the encoding number of pheromone is represented as the integers, instead of using floating points shown in formula (3).

$$\tau(i,j) = m, (m \in integer, 1 \leq m \leq 2^n, n \in integer). \tag{3}$$

In formula (1), $\tau(i,j)$ is the pheromone of the edge (i,j), m is an integer with a fixed range. The size of m have great effects on the optimal solution, this will be discussed section 3.

It is obvious that a numerical variable with a limited representation range will finally be overflowed with the continuous increasing of iterations. After the pheromone variable achieving the maximum range, pheromone value will not be increased, even ants construct the best tour, and pheromone can no longer reflect that whether the corresponding tour is the best or not. With increase of iterations, maybe the new optimal path would be constructed, and pheromone on edges of the new path also have the probability could be got the maximum, that is to say, the pheromone on the optimal path and the pheromone non-optimal path are the same, they may have the same chance to be visited, and it doesn't benefit to fast convergence. Similarly when most of pheromone variable achieving the minimum range, it doesn't benefit to explore new path. In order to solve this problem, a cycling strategy will be introduced. When $\tau(i,j)$ is the maximum, its value will not be added, and if the best solution wasn't changed yet after some iterations, let all $\tau(i,j)$ is initialized to the initialization value τ_0; when $\tau(i,j)$ is the minimum, its value would also no longer be decreased. And if the best solution wasn't changed yet after some iteration, let all $\tau(i,j)$ be initialized to the initialization value τ_0 on all edges.

2.4   Pheromone Management Strategy

In traditional ACO [1, 3, 5, 7], it requires two decay parameters ρ and ε, ρ represents the global decay coefficient and ε represents the local decay coefficient. Both of them are limited in an interval [0, 1]. In the proposed QACS, two decay parameters are eliminated and a new updating rule was presented in this algorithm, the details could be seen in the following section.

2.4.1   Pheromone Updating Rule

In ACS [11], global updating rule and local updating rule are used in the pheromone updating rules. In the proposed QACS, however, considerate both of them only an updating rule consideration global-best tour and iteration-best tour together. According to the research results of Thomas Stützle in reference [6], the global best updating could make solutions are converged fast to T_b with obvious direction and iteration-best updating could enhance search ability of the algorithm. So in order to offset the absent of the local updating rule and reinforce explore probability, the global best updating and iteration-best updating are used together.

In the proposed QACS, it considerate the global-best tour and iteration-best tour together. The details could be seen in formula (4).

$$\tau(i, j) = \tau(i, j) + \Delta\tau_k(i, j) \tag{4}$$

$$\Delta\tau_k = \begin{cases} 3, (i, j) \in C_b \text{ and } T_b \\ 1, (i, j) \in C_b \\ 1, (i, j) \in T_b \\ 0, others \end{cases}$$

Where T_b is the length of the shortest tour from beginning, and C_b is the shortest tour of the current iteration.

2.4.2   The Pheromone Initialization τ_0

The pheromone of all edges should be assigned an initialization τ_0 at the beginning of the ant system algorithm. In QACS, the representation range of the pheromone variable is limited, and if the difference among some edges is already very large after first several iterations, it will not benefit to seek for the global best solution. Refer the pheromone initialization of MMAS [17], it assigned a middle integer 2^{n-1} to τ_0, (Assume the maximum range is set at 2^n.)

2.4.3   Pheromone Decay Rule

In QACS, and the pheromone value, however, must be decayed, it doesn't require any decay parameters. Thus, a new simply decay strategy is introduced. Pheromones on all edges are integers, and the size of decay unit also must be a small integer. According to the reinforcement learning strategy in reference [14], those who constructed the best tour should be encouraged, thus, decay unit is smaller than the increment. And it mean that the integer '1' is only choice. The decay strategy thus could be presented in formula (5) as follows.

$$\tau(i,j) = \tau(i,j) - 1.\tag{5}$$

When ants finished its iteration, the pheromone on each edge will be volatilized according to formula (5), thus, the parts of pheromone on all edges were volatilized, and the optimal path still was kept and reinforced.

2.5   Steps of the Proposed QACS

INPUT: A city set $c = \{c_1, c_2, \ldots . c_n\}$ and weight control parameters between initialization pheromone and path length α and β; the number of m ants; control probability of exploitation and exploration q_0.

OUTPUT: A shortest path of the closed circuit for TSP.

STEP 1: Use greedy algorithm to construct a path and get the length of the path C^{nn}; take C^{nn} as history best tour and the global best tour, and initialize all edges with $\tau_0 = 128$.

STEP 2: Construct the tour C_i of each ant i, and store the best solution.

Substep 2-1: Each ant k, select a city i as an initially randomly.

Substep 2-2: Ant k select the next city j according to the state transition rule shown in formula (1) and formula (2).

Substep 2-3: If ant k doesn't finish constructing the tour go to Substep 2-2.

Substep 2-4: If each ant k finishes constructing the tour use pheromone decay rule to updating the pheromone on each edge, or else go to Substep 2-2.

Substep 2-5: If each ant finishes the constructing the path then using pheromone updating rule to updating pheromone on the iteration-best tour, or else go to Substep 2-2.

Substep 2-6: If the best tour meet the condition or the iteration > MAXiteration, then go to STEP 3, or else go to Substep 2-1.

STEP 3: Output the best solution.

2.6   An Example

In this Section, an example is given to illustrate the proposed QACS. Assume four cities A, B, C, D. The distance matrix is $d_{ij} = \begin{bmatrix} \infty & 3 & 1 & 2 \\ 3 & \infty & 5 & 4 \\ 1 & 5 & \infty & 2 \\ 2 & 4 & 2 & \infty \end{bmatrix}$, and the number of ants is initially set at 3, and $\alpha = 1, \beta = 2, q_0 = 0.5$.

STEP 1: Use greedy algorithm to construct a path $(ACDBA)$, the length

$C^{nn} = f(ACDBA) = 1+2+4 \ \ +3 = 10$, and take it as the best solution.

Initialize all edges with $\tau_0 = 128$.

STEP 2: Construct the tour C_i of each ant i, and store the best solution.

Substep 2-1: Each ant k, select a city i as an initially randomly. Assume that the first ant select city A, the second ant select city B, and the third ant select city C.

Substep 2-2: Ant k select the next city j according to the state transition rule shown in formula (1) and formula (2). It takes the first ant as an example, the current city $i = A$ and unvisited city set $J_1(i) = \{B, C, D\}$. The probability for first ant selects B, C, and D as the next visiting object is:

$$A \Rightarrow \begin{cases} B: \ \tau_{AB} * \eta^2 = 128*1/9 = 128/9 & p(B) = 0.0816 \\ C: \ \ \ \tau_{AC} * \eta^2 = 128*1 = 128 & \Rightarrow p(C) = 0.7346 \\ D: \ \ \tau_{AD} * \eta^2 = 128*1/4 = 32 & p(D) = 0.1837 \end{cases}$$

Next City	Bias exploration Probability
C	0.5--0.7346
B	0.7346--0.8162
D	0.8162--11

Assume   that   random   select   probability   is $q = random(0,1) = 0.85$, the first ant will select city B as the next visiting city. In a similar way, assume the second ant select city D, and the third ant select city A.

Substep 2-3: Ant k finished their tour, assume the first ant construct the path $(ABCDA)$, the second ant constructed the path $(BDCAB)$, and the third ant constructed the path $(DACBD)$

Substep 2-4: If each k complete the tour then updating the pheromone with decay rule shown in formula (5). Pheromones on all edges become 127; Or else go to Substep 2-2.

Substep 2-5: The first ant constructed the best tour $(ABCDA)$ which is both iteration-best and the global best tour, then edges on the best tour are added 3 and they become into 130.

Substep 2-6: If the best solution 10 meets the acquirement or iteration > MAXiteration go to STEP 3, Or else go to Substep 2-1.

STEP 3: Output the best solution $(ABCDA)$.

3   Experimental Results

Experiments were made to show the performance of the proposed QACS, comparing to the traditional QACS in TSP. and experiments were implemented in VC++ 6.0 on a personal computer with an Intel Celeron G530 2.4GHz, 4G RAM; The demo system is set at windows 7 platform.

The eil51 dataset [6] is used in the experiments and the results are shown in Table 1.

Full consideration the optimal solution and corresponding runtime, it was assigned a middle number of the maximal representation range to the pheromone initialization τ_0 ; besides these, by studying some program languages, it shows that an unsigned integer maybe is a good choice, and too larger integer also could cause computational complexity and longer runtime, so it is set the maximal represented range to an integer 255, and all experiments were all run with $\tau_0 = 128$ and $Maxm = 255$.

Table 1. Different m and τ_0 on City set eil51

Max m	τ_0	The best Solution	Runtime	City set
15	1	462	8488ms	eil51
15	8	449	8531 ms	eil51
15	15	449	8405 ms	eil51
255	1	427	8427 ms	eil51
255	128	426	8386ms	eil51
255	255	427	8393ms	eil51
655535	1	427	8455ms	eil51
655535	128	427	8422ms	eil51
65535	255	426	8466ms	eil51
655535	32768	426	8460ms	eil51
655535	65535	427	8456ms	eil51

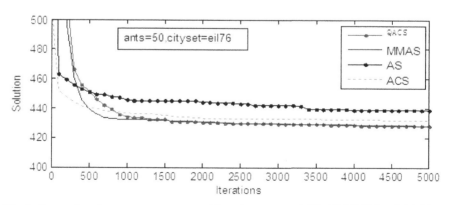

Fig. 1. Comparison the convergence rate and iterations among ACS, AS MMAS and QACS

The convergence rate is also evaluated in the experiments, comparing to the QACS and AS- based [1, 11, 18] algorithm, the data set eil76 [6] used in the experiments and the results are shown in Figure1.

In Figure 1, it is obviously that the proposed QACS has a nearly optimal performance of the convergence rate.

Experiments are also progressed to evaluate the execution time, comparing to the proposed QACS with other Ant-based algorithms.

The data set eil51 is used again in experiments, the optimal solution which was acquired under the same execution time was counted and the results are shown in Figure 2.

In Figure 2, it could be seen that the proposed QACS could get better solutions step by step; it could have the best performance finally.

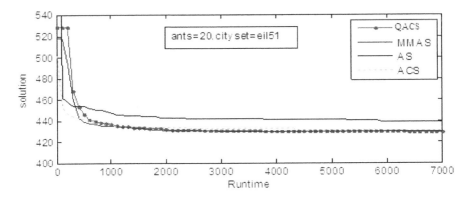

Fig. 2. Comparison based on average solution and iterations

4   Conclusion

In this paper, Qualification-based ant colony algorithm was proposed, it introduces new updating rule of pheromone and eliminate two decay parameters, thus simplifying the process of pheromone updating; and it adopts integers to represent pheromone, then reducing the dependencies of large memory. Experimental results show that it has good performance.

Qualification-based representation of pheromone could reduce the dependencies of memory; it also has some troubles because of limited range. And less updating rule of pheromone also doesn't benefit to explore the global best solution. A more reasonable updating rule of pheromone is the main work in the future.

References

1. Colorni, A., Dorigo, M., Maniezzo, V.: Distributed optimization by ant colonies. In: Proceedings of the European Conference on Artificial Life, Paris, France, pp. 134–142 (1991)

2. Bullnheimer, B., Hartl, R., Strauss, C.: An Improved Ant System Algorithm for the Vehicle Routing Problem. Annals of Operations Research 89, 319–328 (1999)
3. Bullnheimer, B., Hartl, R., Strauss, C.: A new rank-based version of the ant system: a computational study. Central European Journal for Operations Research and Economics 7(1), 25–38 (1999)
4. Nezamabadi-pour, H., Rashedi, S.: Edge detection using ant algorithms. Soft Computing 10(7), 623–628 (2006)
5. Yan, H., Shen, X.Q., Li, X., Wu, M.H.: An improved ant algorithm for job scheduling in grid computing. Machine Learning and Cybernetics, 2957–2961 (2005)
6. http://elib.zib.de/pub/mp-testdata/tsp/tsplib/tsplib.html
7. Zhang, J., Chen, W.-N., Zhong, J.-H., Tan, X., Li, Y.: Continuous Function Optimization Using Hybrid Ant Colony Approach with Orthogonal Design Scheme. In: Wang, T.-D., Li, X., Chen, S.-H., Wang, X., Abbass, H.A., Iba, H., Chen, G.-L., Yao, X. (eds.) SEAL 2006. LNCS, vol. 4247, pp. 126–133. Springer, Heidelberg (2006)
8. Abbaspour, K.C., Schulin, R., Van Genuchten, M.T.: Estimating unsaturated soil hydraulic parameters using ant colony optimization. Advances In Water Resources 24(8), 827–841 (2001)
9. Wang, L., Wu, Q.D.: Linear system parameters identification based on ant system algorithm. In: Proceedings of the IEEE Conference on Control Applications, Mexico, pp. 401–406 (2001)
10. Dorigo, M.: Optimization, learning, and natural algorithms. Ph.D. Thesis, Dip. Elettronica, Politecnico di Milano, Italy (1992)
11. Dorigo, M., Luca Gambardella, M.: Ant colony system: a cooperative learning approach to the traveling salesman problem. IEEE Transactions on Evolutionary Computation 1(1), 53–66 (1997)
12. Dorigo, M., Stutzle, T.: Ant colony optimization. The MIT Press, London (2004)
13. Dorigo, M., Maniezzo, V.: Ant system for job-shop scheduling. Belgian Journal of Operations Research, Statistics and Computer Science 34, 39–53 (1994)
14. Luca Gambardella, M., Dorigo, M.: Ant-Q: A Reinforcement Learning approach to the traveling salesman problem. In: Proceedings of the International Conference on Machine Learning, Tahoe, California, pp. 252–260 (1995)
15. Luca Gambardella, M., Taillard, E., Agazzi, G.: A multiple ant system for vehicle routing problem with time windows. New Ideas on Optimization, 285–296 (1999)
16. David, P., Matthieu, C., Arnaud, R.: Image Retrieval over Networks: Active Learning using Ant Algorithm. IEEE Transactions on Multimedia 10(7), 1356–1365 (2008)
17. Stützle, T., Hoosb, H.: MAX–MIN Ant System and Local search for the traveling salesman problem. In: Proceedings of the IEEE International Conference on Evolutionary Computation, Pistcataway, USA, pp. 309–314 (1997)
18. Stützle, T., Hoosb, H.: MAX–MIN Ant System. Future Generation Computer Systems 16(8), 927–935 (2000)
19. Hu, X.M., Zhang, J., Xiao, J., Li, Y.: Protein Folding in Hydrophobic-Polar Lattice Model: A Flexible Ant-Colony Optimization Approach. Protein and Peptide Letters 15(5), 469–477 (2008)

Directional Discriminant Analysis for Image Feature Extraction

Lijun Yan, Jeng-Shyang Pan, and Xiaorui Zhu

Shenzhen Graduate School, Harbin Institute of Technology, University Town, 518055, Shenzhen, China
yanlijun@126.com, jengshyangpan@gmail.com, xiaoruizhu@hitsz.edu.cn

Abstract. A novel subspace learning algorithm based on nearest feature line and directional derivative gradient is proposed in this paper. The proposed algorithm combines neighborhood discriminant nearest feature line analysis and directional derivative gradient to extract the local discriminant features of the samples. A discriminant power criterion based on nearest feature line is used to find the most discriminant direction in this paper. Some experiments are implemented to evaluate the proposed algorithm and the experimental results demonstrate the effectiveness of the proposed algorithm.

Keywords: Directional derivative gradient, feature extraction, nearest feature line.

1 Introduction

Over the past 20 years, biometric and related technologies have become very popular in person authentication, computer vision and machine learning [1–3]. Many researches on biometric were based on image classification, so a lot of image feature extraction algorithms were proposed. Principal Component Analysis (PCA) [4], Linear Discriminant Analysis (LDA)[5] are some of most popular approaches. However, PCA projects the original samples to a low dimensional space, which is spanned by the eigenvectors associated with the largest eigenvalues of the covariance matrix of all samples. PCA is the optimal representation of the input samples in the sense of minimizing the mean squared error. However, PCA is an unsupervised algorithm, which may lead to a lower recognition accuracy. Linear subspace analysis (LDA) finds a transformation matrix U that linearly maps high-dimensional sample $x \in R^n$ to low-dimension data y by $y = U^T x \in R^m$, where $n > m$. LDA can calculate an optimal discriminant projection by maximizing the ratio of the trace of the between-class scatter matrix to the trace of the within-class scatter matrix. LDA takes consideration of the labels of the input samples and improves the classification ability. However, LDA suffers from the small sample size (SSS) problem. Some algorithms using the kernel trick are developed in recent years [6], such as kernel principal component analysis (KPCA)[7, 8], kernel discriminant analysis (KDA) [9] and Locality Preserving

J.-S. Pan et al. (eds.), *Genetic and Evolutionary Computing,*
Advances in Intelligent Systems and Computing 238,
DOI: 10.1007/978-3-319-01796-9_37, © Springer International Publishing Switzerland 2014

Projection[10] used in many areas[11]. Researchers have developed a series of KDA and related algorithms[12–14].

An alternative way to handle the above problem is to extract features from the face image matrix directly. In resent years, several methods based on matrix are proposed, such as Two-Dimensional Principal Component Analysis (2DPCA) [15] and Two-Dimensional Linear Discriminant Analysis (2DLDA) [16]. 2DPCA and 2DLDA can extract the features in a straightforward manner based on the image matrix projection. And these algorithms, not only greatly reduce the computational complexity, but also enhance the recognition effectiveness. Many works based on matrix were presented in these years [17].

The above algorithms are based on Euclidean Distance. Nearest feature line (NFL) [18] is a classifier, proposed by Li in 1998, firstly. In particular, it performs better when only limited samples are available for training. The basic idea underlying the NFL approach is to use all the possible lines through every pair of feature vectors in the training set to encode the feature space in terms of the ensemble characteristics and the geometric relationship. As a simple yet effective algorithm, the NFL has shown its good performance in face recognition, audio classification, image classification, and retrieval. The NFL takes advantage of both the ensemble and the geometric features of samples for pattern classification. Some improved algorithms were proposed in the recent years [19, 20].

While NFL has achieved reasonable performance in data classification, most existing NFL-based algorithms just use the NFL metric for classification and not in the learning phase. While classification can be enhanced by NFL to a certain extent, the learning ability of existing subspace learning methods remains to be poor when the number of training samples is limited. To address this issue, a number of enhanced subspace learning algorithms based on the NFL metric have been proposed, recently. For example, Zheng et al. proposed a Nearest Neighbour Line Nonparametric Discriminant Analysis (NFL-NDA) [21] algorithm, Pang et al. presented a Nearest Feature Line-based Space (NFLS)[22] method, Lu et al. proposed the Uncorrelated Discriminant Nearest Feature Line Analysis (UDNFLA) [23], Yan et al. proposed the Neighborhood Discriminant Nearest Feature Line Analysis (NDNFLA) [24] and some improve algorithms based on NFL [25].

However, the most subspace learning based image feature extraction algorithms only extract the statistical features of images and ignore the features of images as two-dimensional signals. In this paper, a novel image feature extraction algorithm, named Directional Discriminant Analysis (DDA), is proposed. Its effectiveness of the proposed method is verified by some experiments on AR face database. The rest of the paper is organized as follows. In section 2 introduces some preliminaries. In section 3, we give the presentation of the proposed method. In section 4, the experiments are implemented to justify the superiority of the proposed algorithm. And conclusions are made in section 5.

2 Preliminaries

2.1 Directional Derivative Gradient

Given a signal f and a direction vector $(\sin\theta, \cos\theta)$, the first directional derivative [26] $f'_\theta(r,c)$ of f in the direction θ can be treated as the component of the
gradient ∇f along the direction vector, that is,

$$f'_\theta(r,c) = \frac{\partial f}{\partial r}\sin\theta + \frac{\partial f}{\partial c}\cos\theta \qquad (1)$$

Then the points in $I_{\alpha,\beta}$ a digital line with slope α and intercept β. Given an α,
the digital lines with different β can cover the plane.

2.2 Nearest Feature Line

Nearest feature line is a classifier [18]. It is first presented by Stan Z. Li and
Juwei Lu. Given a training samples set, $X = \{x_n \in R^M : n = 1, 2, \cdots, N\}$,
denote the class label of x_i by $l(x_i)$, the training samples sharing the same
class label with x_i by $P(i)$, and the training samples with different label with
x_i by $R(i)$. NFL generalizes each pair of prototype feature points belonging to
the same class: $\{x_m, x_n\}$ by a linear function $L_{m,n}$, which is called the feature
line. The line $L_{m,n}$ is expressed by the span $L_{m,n} = sp(x_m, x_n)$. The query x_i
is projected onto $L_{m,n}$ as a point $x^i_{m.n}$. This projection can be computed as

$$x^i_{m,n} = x_m + t(x_n - x_m) \qquad (2)$$

where $t = [(x_i - x_n)(x_m - x_n)]/[(x_m - x_n)^T(x_m - x_n)]$.

The Euclidean distance of x_i and $x^i_{m,n}$ is termed as FL distance. The less
the FL distance is, the bigger probability that x_i belongs to the same class as
x_m and x_n is. Fig. 1 shows a sample of FL distance. In Fig. 1, the distance
between y_p and the feature line $L_{m,n}$ equals to the distance between y_q and
y_p, where y_p is the projection point of y_q to the feature line $L_{m,n}$.

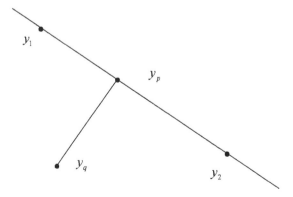

Fig. 1. Feature Line Distance

2.3 NDNFLA

Let's introduce two definitions firstly.

Definition 1. *Homogeneous neighborhoods: For a sample x_i , its k nearest homogeneous neighborhood N_i^o is the set of k most similar data which are in the same class with x_i .*

Definition 2. *Heterogeneous neighborhoods: For a sample x_i , its k nearest Heterogeneous neighborhoods N_i^e is the set of k most similar data which are not in the same class with x_i .*

In NDNFLA approach, the optimization problem is as follows:

$$
\max J(W) = \left(\sum_{i=1}^{N} \frac{1}{NC_{|N_i^e|}^2} \sum_{x_m, x_n \in N_i^e} \left\| W^T x_i - W^T x_{m,n}^i \right\|^2 \right.
$$
$$
\left. - \sum_{i=1}^{N} \frac{1}{NC_{|N_i^o|}^2} \sum_{x_m, x_n \in N_i^o} \left\| W^T x_i - W^T x_{m,n}^i \right\|^2 \right)
$$
(3)

Using matrix computation,

$$
\sum_{i=1}^{N} \frac{1}{NC_{|N_i^e|}^2} \sum_{x_m, x_n \in N_i^e} \left\| W^T x_i - W^T x_{m,n}^i \right\|^2
$$
$$
= \sum_{i=1}^{N} \frac{1}{NC_{|N_i^e|}^2} \sum_{x_m, x_n \in N_i^e} \operatorname{tr}[W^T (x_i - x_{m,n}^i)(x_i - x_{m,n}^i)^T W]
$$
(4)
$$
= \operatorname{tr}\{W^T \sum_{i=1}^{N} \frac{1}{NC_{|N_i^e|}^2} \sum_{x_m, x_n \in N_i^e} [(x_i - x_{m,n}^i)(x_i - x_{m,n}^i)^T] W\}
$$

where tr denotes the trace of a matrix. Similar with the above,

$$
\sum_{i=1}^{N} \frac{1}{NC_{|N_i^o|}^2} \sum_{x_m, x_n \in N_i^o} \left\| W^T x_i - W^T x_{m,n}^i \right\|^2
$$
$$
= \operatorname{tr}\{W^T \sum_{i=1}^{N} \frac{1}{NC_{|N_i^o|}^2} \sum_{x_m, x_n \in N_i^o} [(x_i - x_{m,n}^i)(x_i - x_{m,n}^i)^T] W\}
$$
(5)

Then the problem becomes

$$
\max J(W) = \operatorname{tr}[W^T (A - B)W]
$$
(6)

where

$$
A = \sum_{i=1}^{N} \frac{1}{NC_{|N_i^e|}^2} \sum_{x_m, x_n \in N_i^e} [(x_i - x_{m,n}^i)(x_i - x_{m,n}^i)^T]
$$
(7)

$$
B = \sum_{i=1}^{N} \frac{1}{NC_{|N_i^o|}^2} \sum_{x_m, x_n \in N_i^o} [(x_i - x_{m,n}^i)(x_i - x_{m,n}^i)^T]
$$
(8)

A length constraint $w^T w = 1$ is imposed on the proposed NDNFLA. Then, the optimal projection W of NDNFLA can be obtained by solving the following eigenvalue problem.

$$(A - B)w = \lambda w \tag{9}$$

Let $w_1, w_2, \cdots, w_q$ be the eigenvectors of formula(9) corresponding to the q largest eigenvalues ordered according to $\lambda_1 \geq \lambda_2 \geq \cdots \geq \lambda_q$. An $M \times q$ transformation matrix $W = [w_1, w_2, \cdots, w_q]$ can be obtained to project each sample $M \times 1$ x_i into a feature vector $q \times 1$ y_i as follows:

$$y_i = W^T x_i, \qquad i = 1, 2, \cdots, N \tag{10}$$

3 The Proposed Algorithm

In this paper, a Directional Discriminant Analysis (DDA) is proposed to extract the directional features of the images. In DDA approach, the directional features with a most discriminant direction are extracted based on directional derivative gradient of images.

Firstly, discriminant power criterion based on NFL is proposed in this section. Let $X = \{X_1, X_2, \cdots, X_N\} \subset R^{d_1 \times d_2}$ denote the prototype sample set. X_i^θ denotes the directional derivative gradient of images X_i with the direction θ. Then transform the matrix X_i^θ to a vector $x_i^\theta \in R^D$, where $D = d_1 \times d_2$. Denote $X_\theta = \{x_1^\theta, x_2^\theta, \cdots, x_N^\theta\} \subset R^D$.

Let $l_i(\theta)$ denote the number of FLs in the same class with x_i^θ among its k nearest feature lines in X_θ. Then, let $L(\theta) = \sum_{i=1}^{N} l_i(\theta)$. At last, let

$$J_{DP}(\theta) = \frac{L(\theta)}{k * N} \tag{11}$$

According to the formula(11), it is clear that the bigger $J_{DP}(\theta)$ is, the more discriminant features are. So the most discriminant direction can be find by maximizing J_{DP}.

The main idea of the proposed feature extraction algorithm, is to extract the local discriminant features from the most discriminant direction. The detailed procedure of proposed method is as follows:

Training stage:

Step 1, using the discriminant power criterion based on NFL, find the most discriminant direction θ_0;

Step 2, perform the directional derivative gradient operator with direction θ_0 to all the prototype samples to get a new prototype samples set;

Step 3, apply NDNFLA to find the optimal transformation matrix W on the new prototype samples set;

Step 4, Extract the features of prototype samples following formula(10).

Classification stage:

Step 1, perform the directional derivative gradient operator with direction θ_0 to the query sample;

Step 2, Extract the features of query following formula(10);

Step 3, Classify with NFL.

4   Experimental Results

AR face database [27] was created by Aleix Martinez and Robert Benavente in the Computer Vision Center (CVC) at the U.A.B. It contains over 4,000 color images corresponding to 126 people's faces (70 men and 56 women). Images feature frontal view faces with different illumination conditions, facial expressions, and occlusions (sun glasses and scarf). The pictures were taken at the CVC under strictly controlled conditions. Each person participated in two sessions, separated by two weeks (14 days) time. The same pictures were taken in both sessions. In the following experiments, only nonoccluded images of 120 people in AR face database are selected. Five images per person are randomly selected for training and the other images are for testing. This system also runs 20 times. Some samples of AR face database are shown in Fig. 2. Table 1 tabulates the maximum average recognition rate (MARR) of these algorithms on AR face database. Clearly, MARR of the proposed algorithm is higher than other approaches.

Fig. 2. Some samples of AR face database

Table 1. MARR of different algorithms on AR face database

Algorithms	MARR	Feature dimension
fisherface	0.9481	120
PCA+NN	0.7604	120
PCA+NFL	0.8521	190
UDNFLA	0.9353	120
NFLS	0.9126	190
NDNFLA	0.9690	150
Proposed algorithm	0.9735	130

5   Conclusions

In this paper, a novel image feature extraction algorithm, called Directional Discriminant Analysis is proposed. DDA is based on the directional derivative gradient and nearest feature line. It can find the optimal direction adaptively to

Acknowledgments. This work was supported in part by the Peacock Project of Shenzhen, Intelligent Vehicle Cloud Client System, under Project NO. KQC20110 9020055A, Shenzhen Strategic Emerging Industries Program under Grants No. ZDSY20120613125016389.

References

1. Li, J.-B., Chu, S.-C., Pan, J.-S., Jain, L.C.: Multiple Viewpoints Based Overview for Face Recognition. Journal of Information Hiding and Multimedia Signal Processing 3, 352–369 (2012)
2. Krinidis, S., Pitas, I.: Statistical Analysis of Human Facial Expressions. Journal of Information Hiding and Multimedia Signal Processing 1, 241–260 (2010)
3. Zhang, L., Zhang, L., Zhang, D., Guo, Z.: Phase Congruency Induced Local Features for Finger-Knuckle-Print Recognition. Pattern Recognition 45, 2522–2531 (2012)
4. Wu, J., Zhou, Z.H.: Face Recognition with One Training Image Per Person. Pattern Recognition Letters 23, 1711–1719 (2002)
5. Belhumenur, P.N., Hepanha, J.P., Kriegman, D.J.: Eigenfaces vs Fisherfaces: Recognition Using Class Specific Linear Projection. IEEE Trans. Pattern Analysis Machine Intelligence 19, 711–720 (1997)
6. Li, J.-B., Pan, J.-S., Lu, Z.-M.: Kernel Optimization-Based Discriminant Analysis for Face Recognition. Neural Computing and Applications 18, 603–612 (2009)
7. Yang, J., Frangi, A.F., Yang, J.-Y., Zhang, D., Jin, Z.: KPCA Plus LDA: A Complete Kernel Fisher Discriminant Framework for Feature Extraction and Recognition. IEEE Trans. Pattern Analysis and Machine Intelligence 27, 230–244 (2005)
8. Li, J.-B., Yu, L.-J., Sun, S.-H.: Refined Kernel Principal Component Analysis Based Feature Extraction. Chinese Journal of Electronics 20, 467–470 (2011)
9. Pan, J.-S., Li, J.-B., Lu, Z.-M.: Adaptive Quasiconformal Kernel Discriminant Analysis. Neurocomputing 71, 2754–2760 (2008)
10. Li, J.-B., Pan, J.-S., Chen, S.-M.: Kernel Self-Optimized Locality Preserving Discriminant Analysis for Feature Extraction and Recognition. Neurocomputing 74, 3019–3027 (2011)
11. Li, J.-B., Pan, J.-S., Chu, S.-C.: Kernel Class-wise Locality Preserving Projection. Information Sciences 178, 1825–1835 (2008)
12. Li, J.-B., Gao, H., Pan, J.-S.: Common Vector Analysis of Gabor Features with Kernel Space Isomorphic Mapping for Face Recognition. International Journal of Innovative Computing, Information and Contro 6(10) (October 2010)
13. Xu, Y., Zhang, D., Jin, Z., Li, M., Yang, J.Y.: A Fast Kernel-Based Nonlinear Discriminant Analysis for Multi-Class Problems. Pattern Recognition 39, 1026–1033 (2006)
14. Xu, Y., Zhang, D.: Represent and Fuse Bimodal Biometric Images at the Feature Fevel: Complex-Matrix-Based Fusion Scheme. Opt. Eng. 49, 037002 (2010)
15. Yang, J., Zhang, D., Frangi, A.F., Yang, J.: Two-dimensional PCA: a new approach to appearance-based face representation and recognition. IEEE Trans. Pattern Analysis and Machine Intelligence 26, 131–137 (2004)
16. Yang, J., Zhang, D., Yong, X., Yang, J.Y.: Two-dimensional discriminant transform for face recognition. Pattern Recognition 37, 1125–1129 (2005)
17. Yan, L., Pan, J.-S., Chu, S.-C., Muhammad, K.K.: Adaptively weighted sub-directional two-dimensional linear discriminant analysis for face recognition. Future Generation Computer Systems 28, 232–235 (2012)

18. Li, S.Z., Lu, J.: Face Recognition Using the Nearest Feature Line Method. IEEE Trans. Neural Networks 10, 439–443 (1999)

19. Feng, Q., Pan, J.-S., Yan, L.: Restricted Nearest Feature Line with Ellipse for Face Recognition. Journal of Information Hiding and Multimedia Signal Processing 3, 297–305 (2012)

20. Feng, Q., Pan, J.-S., Yan, L.: Nearest Feature Centre Classifier for Face Recognition. Ellectronics Letters 48, 1120–1122 (2012)

21. Zheng, Y.-J., Yang, J.-Y., Yang, J., Wu, X.-J., Jin, Z.: Nearest Neighbour Line Nonparametric Discriminant Analysis for Feature Extraction. Electronics Letters 42, 679–680 (2006)

22. Yang, Y., Yuan, Y., Li, X.: Generalised Nearest Feature Line for Subspace Learning. Electronics Letters 43, 1079–1080 (2007)

23. Lu, J., Tan, Y.P.: Uncorrelated Discriminant Nearest Feature Line Analysis for Face Recognition. IEEE Signal Processing Letter 17, 185–188 (2010)

24. Yan, L., Pan, J.-S., Zheng, W., Chu, S.-C., Roddick, J.F.: Neighborhood Discriminant Nearest Feature Line Analysis for Face Recognition. Journal of Internet Technology 14, 344–347 (2013)

25. Yan, L., Wang, C., Chu, S.-C., Pan, J.-S.: Discriminant Analysis Based on Nearest Feature Line. In: Yang, J., Fang, F., Sun, C. (eds.) IScIDE 2012. LNCS, vol. 7751, pp. 356–363. Springer, Heidelberg (2013)

26. Zuniga, O.A., Haralick, R.M.: Integrated Directional Derivative Gradient Operator. IEEE Trans. Systems, Man, and Cybernetics 17, 508–517 (1987)

27. Martinez, A.M., Benavente, R.: The AR Face Database. CVC Technical Report 24 (1998)

Novel Matrix Based Feature Extraction Method for Face Recognition Using Gaborface Features

Qi Zhu[1,2], Yong Xu[1,2,*], Yuwu Lu[1,2], Jiajun Wen[1,2], Zizhu Fan[3], and Zhengming Li[4]

[1] Key Laboratory of Network Oriented Intelligent Computation, Shenzhen China
[2] Harbin Institute of Technology, Shenzhen Graduate School, Shenzhen China
[3] School of Basic Science, East China Jiaotong University, Nanchang, China
[4] Guangdong Industrial Training Center, Guangdong Polytechnic Normal University, Guangzhou, China
yongxuhitsz@163.com

Abstract. This study proposes a framework to integrate the Gaborface features and the matrix based feature extraction method for face recognition. In this framework, we first select a subset of Gaborfaces to construct the optimal ensemble Gaborface. Then, a two-phase matrix based feature extraction method, i.e.: two-dimensional linear discriminant analysis (2DLDA) plus multi-subspaces principle component analysis (MSPCA), is developed to directly and effectively extract features from the optimal ensemble Gaborface matrixes. Experiment results on ORL and AR face datasets demonstrate the effectiveness of our method.

Keywords: face recognition, Gaborface, Gabor filter optimization, matrix based feature extraction.

1 Introduction

Many subspace analysis based feature extraction methods have been proposed for face recognition in the past few decades [1-7]. Eigenface [5], Fisherface [7] are two of the most well-known and widely used methods in this filed. In the two methods, the 2D facial image matrix must be previously transformed into high-dimensional 1D vector. The matrix-to-vector operation will induce a large scale covariance matrix. As a result, the computation of the eigen-vectors of the two methods is very time-consuming. More importantly, this operation destroys the structural information embedding in the image matrix. In 2003, an improved PCA technique named two-dimensional PCA (2DPCA) was proposed by Yang et al. [8]. The 2DPCA directly extracts the features from the image matrix by projecting the image matrix along the projection axes that are the eigen-vectors of the 2D images covariance matrix. As the covariance matrix of 2DPCA has a lower dimensionality than that of PCA, 2DPCA is computationally more efficient than PCA. Motivated by 2DPCA, Xu et al. proposed

---

* Corresponding author.

J.-S. Pan et al. (eds.), *Genetic and Evolutionary Computing,*
Advances in Intelligent Systems and Computing 238,
DOI: 10.1007/978-3-319-01796-9_38, © Springer International Publishing Switzerland 2014

to combine two solution schemes of 2DPCA to extract features from matrixes [9]. Gao and Zhang et al. propose the two-dimensional independent component analysis (2DICA) [10] that directly evaluates the two correlated demixing matrices from the image matrix without matrix-to-vector transformation. The 2DICA greatly alleviates the SSS problem [11] and the curse of dimensionality, reduces the computational complexity, and simultaneously exploits the spatial and structural information embedded in image. Li et al. combined the idea of matrix based feature extraction and Fisher Linear Discriminant Analysis (FLDA) [12] for face recognition and proposed the two-dimensional LDA (2DLDA) [13]. In general, 2DLDA does not suffer from the SSS problem in face recognition, whereas FLDA method encounters. In 2010, Xu et al. proposed the complex matrix based LDA for bimodal biometrics [14]. 2DPCA, 2DICA, 2DLDA and the complex matrix based LDA can be all referred to as matrix based feature extraction methods.

We propose a new matrix based feature extraction method named MSPCA. In MSPCA, the feature space is divided into several subspaces without overlap, and Karhunen-Loeve (K-L) transform is conducted in each subspace. MSPCA has complete theoretical basis and some advantages over the other PCA based methods, e.g.: we have proven that the MSPCA can achieve larger divergence than that of 2DPCA. Motivated by the bidirectional matrix based feature extraction, we present a two-phase method, i.e.: 2DLDA + MSPCA, to extract the most representative and discriminative information from samples. For the first phase, we use the 2DLDA working in row direction to extract the discriminant information from the matrixes. We get the feature matrix of 2DLDA by projecting the training samples $\{A_1, A_2, ..., A_l\}$, where each sample is the $m \times n$ matrix, onto d projection axes $x_1, x_2, ..., x_d$. For the second phase, we use the MSPCA proposed in this paper to extract the most representative information from the feature matrixes in the column direction. For the details: by the projection axes x_j, the projection results from training samples will span a subspace $S_j = span\{A_1 x_j, A_2 x_j, ..., A_l x_j\}$ ($j = 1, 2, ..., d$). Then, K-L transform is implemented in each subspace ϕ_j ($j = 1, 2, ..., d$). If r primary principle components are included ($r < m$), the size of the final feature matrix extracted by our method, i.e.:($r \times d$), is much smaller than that of 2DLDA, i.e.: ($m \times d$).

2   Optimal Ensemble Gaborface Selection

One of the potential problems of the ensemble Gaborface [15] is that it is redundant and too high-dimensional. For example, if the size of the facial image is 100×80, the size of the ensemble Gaborface will reach (100×5)×(80×8), which is incredibly large for the matrix based feature extraction methods. Additionally, no evidence is found that every Gaborface is helpful for improving the classification accuracy. It is meaningful to conduct feature dimension reduction on the ensemble Gaborface. Dozens of Gabor filters are usually adopted for constructing the ensemble Gaborface, so exhaustive search is too time-consuming to get the solution. We design a heuristic search algorithm with forward selection. In our algorithm, the sum of the absolute values of the eigen-values of $G_w^{-1}G_b$ is used to evaluate the quality of the selected subset, where G_w, G_b are the between-class and within-class scatter matrices of 2DLDA respectively. Employing this criterion ensures that the new ensemble Gaborface has the maximum Fisher's ratio, which is favorable for improving the classification accuracy of the training samples. The Gabor filter selection algorithm follows these steps:

Step 1. Initialize the parameter v that is the number of the filters to be selected.

Step 2. According to the above criterion, select the first Gabor filter ψ_{s_1} from the Gabor filter set { $\psi_1, \psi_2, ..., \psi_k$ }.

Step 3. There are $k-1$ choices for selecting the second Gabor filter. Denote the Gaborface corresponding to ψ_i as O_i. For each choice, such as ψ_i, we ensemble the two Gaborfaces of each training sample as the matrix [O_{s_1}, O_i]. Then, we calculate the value of the criterion function with the choice of ψ_i. ψ_{s_2} is selected as the second optimal Gabor filter, which has the maximum criterion function value among the $k-1$ choices.

Step 4. When the number of the Gabor filters selected reaches v, the algorithm terminates, otherwise, go to Step 5.

Step 5. Supposing t Gabor filters has been selected, there are $k-t$ choices for selecting the $(t+1)$th Gabor filter. For each choice, such as ψ_i, we ensemble the $t+1$ Gaborfaces of each training sample as the matrix [$O_{s_1}, O_{s_2}, ..., O_{s_t}, O_i$]. Then, we calculate the criterion function with the choice of ψ_i. $\psi_{s_{(t+1)}}$ is selected as the $(t+1)$th Gabor filter, which has the maximum criterion function value among the $k-t$ choices.

By the above algorithm, v optimal Gabor filters ($\psi_{s_1}, \psi_{s_2}, ..., \psi_{s_v}$) are selected. Our optimal ensemble Gaborface is constructed by the form of [$O_{s_1}, O_{s_2}, ..., O_{s_v}$], whose size is much smaller than that of the naïve ensemble Gaborface.

3   The Description of Our Feature Extraction Method

In this section, we propose a two-phase matrix based feature extraction method, i.e.: 2DLDA +MSPCA, for directly extracting features from the optimal ensemble Gaborfaces.

3.1   Overview of 2DLDA

Suppose the l training samples $(A_1, A_2, ..., A_l)$, mentioned in section 1, belong to c classes, and each class includes l_c samples. Let y_i be the feature vector extracted from the training sample A_i. y_i is obtained by projecting A_i onto x using:

$$y_i = A_i x \quad (i = 1, 2, \ldots, l) \tag{1}$$

The between-class and within-class scatter matrices G_b and G_w are defined as follows:

$$G_b = \sum_{j=1}^{c} l_j (u_j - \overline{u})^T (u_j - \overline{u}) \tag{2}$$

$$G_w = \sum_{j=1}^{c} \sum_{k \in class\, c}^{l_c} (A_k - \overline{u_j})^T (A_k - \overline{u_j}) \tag{3}$$

$\overline{u_j}$ and $\overline{u}$ denote the mean of j th class and the mean of whole training set, respectively. 2DLDA uses the traces of G_b and G_w to characterize the Fisher's criterion $J(x)$ by:

$$J(x) = \frac{x^T G_b x}{x^T G_w x} \tag{4}$$

The optimal projection x_{opt} is chosen, when the criterion is maximized. We have:

$$x_{opt} = \arg \max_{x} J(x) \tag{5}$$

For face recognition problems, G_w in 2DLDA is always nonsingular. $J(x)$ is maximized when the projection vector is the eigen-vector of $G_w^{-1} G_b$. Generally, it is not enough to have only one Fisher optimal projection axis. We usually need to select a set of projection axes, $x_1, x_2, \ldots, x_d$, satisfying:

$$\begin{cases} x_1, x_2, \ldots, x_d = \arg \max_{x} J(x) \\ x_i^T x_j = 0 \ (i, j = 1, 2, \ldots, d \text{ and } i \neq j) \end{cases} \tag{6}$$

2DLDA uses the feature matrix to present the feature extract result. For an image sample A, the feature matrix Y obtained by 2DLDA can be computed as following equation:

$$Y = AX \tag{7}$$

Where $Y = [y_1, y_2, \ldots, y_d]$ and $X = [x_1, x_2, \ldots, x_d]$.

3.2   MSPCA

We propose MSPCA to extract the representative information from each discriminant feature vector of 2DLDA. It works as follows: When we have obtained the projection axes $(x_1, x_2, ..., x_d)$ of 2DLDA, we projected all the training samples onto $(x_1, x_2, ..., x_d)$ by Eq. (1). For the subspace S_c mentioned in Section 1, it is spanned by $\{A_1 x_c, A_2 x_c, ..., A_l x_c\}$. The covariance matrix for the l vectors in S_c is:

$$G_c = \frac{1}{l} \sum_i^l (y_c(i) - \overline{y_c})(y_c(i) - \overline{y_c})^T, \tag{8}$$

where $y_c(i)$ is the cth Fisher discriminant vector of $y(i)$, and $\overline{y_c} = \sum_i^l y_c(i)$.

Let $W_c = (w_c^1, w_c^2, ..., w_c^r)$ be the K-L transform matrix formed by r eigen-vectors corresponding to the r largest eigen-values of G_c. For a feature vector y_c obtained by 2DLDA, its reduced feature vectors can be obtained by projecting it onto W_c^T. The new feature matrix obtained by our method is:

$$Y' = [y'_1, y'_2, ..., y'_d] = [W_1^T A x_1, W_2^T A x_2, ..., W_d^T A x_d] \tag{9}$$

Clearly, the size of the feature matrix obtained by our method is much smaller than that of 2DLDA $(r \times d < m \times d)$.

It is easy to reconstruct the image A by:

$$\tilde{A} = [W_1 y'_1, W_2 y'_2, ..., W_d y'_d] X^T \tag{10}$$

If $d < n$ and $r < m$, the reconstructed image $\tilde{A}$ is an approximation for the training image A. If $d = n$ and $r = m$, the training image can be completely reconstructed by Eq. (10).

The nearest neighbor classifier is employed for classification. Here, the distance between two arbitrary feature matrixes $Y'(j)$ and $Y'(k)$ extracted from samples A_j and A_k respectively can be calculated by:

$$dis(Y'(j), Y'(k)) = \sum_{c=1}^d \| y'_c(j) - y'_c(k) \|^2 \tag{11}$$

Where $y'_c(j)$ is the cth column vector of $Y'(j)$ and $\| y'_c(j) - y'_c(k) \|^2$ denotes the Euclidean distance between the two vectors $y'_c(j)$ and $y'_c(k)$.

4   Experiments

In this section, we assess the performance of our method on two well-known datasets: AR and ORL. As the comparisons, five state of the art face recognition methods including Fisherface, Gaborface+PCA, Gaborface+LDA, 2DPCA and 2DLDA also perform on the datasets.

The first face image dataset we used in experiments is the ORL dataset (http://www.uk.research.att.com/facedataset.html). ORL dataset consists of 400 frontal faces: 10 tightly cropped images of 40 individuals with variation in pose, illumination, facial expression (open/closed eyes, smiling/not smiling) and facial details (glasses/no glasses). The size of each image is 112×92. Figure 1 depicts the images of a subject in ORL dataset. We choose the following 5 images of each subject, i.e. (b) (d) (f) (h) (j), as the training sample, and the remainder is used as testing samples.

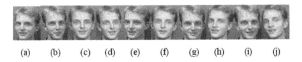

(a)   (b)   (c)   (d)   (e)   (f)   (g)   (h)   (i)   (j)

Fig. 1. Images of a subject in ORL dataset

Comparisons of recognition rates of our method and the other five methods with different number of eigen-vectors (p =5,6,...,25) are shown in Figure 2. For our method, the value of horizontal axis in Figure 2 denotes the discriminant vector number in 2DLDA phase and we choose 10 primary components in MSPCA phase. So, every training sample is transformed to $10 \times p$ matrix in our method, where p is the discriminant vector number we used. As we can see from Figure 2, our method performs better than the others in all cases. Table 1 shows the average and top recognition rates of all methods. Compared with the results from Table 1, the average recognition rate of 97.73% achieved by our method has 3.44 percentage points higher than the second highest average recognition rate achieved by Fisherface method. The top average recognition rate of 99% obtained by our method is also significantly higher than that of the others.

Table 1. Comparison of average and top recognition rate on ORL dataset

method	Average recognition rate (%)	Top recognition rate (%) (feature extracted number)
Our method	97.73	99.00 (10×13)
2DLDA	91.21	92.50 (112×9)
2DPCA	88.00	88.50 (112×8)
Fisherface	94.29	97.00 (14, 16 or 17)
Gaborface + LDA	93.60	96.00 (23)
Gaborface +PCA	90.23	93.50 (30)

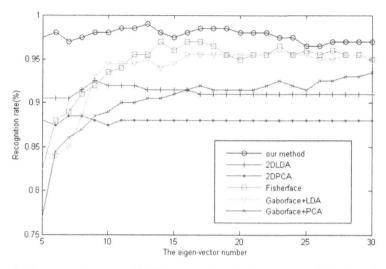

Fig. 2. The recognition rate with different number of eigen-vectors of all the methods on ORL dataset

The public face image dataset, i.e. the AR dataset, is also used in our experiments. It contains over 4000 face images of 126 people (70 men and 56 women) including frontal views of faces with different facial expressions, lighting conditions and occlusions. The face portion of each image is manually cropped and then normalized to the size of 50×40. We used only the images of these 120 subjects and the 14 non-occluded face images of each subject to test different methods in our experiment. Figure 3 shows the images of one subject we used.

Fig. 3. The images of one subject in AR dataset

Table 2. Average recognition rate of different methods on AR dataset

Method	2 training samples per class (%)	3 training samples per class (%)	4 training samples per class (%)	5 training samples per class (%)
Our method	84.72	87.95	92.69	94.14
2DLDA	72.43	72.95	80.56	92.96
2DPCA	71.74	72.05	78.38	81.94
Fisherface	83.54	87.12	93.36	92.42
Gaborface + LDA	81.25	85.23	92.33	93.11
Gaborface +PCA	82.50	87.58	89.89	90.20

To examine how the recognition rate changes depending on the number of training samples of each class, we pick k (k =2,3,4,5) images of each subject at random for training. Remaining $14-k$ images are employed for testing. 10 possible selections of k training images per class are chosen in the experiments, and the experiments are repeated 10 times with these selections. We use 30 eigen-vectors in all methods. Table 2 shows us the comparisons of our method and the other five methods on AR dataset. From Table 2, we can find our method has the highest average recognition rates of 84.72%, 87.95% and 94.14% with 2, 3 and 5 training samples per class among all methods. With 4 training samples per class, our method achieves the second highest average recognition rate of 92.69% that is slightly lower than that of Fisherface.

5   Conclusions

This paper attempts to use matrix based feature extraction method with Gaborface features for face recognition. For achieving this aim, we proposed an optimal ensemble Gaborface representation method and a two-phase matrix based feature extraction method, and then we integrate them to present a novel framework for face recognition. This framework offers a feasible 2D Gaborface representation for directly applying matrix based feature extraction method, and extracts the most representative and discriminative information from Gaborface features with a small scale feature matrix. The results of the experiments carried on ORL and AR demonstrate that our method is effective.

Acknowledgments. This paper is partially supported by Shenzhen Municipal Science and Technology Innovation Council (Nos. JC201005260122A, JCYJ2012061315 3352732 and CXZZ20120613141657279), and the National Natural Science Foundation of China under Grant No. 61263032.

References

1. Xu, Y., Zhong, A., Yang, J., Zhang, D.: LPP solution schemes for use with face recognition. Pattern Recognition 43(12), 4165–41761 (2010)
2. Fan, Z., Xu, Y., Zhang, D.: Local linear discriminant analysis framework using sample neighbors. IEEE Transactions on Neural Networks 22(7), 1119–11321 (2011)
3. Chen, Y.-W., Xu, R., Ushikome, A.: Serially-connected dual 2d pca for efficient face representation and face recognition. International Journal of Innovative Computing, Information and Control 5(11), 4367–4372
4. Xu, Y., Zhang, D., Yang, J., Jin, Z., Yang, J.: Evaluate dissimilarity of samples in feature space for improving KPCA. International Journal of Information Technology & Decision Making 10(3), 479–495 (2011)
5. Turk, M., Pentland, A.: Eigenfaces for recognition. J. Cognitive Neurosci. 3(1), 71–86 (1991)

6. Xu, Y., Zhang, D., Jin, Z., Li, M., Yang, J.-Y.: A fast kernel-based nonlinear discriminant analysis for multi-class problems. Pattern Recognition 39(6), 1026–1033 (2006)
7. Belhumeur, P.N., Hespanha, J.P., Kriegman, D.J.: Eigenfaces vs. Fisherfaces: Recognition using class specific linear projection. IEEE Trans. Pattern Anal. Machine Intell. 19(7), 711–720 (1997)
8. Gao, Q., Zhang, L., Zhang, D.: Face Recognition using FLDA with Single Training Image Per-person. Applied Mathematics and Computation 205, 726–734 (2008)
9. Xu, Y., Zhang, D., Yang, J., Yang, J.Y.: An approach for directly extracting features from matrix data and its application in face recognition. Neurocomputing 71, 1857–1865 (2008)
10. Gao, Q., Zhang, L., Zhang, D., Xu, H.: Independent components extraction from image matrix. Pattern Recognition Letters 31(3), 171–178 (2010)
11. Yang, W., Wang, J., Ren, M., Zhang, L., Yang, J.: Feature extraction using fuzzy inverse FDA. Neurocomputing 72(13-15), 3384–3390 (2009)
12. Zhang, Q., Zhou, C.J., Zhao, J.: Face Recognition Based on FLDA, CPCA and Improved HMM. International Journal of Innovative Computing, Information and Control 6(2), 801–808 (2010)
13. Li, M., Yuan, B.Z.: 2D-LDA: A statistical linear discriminant analysis for image matrix. Pattern Recognition Letters 26, 527–532 (2005)
14. Xu, Y., Zhang, D.: Represent and fuse bimodal biometric images at the feature level: complex-matrixbased fusion scheme. Optical Engineering 49(3) (2010)
15. Wang, L., et al.: 2D Gaborface representation method for face recognition with ensemble and multichannel model. Image and Vision Computing 26, 820–828 (2008)
16. Lee, Y.-C., Chen, C.-H.: A Gabor Feature Based Horizontal and Vertical Discriminant for Face Verification. International Journal of Innovative Computing, Information and Control 9(5), 2111–2123 (2013)

Hybrid Digit-Serial Multiplier for Shifted Polynomial Basis of $GF(2^m)$

Chiou-Yng Lee[1], Wen-Yo Lee[1], Che Wun Chiou[2,*],
Jeng-Shyang Pan[3], and Cheng-Huai Ni[1]

[1] Lunghwa University of Science and Technology, Taiwan
[2] Chien Hsin University of Science and Technology
[3] Shenzhen Graduate School, Harbin Institute of Technology, China
pp010@mail.lhu.edu.tw, cwchiou@uch.edu.tw

Abstract. Recently, a shifted polynomial basis is a variation of polynomial basis representation. Such kind basis provides better performance in designing bit-parallel and subquadratic space complexity multipliers over binary extension fields. In this paper, we study a new shifted polynomial basis multiplication algorithm to implement a hybrid digit-serial multiplier. The proposed algorithm effectively integrates classic schoolbook multiplication, Karatsuba multiplication algorithms to reduce computational complexity, and the modular multiplication with the shifted polynomial basis reduction. We note that, comparably, the proposed architecture achieves lower computation time and higher bit-throughput compared to the best known digit-serial multipliers. Our proposed multipliers can be modular, regular, and suitable for very-large-scale integration (VLSI) implementations. The proposed digit-serial architecture makes the hardware implementations of cryptographic systems more high-performance, and are thus much suitable for efficient applications such as the elliptic curve cryptography (ECC) and pairing computation.

1 Introduction

Finite field arithmetic has many important applications in cryptography, such as National Institute of Standards and Technology (NIST) Digital Signature Standard [1] and the IEEE Standard Specifications for Public-Key Cryptography [2]. Arithmetic operations in finite fields have been utilized in the literature for computing the primitives of cryptographic algorithms such as the point multiplication of the elliptic curve cryptography (ECC), see, for instance, [3]. Many research works have been focused on the efficient and low-complexity hardware architectures of the arithmetic units for the ECC. Additionally, cryptographic pairing has been used extensively to derive various security protocols, such as identity based cryptography. For the tradeoff of the time and space complexities between bit-parallel and bit-serial multipliers,[4]. For such protocols, the Weil and Tate pairings based on elliptic curve arithmetic require thousands of additions and multiplications over large finite fields, and have drawn the attention

---

* Corresponding author.

J.-S. Pan et al. (eds.), *Genetic and Evolutionary Computing,*
Advances in Intelligent Systems and Computing 238,
DOI: 10.1007/978-3-319-01796-9_39, © Springer International Publishing Switzerland 2014

of many researchers [5][6]. In VLSI implementations, based on the requirements of the applications in resource-constrained environments, a trade-off is needed to be reached between the speed of computations and the area complexity.

The shifted polynomial basis (SPB) [7] is a variation of polynomial basis representation. This basis multiplication is depended on the factor x^{-v}, where v is a positive integer, and x is the root of irreducible polynomial. In [8], bit-parallel SPB multiplier architectures for trinomials and type-II pentanomials are faster than the well-known polynomial basis and dual basis multipliers. Employing a subquadratic Toeplitz matrix-vector scheme, bit-parallel SPB multiplier is proposed in [9]. Later, based on the SPB representation, different bit-parallel multiplier designs are seen in [10][11].

For trade-off between bit-parallel and bit-serial multipliers, digit-serial non-systolic multipliers with digit-in parallel-out (DIPO) structures are presented in the literature, see, for instance, [12][13][14]. These multipliers have typically the latency of $O(\frac{m}{d})$ clock cycles, where d is the selected digit-size. Digit-serial systolic multipliers with digit-in digit-out (DIDO) structures are also presented in the previous research works such as [15]. Using hybrid architecture, digit-serial SPB multiplier is outlined by [16].

In [17], low-complexity bit-parallel finite field multiplier architectures using Karatsuba-Ofman algorithm (KOA) is presented. Montgomery [18] and Fan-Hasan [19] presented some improved t-term ($4 \leq t \leq 19$) Karatsuba-like formulae. Lately, some finite field multipliers have mixed two-term and three-term formulae within iterative evaluation-multiply-reconstruction (EMR) method to achieve low-complexity architectures, as seen in [20][21][22]. By using recursive two-term formulae, it is shown that the space complexity of bit-parallel multipliers can be reduced from $O(m^2)$ to $O(m^{log_2 3})$.

In this paper, we choose efficient values to construct the generalized formulation of SPB multiplication. Applying the decomposition method of KOM scheme, the proposed algorithm can achieve a novel hybrid digit-serial SPB multiplier over $GF(2^m)$. The latency of the derived multiplier can be reduced from $O(\frac{m}{d})$ into $O(\frac{m}{dn})$, where d is the selected digit-size and n is the number term of Karatsuba formulae. Our proposed multiplier architecture can achieve lower computation time and higher bit-throughput as compared to the best known digit-serial architectures [13][14][23]. It is noted that the proposed multiplier effectively integrates classic schoolbook multiplication, Karatsuba multiplication algorithms to reduce computational complexity, and the modular multiplication with the shifted polynomial basis reduction.

2   Reviewing Hybrid Digit-Serial Multiplier for Shifted Polynomial Basis of GF(2^m)

The hybrid modular multiplication was introduced for the propose of two-way parallel computation [14]. This way is combined by a classic modular multiplication and a SPB multiplication to improve the speed. It is based on the split operand multiplier into two parts within parallel computations, and increases the calculation speed. We outline the main idea of hybrid multiplication as follows.

Let the field GF(2^m) be constructed from an irreducible polynomial $F(x)$ of degree m. And let the field element A be represented by the polynomial basis of GF(2^m), i.e., $A = a_0 + a_1 x + \cdots + a_{m-1} x^{m-1}$, where a_i and b_i for $0 \le i \le m-1$ are 0 or 1. The set $\{1, x, x^2, \cdots, x^{m-1}\}$ is well-known as the polynomial basis (PB) of $GF(2^m)$.

Let v be a positive integer with $0 \le v \le m-1$. The shifted polynomial basis (SPB) of $GF(2^m)$ is defined by the set $\{x^{-v}, x^{1-v}, x^{2-v}, \cdots, x^{m-1-v}\}$ [8]. This basis representation is similar to the polynomial basis, it is possible to represent each field element using the SPB. For example, if A and B are two elements in $GF(2^m)$, one can write

$$\widehat{A} = x^{-v}A = \sum_{i=0}^{m-1} a_i x^{i-v}, \ \widehat{B} = x^{-v}B = \sum_{i=0}^{m-1} b_i x^{i-v}$$

Given the SPB representation, the multiplication C is represented by

$$\widehat{C} = \widehat{A}\widehat{B} \bmod : F(x) = x^{-v}AB \bmod : F(x) = \sum_{i=0}^{m-1} c_i x^{i-v} \qquad (1)$$

The product $\widehat{C}$ is also a field element of degree $m-1$. Let us assume that $q = \lceil \frac{m}{d} \rceil$, where d is the selected digit-size. If m is not a multiple of dq, then a field element can be padded by $(qd-m)$-bit zeros as $A = (a_0, a_1, \cdots, a_{m-1}, \underbrace{0, \ldots, 0}_{qd-m \text{ bits}})$.

Accordingly, an element A can be represented by $A = \sum_{i=0}^{q-1} A_i x^{id}$, where $A_i = a_{id} + a_{id+1}x + \cdots + a_{id+d-1}x^{d-1}$. By using LSD-first multiplication scheme, the product $\widehat{C}$ can be rewritten as

$$\widehat{C} = x^{-v}AB \quad \bmod F(x) = B(A_0 x^{-v} + A_1 x^{d-v} +$$

$$\cdots + A_{q-1} x^{(q-1)d-v}) \bmod F(x) \qquad (2)$$

Consider the factor value v in (2), we choose $v = \lfloor \frac{q}{2} \rfloor d$ to write (2) as

$$\widehat{C} = BA_0 x^{-\lfloor \frac{q}{2} \rfloor d} + BA_1 x^{d-\lfloor \frac{q}{2} \rfloor d} + \cdots + BA_{\lfloor \frac{q}{2} \rfloor - 1} x^{-d}$$

$$+BA_{\lfloor \frac{q}{2} \rfloor} + BA_{\lfloor \frac{q}{2} \rfloor +1} x^d + \cdots + BA_{q-1} x^{\lfloor \frac{q-2}{2} \rfloor d} \bmod \ F(x) \qquad (3)$$

$$= \widehat{C}_0 + \widehat{C}_1$$

From (3), the product C is included by two parts. $\widehat{C}_0 = BA_0 x^{-\lfloor \frac{q}{2} \rfloor d} + BA_1 x^{d-\lfloor \frac{q}{2} \rfloor d} + \cdots + BA_{\lfloor \frac{q}{2} \rfloor -1} x^{-d}$ is the negative powers of x; and $\widehat{C}_1 = BA_{\lfloor \frac{q}{2} \rfloor} + BA_{\lfloor \frac{q}{2} \rfloor +1} x^d + \cdots + BA_{q-1} x^{\lfloor \frac{q-2}{2} \rfloor d}$ is the positive powers of x. We use a similar approach for

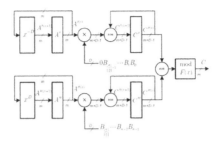

Fig. 1. Tradtional Hybrid digit-serial SPB multiplier [14]

digit-serial multiplication algorithm in [13], Algorithm 1 shows the hybrid digit-serial SPB multiplication. Fig. 1 shows a digit-serial multiplier over $GF(2^m)$ based on Algorithm 1. It consists of two multiplier cores, three reduction operations ($Bx^d \bmod F(x)$, $Bx^{-d} \bmod F(x)$ and $\overline{C} \bmod F(x)$), and one $(m + d)$-bit adder. Two multiplier cores computes two-steps 3 and 4 computations. In the initial step, two registers $< B' >$ and $< B'' >$ are initialized by the element B, and two registers $< \widehat{C}_0 >$ and $< \widehat{C}_1 >$ are initialized by zero. According to SPB multiplication of (3), after $\lceil m/2d \rceil + 1$ clock cycles, two registers $< \widehat{C}_0 >$ and $< \widehat{C}_1 >$ provide the calculation of a classic modular multiplication and a SPB multiplication, respectively. And the final reduction in Step 6 is performed by computing $\widehat{C} = \widehat{C}_0 + \widehat{C}_1 \bmod F(x)$ to obtain the complete multiplication. Thus, the architecture of Fig.1 for the hybrid digit-serial SPB multiplier requires $\lceil m/2d \rceil + 2$ clock cycles.

Algorithm 1: Traditional hybrid digit-serial SPB multiplication scheme [14]

Input: $A, B, F(x)$, $q = \lceil \frac{m}{d} \rceil$ and $v = \lfloor \frac{q}{2} \rfloor d$
Output: $\widehat{C} = x^{-v} A B \bmod F(x)$
1. $B' = B$, $B'' = B, C' = 0$, $C'' = 0$.
2. for i=0 to $\lfloor \frac{q}{2} \rfloor$
3. $B' = B' x^{-d} \bmod F(x)$.
4. $\widehat{C}_0 = \widehat{C}_0 + B' A_{\lfloor \frac{q}{2} \rfloor - 1 - i}$
5. $\widehat{C}_1 = \widehat{C}_1 + B'' A_{\lfloor \frac{q}{2} \rfloor + i}$
6. $B'' = B'' x^d \bmod F(x)$
7. endfor
8. $\widehat{C} = \widehat{C}_0 + \widehat{C}_1 \bmod F(x)$

3   Modified Hybrid Digit-Serial SPB Multiplier

In this section, we present a new way for deriving a hybrid digit-serial multiplication algorithm to achieve a higher speed computation compared to the best known digit-serial multiplier. The proposed architecture is suitable for implementing multicore platforms to explore ample parallelism computations.

The divide-and-conquer algorithm for high-precision multiplication was introduced by Karatsuba and Ofman. This algorithm can reduce the number of single-precision multiplications by reusing the intermediate partial products. By using recursively Karatsuba's algorithm, the multiplication of two m-digit polynomials have $O(m^{log_2 3})$ space complexity, while normal schoolbook multiplication requires $O(m^2)$ space complexity. For example, based on two-term Karatsuba's algorithm, $A = A_0 + x^{\frac{m}{2}} A_1$ and $B = B_0 + x^{\frac{m}{2}} B_1$ are two polynomials of degree m, where A_0, A_1, B_0, B_1 are $(m/2)$-bit polynomials. The product of A and B can be rewritten as

$$
\begin{aligned}
C = AB &= (A_0 + x^{\frac{m}{2}} A_1)(B_0 + x^{\frac{m}{2}} B_1) \\
&= A_0 B_0 + x^{\frac{m}{2}}(A_0 B_1 + A_1 B_0) + A_1 B_1 x^m \\
&= A_0 B_0(1 + x^{\frac{m}{2}}) + (A_0 + A_1)(B_0 + B_1)x^{\frac{m}{2}} + A_1 B_1(x^m + x^{\frac{m}{2}})
\end{aligned}
$$

$$
= C_0(1 + x^{\frac{m}{2}}) + C_1(x^m + x^{\frac{m}{2}}) + C_2 x^{\frac{m}{2}}, \tag{4}
$$

where

$$
C_0 = A_0 B_0, C_1 = A_1 B_1, C_2 = (A_0 + A_1)(B_0 + B_1) = A_2 B_2.
$$

We employ the Karatsuba scheme in (4) to compute the SPB multiplication. When $v = \frac{m}{2}$, the SPB product can obtain through the following formulation:

$$
\widehat{C} = x^{-\frac{m}{2}} AB \bmod F(x) = C_0(1 + x^{-\frac{m}{2}}) + C_1(1 + x^{\frac{m}{2}}) + C_2 \bmod F(x) \tag{5}
$$

$$
= x^{-\frac{m}{2}} C_0 R + C_1 R + C_2 \bmod F(x).
$$

where

$$
R = 1 + x^{\frac{m}{2}}.
$$

In (5), the product $\widehat{C}$ includes three partial products, such as $x^{-\frac{m}{2}} C_0 R \bmod F(x)$, $C_1 R \bmod F(x)$, and C_2. The multiplication includes three-type multiplications. $x^{-\frac{m}{2}} C_0 R$ is a multiplication with a shifted polynomial basis reduction; $C_1 R \bmod F(x)$ is a modular multiplication; and C_2 is schoolbook multiplication. For realizing a digit-serial multiplication structure, assume that d is the selected digit size, each subword B_i can be rewritten as $B_i = \sum_{j=0}^{\lceil \frac{m}{2d} \rceil - 1} B_{i,j} x^{jd}$, where $B_{i,j} = \sum_{l=0}^{d-1} b_{i,jd+l} x^l$. In the following, we discuss three block partial products.

For computing $x^{-\frac{m}{2}} C_0 R \bmod F(x)$, let we choose $n = \lfloor \frac{m}{2d} \rfloor$, then this term can be rewritten as

$$
\begin{aligned}
&x^{-\frac{m}{2}} C_0 R \bmod F(x) \\
&= x^{-(n-1)d}(A_0 R B_{0,0} + A_0 R B_{0,1} x^d + \cdots + A_0 R B_{0,n-1} x^{(n-1)d}) \bmod F(x)
\end{aligned}
$$

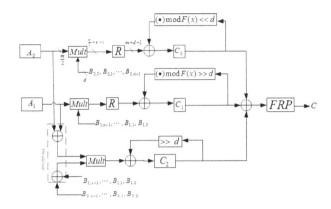

Fig. 2. Proposed digit-serial SPB multiplier over $GF(2^m)$

$$= A_0 R B_{0,0} x^{-(n-1)d} + A_0 R B_{0,1} x^{-(n-2)d} + \cdots + A_0 R B_{0,n-1} \bmod F(x) \quad (6)$$

For computing $C_1 R \bmod F(x)$, we have

$$A_1 B_1 R \bmod F(x) = A_1 R B_{1,0} + A_1 R B_{1,1} x^d + \cdots$$

$$+ A_1 R B_{1,n-1} x^{(n-1)d} \bmod F(x) \quad (7)$$

For computing C_2, we directly obtain

$$C_2 = A_2 B_2 = A_2 B_{2,0} + A_2 B_{2,1} x^d + \cdots + A_2 B_{2,n-1} x^{(n-1)d} \quad (8)$$

Based on (6)-(8), we can derive the hybrid digit-parallel multiplication scheme as stated in Algorithm 2.

Algorithm 2: Hybrid digit-serial SPB multiplication algorithm

Inputs: $A = A_0 + A_1 x^{\frac{m}{2}}$, $B = B_0 + B_1 x^{\frac{m}{2}}$, $R = 1 + x^{\frac{m}{2}}$, $n = \left\lceil \frac{m}{2d} \right\rceil$
Output: $\widehat{C} = x^{-\frac{m}{2}} AB : \bmod : F(x)$
1. $C_0 = 0$, $C_1 = 0$, $C_2 = 0$
2. for $j = n - 1$ to 0
3. $A_2 = A_1 + A_0$, $B_{2,j} = B_{1,j} + B_{0,j}$
4. $C_0 = (C_0 \bmod F(x)) x^{-d} + A_0 B_{0,j} R$
5. $C_1 = (C_1 \bmod F(x)) x^d + A_1 B_{1,n-1-j} R$
6. $C_2 = C_2 x^d + A_2 B_{2,j}$
7. endfor
8. $\widehat{C} = C_0 + C_1 + C_2 \bmod F(x)$

According to Algorithm 2, Fig.2 shows the proposed SPB multiplication architecture. In Fig. 2, The module labeled $Mult$ performs the multiplication of

A_i and $B_{i,j}$. The final reduction polynomial (FRP) module is carried out by $\widehat{C} = C_0 + C_1 + C_2 \mod F(x)$. The pre-addition module calculates $A_2 = A_1 + A_0$, $B_{2,j} = B_{1,j} + B_{0,j}$. The module labeled "($\bullet$)mod $F(x) \gg d$" calculates a reduction polynomial $F(x)$ after shift to right operation. The module labeled R performs a multiplication by $1 + x^{m/2}$ to produce $(m+d-1)$-bit results. Also, A_0 and A_1 are $(m/2)$-bit registers whereas C_0 and C_1 are $(m+d-1)$-bit registers, and C_2 is m-bit registers.

It is to note that (6)-(8) are $n(= \lceil \frac{m}{2d} \rceil)$-term partial products. Obtaining three results C_0, C_1, and C_2 require n clock cycles. Moreover, Fig. 2 needs one extra clock cycle for computing the FRP module to obtain the final multiplication. Thus, the total latency is $n+1 = \lceil \frac{m}{2d} \rceil + 1$ clock cycles. We explain the complexity of this algorithm in the next section.

Furthermore, we can also use multi-term Karatsuba algorithms to construct the hybrid digit-serial SPB multiplier. For example, we use three-term Karatsuba algorithm, two polynomials are partitioned as $A = A_0 + x^{m/3} A_1 + x^{2m/3} A_2$ and $B = B_0 + x^{m/3} B_1 + x^{2m/3} B_2$. We select $v = \frac{2m}{3}$, the SPB product of A and B can then be obtained as

$$\widehat{C} = x^{-\frac{2m}{3}} AB = (1 + x^{-\frac{m}{3}} + x^{-\frac{2m}{3}})C_0 + (1 + x^{-\frac{m}{3}} + x^{\frac{m}{3}})C_1$$

$$+(x^{\frac{m}{3}} + x^{\frac{2m}{3}} + x^m)C_2 + C_3 + x^{\frac{m}{3}}C_4 + x^{\frac{2m}{3}}C_5 \qquad (9)$$

$$= x^{-\frac{2m}{3}} C_0 R + x^{-\frac{m}{3}} C_1 R + x^{\frac{m}{3}} C_2 R + C_3 + x^{\frac{m}{3}} C_4 + x^{\frac{2m}{3}} C_5$$

where

$$C_0 = A_0 B_0, C_1 = A_1 B_1, C_2 = A_2 B_2,$$
$$C_3 = (A_0 + A_1)(B_0 + B_1), C_4 = (A_0 + A_2)(B_0 + B_2),$$
$$C_5 = (A_1 + A_2)(B_1 + B_2), R = 1 + x^{\frac{m}{3}} + x^{\frac{2m}{3}}.$$

Applying the structure of (9), this case SPB multiplier allows us to perform six-type parallelism sub-products. Moreover, each term can be computed independently by using digit-serial multiplication scheme. In general, based on k-term Karatsuba algorithm, the latency of the proposed architecture can be reduced from $\frac{m}{d} + 1$ into $\frac{m}{dk} + 1$ as compared with the best known digit-serial multipliers [13][14].

4   Performance Analysis

Table 1 lists the hardware components used by our proposed multiplier and the existing digit-serial multipliers [13][14]. From this table we can find that the proposed digit-serial multiplier has the latency of $\frac{m}{dn} + 1$ clock cycles, where n is the number term of Karatsuba algorithm, while MSD-first/LSD-first and hybrid digit-serial multipliers involve latencies of $\lceil \frac{m}{d} \rceil + 1$ and $\lceil \frac{m}{2n} \rceil + 2$ clock cycles, respectively. When we use two-term KA to develop the proposed multiplier,

Table 1. Comparison of two digit-serial multipliers over $GF(2^m)$

Multipliers	Basis	#XOR	#AND	#FF
Hybrid [14]	SPB	$d(2m+3k+1)+m-1$	$2md$	$4m+2d-2$
MSD-first 14]	SPB	$d(3m-2)-m+1$	$d(3m-2)-m+1$	$2m+d-1$
LSD-first[13]	PB	$(m+k)d+(k+1)(d-1)$	$(m+k)d+(k+1)(d-1)$	$2m+d+k$
Fig. 2	SPB	$\frac{md(n+1)}{2}+\frac{m}{n}$ $+d(\frac{n^2+n}{2}(1+k)+1)$	$\frac{md(n+1)}{2}$	$2m+(m+d)\frac{n^2+n-2}{2}$

Note: (1) n is the number of terms KA. (2) k is the Hamming weight of irreducible polynomial

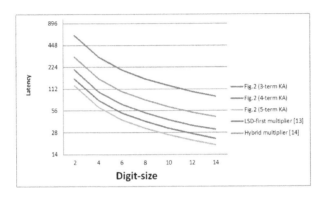

Fig. 3. Comparison of the latency for various digit-serial multipliers over $GF(2^{1223})$

our derived latency is the same of hybrid digit-serial multiplier [14]. Moreover, if we use increasing the number of term KA, the proposed architecture has lesser latency compared to the best known digit-serial multipliers. In $GF(2^{1223})$, Fig.3 shows the comparison of the latency for various digit-serial multipliers. In this figure, the proposed architecture for using 5-term KA can save about 59% and 78.6% latency complexity compared to hybrid multiplier[14] and LSD-first multiplier[13], respectively.

We select the field $GF(2^{1223})$ constructed by the trinomial $x^{1223}+x^{155}+1$ to estimate critical-path, area complexity, and area-delay product for various digit-serial multipliers. We have used the NanGate's Library Creator and the 45-nm FreePDK Base Kit from North Carolina State University (NCSU) [24] to synthesize the proposed and the corresponding existing digit-serial multipliers and obtained time and area complexities. Amongst all the existing digit-serial multipliers, the multiplier of [14] has the minimum time-complexity. But as shown in Table 2, the proposed multiplier using 3-term KA involves lesser area-delay product (ADP), higher bit-throughput (BT), and energy per output bit (EOB) compared with those of two existing multipliers [13][14] as the digit-size $d=8$.

Table 2. Comparison of various digit-serial multipliers over $GF(2^{1223})$ in the terms of latency, ADP, area, power, EOB, and BT for digit-size d=8

multipliers	Latency	area (μm^2)	ADP (μm^2)ns	Power($\mu W/GHz$)	BT	EOB(pJ)
Fig.2	52	93,764.5	1,023,908	160,421.2	23.52	6.82
Hybrid [14]	77	85875.71	1,388,610	163,265.4	15.88	10.28
LSD-first [13]	154	44,034.97	1,424,090.9	74,756.56	7.94	9.413

5   Conclusions

We have presented a novel hybrid digit-serial multiplier over $GF(2^m)$. The proposed digit-serial architecture for using n-term KA scheme has latency of $\lceil \frac{m}{dn} \rceil + 1$ clock cycles, which is less than the best of existing digit-serial architectures. For exploring the area-time trade-off for large field arithmetic architectures, we have used Karatsuba schemes to implement the digit-serial systolic multiplier over $GF(2^{1223})$. As the number of terms in the Karatsuba method increases, it provides significantly higher bit-throughput and less energy per output bit and area-delay product. Therefore, the proposed digit-serial multiplier using KA with different number of terms could be used to have the desired trade-off among speed, ADP/EOB, and BT of digit-wise multipliers. The analytical results provide a valuable reference for implementing pairing algorithm and elliptic curve digital signature algorithm (ECDSA) in resource-constrained embedded systems and smart phones. Moreover, our proposed architecture has the features of regularity, modularity, and concurrency, and are suitable for VLSI chip designs.

References

1. Digital Signature Standard, National Institute of Standards and Technology, 186-2 (January 2000)
2. IEEE Std 1363-2000, IEEE Standard Specifications for Public-Key Cryptography (January 2000)
3. Adikari, J., Dimitrov, V., Cintra, R.: A New algorithm for double scalar multiplication over Koblitz curves. In: IEEE Intl Symp. Circuits and Systems (ISCAS 2011), pp. 709–712 (2011)
4. Boneh, D., Franklin, M.K.: Identity-based encryption from the weil pairing. SIAM Journal on Computing 32(3), 586–615 (2003)
5. Aranha, D.F., Beuchat, J.-L., Detrey, J., Estibals, N.: Optimal eta pairing on supersingular genus-2 binary hyperelliptic curves. In: Dunkelman, O. (ed.) CT-RSA 2012. LNCS, vol. 7178, pp. 98–115. Springer, Heidelberg (2012)
6. Beuchat, J.-L., Detrey, J., Estibals, N., Okamoto, E., Rodriguez-Henriquez, F.: Fast architectures for the T_η pairing over smallcharacteristic supersingular elliptic curves. computers. IEEE Trans. Computers 60(2), 266–281 (2011)
7. Fan, H., Dai, Y.: Fast bit-parallel GF(2^n) multiplier for all trinomials. IEEE Trans. Computers 54(4), 485–490 (2005)
8. Fan, Hasan, M.: Fast bit parallel shifted polynomial basis multipliers in GF(2^n). IEEE Trans. Circuits and Systems I: Regular Papers 53(12), 2606–2615 (2006)

9. Fan, Hasan, M.: Subquadratic computational complexity schemes for extended binary field multiplication using optimal normal bases. IEEE Trans. Computers 56(10), 1435 (2007)

10. Park, S.-M., Chang, K.-Y.: Fast bit-parallel shifted polynomial basis multiplier using weakly dual basis over $GF(2^m)$. IEEE Trans. Very Large Scale Integration (VLSI) Systems 19(12), 2317–2321 (2011)

11. Negre, C.: Efficient parallel multiplier in shifted polynomial basis. Journal of Systems Architecture 53(2-3), 109–116 (2007)

12. Morales-Sandoval, M., Feregrino-Uribe, C., Kitsos, P.: Bit-serial and digit-serial $GF(2^m)$ Montgomery multipliers using linear feedback shift registers. IET Computers and Digital Techniques 5(2), 86–94 (2011)

13. Kumar, S., Wollinger, T., Paar, C.: Optimum digit serial $GF(2^m)$ multipliers for curve-based cryptography. IEEE Trans. Computers 55(10), 1306–1311 (2006)

14. Hariri, A., Reyhani-Masoleh, A.: Digit-Serial Structures for the Shifted Polynomial Basis Multiplication over Binary Extension Fields. In: von zur Gathen, J., Imaña, J.L., Koç, Ç.K. (eds.) WAIFI 2008. LNCS, vol. 5130, pp. 103–116. Springer, Heidelberg (2008)

15. Talapatra, S., Rahaman, H., Mathew, J.: Low complexity digit serial systolic Montgomery multipliers for special class of $GF(2^m)$. IEEE Trans. Very Large Scale Integration (VLSI) Systems 18(5), 487–852 (2010)

16. Hariri, A., Reyhani-Masoleh, A.: Digit-level semi-systolic and systolic structures for the shifted polynomial basis multiplication over binary extension fields. IEEE Trans. VLSI, 8 Transaction 19(11), 2125–2129 (2011)

17. Paar, C.: A new architecture for a parallel finite field multiplier with low complexity based on composite fields. IEEE Trans. Computers 45(7), 856–861 (1996)

18. Montgomery, P.: Five, six, and seven-term karatsuba-like formulae. IEEE Trans. Computers 54(3), 362–369 (2005)

19. Fan, H., Gu, M., Sun, J., Lam, K.-Y.: Obtaining more karatsuba-like formulae over the binary field. IET Information Security 6(1), 434–437 (2012)

20. Zhou, G., Michalik, H., Hinsenkamp, L.: Complexity analysis and efficient implementations of bit parallel finite field multipliers based on Karatsuba-Ofman algorithm on FPGAs. IEEE Trans. Very Large Scale Integr. 18(7), 1057–1066 (2010)

21. Juliano, D.P., Lima, B., Wang, Q.: A karatsuba-based algorithm for polynomial multiplication in chebyshev form. To appear in IEEE Trans. Computers (2013)

22. Fan, H., Sun, J., Gu, M., Lam, K.-Y.: Overlap-free karatsuba-ofman polynomial multiplication algorithms. In: 3rd International Conference on Design and Technology of Integrated Systems in Nanoscale Era, DTIS 2008, vol. 4(1), pp. 8–14 (2010)

23. Lee, C.-Y.: Digit-serial Gaussian normal basis multiplier over $GF(2^m)$ using Toeplitz matrix-approach. In: The 20th VLSI Design/CAD Symposium 2009, Hualien, Taiwan, August 4-7, pp. 1–4 (2009)

24. Nangate standard cell library, http://www.si2.org/openeda.si2.org/projects/nangatelib/

Pipeline Design of Bit-Parallel Gaussian Normal Basis Multiplier over GF(2^m)

Che Wun Chiou[1], Jim-Min Lin[2], Yu-Ku Li[1], Chiou-Yng Lee[3,*],
Tai-Pao Chuang[1], and Yun-Chi Yeh[4]

[1] Department of Computer Science and Information Engineering,
Chien Hsin University of Science and Technology, Jhong-Li 320, Taiwan
{cwchiou,m10113003,tpchuang}@uch.edu.tw
[2] Department of Information Engineering and Computer Science,
Feng Chia University, Taichung City 407, Taiwan
jimmy@fcu.edu.tw
[3] Department of Computer Information and Network Engineering,
Lunghwa University of Science and Technology, Taoyuan County 333, Taiwan
PP010@mail.lhu.edu.tw
[4] Department of Electronic Engineering,
Chien Hsin University of Science and Technology, Jhong-Li 320, Taiwan
yunchi@uch.edu.tw

Abstract. The finite field multiplication over GF(2^m) is the most important arithmetic operation for performing the elliptic curve cryptosystem which is very attractive in portable devices due to small key size. Design of finite field multiplier with low space complexity for elliptic curve cryptosystem is needed. The proposed bit-parallel GNB multiplier using pipeline XOR tree rather than XOR tree in traditional bit-parallel GNB multipliers. The proposed one can save about 99% number of both AND and XOR gates while comparing with existing bit-parallel GNB multipliers.

Keywords: Elliptic curve cryptosystem, finite field arithmeti, multiplier, public key cryptosystem, information security.

1 Introduction

Elliptic Curve Cryptosystem (ECC) [1,2] is attractive for information security of e-commerce. ECC depends on finite field arithmetic operations such as GF(p) and GF(2^m) [3]. GF(2^m) arithmetic operations include addition, multiplication, multiplicative inversion, and exponentiation. Addition in GF(2^m) is easily conducted using only XOR gates. Multiplicative inversion and exponentiation are time-consuming operations. Fortunately, they can be carried out by repeated multiplications. Accordingly, efficient finite field multiplication over GF(2^m) is crucial to cryptographic applications. The efficiency of finite field multiplication highly depends on the field element

---

* Corresponding author.

J.-S. Pan et al. (eds.), *Genetic and Evolutionary Computing*,
Advances in Intelligent Systems and Computing 238,
DOI: 10.1007/978-3-319-01796-9_40, © Springer International Publishing Switzerland 2014

representation. Three popular bases for representing elements in GF(2^m) are polynomial basis (PB) [3-14], dual basis (DB) [15-21], and normal basis (NB) [6, 21-36]. Each basis has its own advantages. PB multipliers are suitable for VLSI implementation due to their advantages of simplicity, regularity, and modularity and. DB multipliers require smaller area comparing with the other two bases multipliers. The NB multipliers can perform the squaring operation only by cyclically shifting its binary form. Hence, NB multipliers are very efficient in performing square operations of multiplicative inversion, squaring, and exponentiation operations. In 1986, Massey and Omura first developed NB multiplication algorithm [22]. Wang *et al.* [23] developed the VLSI architecture for Massey-Omura multiplier. However, the VLSI multiplier by Wang *et al.* [23] lacks modularity. Kwon [28] then proposed a systolic multiplier for an optimal normal basis of type-2 which is a special case of NB. Reyhani-Masoleh [24] developed a new non-systolic architecture for the Gaussian normal basis multiplication of type-t which is also a special case of NB. Lidl and Niederreiter [30] stated that there exists a corresponding normal basis for any positive integer m. Gaussian normal basis (GNB) is a special case of NB for which low space complexity feature is held. All positive integers, except for those are divisible by eight, have GNB [31]. Several information security standards exploit GNB, such as ANSI X9.62 [37], FIPS 186-2 [38] and IEEE Standard 1363-2000 [39]. Since the binary ECC is advantageous for hardware implementation of smart phones [40], the design of fast GNB multipliers with low hardware cost is desirable. Therefore, this study will present a simple means of designing a low-complexity bit-parallel GNB multiplier. The study presents a pipeline XOR-tree architecture for bit-parallel GNB multiplier with type t, while traditional XOR-tree is not pipeline architecture. The proposed GNB multiplier can drastically reduce numbers of both AND and XOR gates in existing bit-parallel GNB multipliers.

2   Preliminaries

The results of the NB multiplication are briefly reviewed in the following paragraphs. For more detail, readers can refer to [3].

Let $\left\{ \alpha^{2^0}, \alpha^{2^1}, \alpha^{2^2}, ..., \alpha^{2^{m-1}} \right\}$ be a normal basis of GF(2^m) for $\alpha \in$ GF(2^m). Each element $A, B \in$ GF(2^m) can be uniquely represented as

$$A = \sum_{i=0}^{m-1} a_i \alpha^{2^i},$$

$$B = \sum_{i=0}^{m-1} b_i \alpha^{2^i},$$

where $a_i, b_i \in \{0,1\}$ for i=0,1,2,..., m-1.

The following features hold for the normal basis.

(1) $A^2 = a_{m-1} \alpha^{2^0} + a_0 \alpha^{2^1} + a_1 \alpha^{2^2} + ... + a_{m-2} \alpha^{2^{m-1}}$,

(2) $(A+B)^2 = A^2 + B^2$.

Let m, t be positive integers such that mt+1 is a prime not dividing 2. Let γ be the primitive $(mt+1)^{th}$ root of unity in GF(2^{mt}). Then for any primitive t^{th} root of unity τ in GF(2^{mt+1}), the element $\alpha = \sum_{i=0}^{t-1} \gamma^{\tau^i}$ is called a Gauss period of type (m,t) over GF(2). It is well known that α is a normal element in GF(2^m). The $\left\{ \alpha^{2^0}, \alpha^{2^1}, \alpha^{2^2}, ..., \alpha^{2^{m-1}} \right\}$ is a normal basis for GF(2^m). The GNB of a type-t presented by the Gaussian period of type (m,t) has the following properties.

Property 1: $\alpha = \sum_{i=0}^{t-1} \gamma^{\tau^i}$,

Property 2: $\tau^t = 1 \mod mt+1$,

Property 3: $\gamma^{mt+1} = \gamma^{(mt+1) \mod (mt+1)} = 1$.

Based on the definition of Gauss period of type (m,t), Property 1 holds. Due to τ is the primitive t^{th} root of unity, Property 2 holds. The Property 3 also holds because γ is the $(mt+1)^{th}$ root of unity.

3   Traditional Bit-Parallel GNB Multiplier

Let the product C of A and B ($C = A \times B$) be represented as

$$C = \sum_{i=0}^{m-1} c_i \alpha^{2^i} ,$$

where finite field elements A,B, and C are represented in GNB of type-t. Let the GNB $\left\{ \alpha^{2^0}, \alpha^{2^1}, \alpha^{2^2}, ..., \alpha^{2^{m-1}} \right\}$ be Ψ . The basis Ψ can then be transformed to the extended polynomial basis $\Psi' = \left\{ \gamma^0, \gamma^1, ..., \gamma^{mt} \right\}$, where $\gamma^i = \gamma^{(2^j \tau^k \mod mt+1)}$ for $1 \le i \le mt$, $0 \le j \le m-1$, and $0 \le k \le t-1$. The elements A, B, and C in Ψ can be rewritten as $A', B',$ and C' in Ψ' , as follows.

$$A' = \sum_{i=0}^{mt} a_i' \gamma^i , \quad B' = \sum_{i=0}^{mt} b_i' \gamma^i , \quad C' = \sum_{i=0}^{mt} c_0' \gamma^i ,$$

where $a_0' = b_0' = 0$, $a_i' = a_j, b_i' = b_j$ for $\gamma^i = \gamma^{(2^j \tau^k \mod mt+1)}$ ($1 \le i \le mt$, $0 \le j \le m-1$, and $0 \le k \le t-1$).

$$C' = \sum_{i=0}^{mt} c_i' \gamma^i = A' \times B'$$

$$= \begin{cases} \left(a_0' b_0' + a_1' b_{mt}' + a_2' b_{mt-1}' + ... + a_{mt}' b_1' \right) + \\ \left(a_0' b_1' + a_1' b_0' + a_2' b_{mt}' + ... + a_{mt}' b_2' \right) \gamma + \\ .... + \\ \left(a_0' b_{mt}' + a_1' b_{mt-1}' + a_2' b_{mt-2}' + ... + a_{mt}' b_0' \right) \gamma^{mt} \end{cases} \quad (1)$$

where $c_i' = \sum_{j=0}^{mt} a_j' b_k'$, $k = mt+1+i-j \mod mt+1$.

The traditional bit-parallel GNB multiplier using XOR tree is shown in Fig.1 and the detail circuit of XOR tree with m=4 and t=3 is drawn in Fig.2.

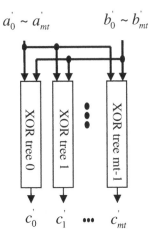

Fig. 1. Traditional bit-parallel GNB multiplier

$a_0'b_0'\ a_1'b_{12}'\ a_2'b_{11}'\ a_3'b_{10}'\ a_4'b_9'\ a_5'b_8'\ a_6'b_7'\ a_7'b_6'\ a_8'b_5'\ a_9'b_4'\ a_{10}'b_3'\ a_{11}'b_2'\ a_{12}'b_1'$

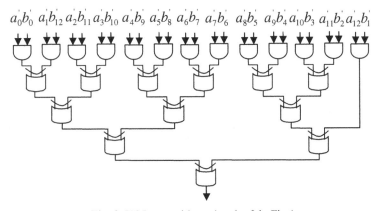

Fig. 2. XOR tree with m=4 and t=3 in Fig.1

4   The Proposed Pipeline Bit-Parallel GNB Multiplier

The XOR tree in existing bit-parallel GNB multipliers is a pure combinational circuit. Thus, the bit-parallel GNB multiplier with type t in [35] requires mt units each with a XOR-tree for computing one result bit c_i. If such a XOR-tree belongs to pipeline architecture, only one unit is required. Therefore, the space complexity will be reduced drastically. The proposed pipeline XOR tree is shown in Fig.3 for m=4 and t=3. In Fig.3, 2mt 1-bit latches are added. The proposed pipeline bit-parallel GNB multiplier is depicted in Fig.4. The latency of the proposed bit-parallel GNB multiplier is $\lceil \log_2(mt+1) \rceil + mt$. Table 1 lists the comparison results of various bit-parallel GNB multipliers. Table 2 shows the comparing results for NIST suggested values. The proposed multiplier can save at least 99% space complexity in both AND and XOR gates and requires 420% space complexity in 1-bit latches while comparing with the one by Azarderakhsh and Reyhani-Masoleh in [36].

$a_0'b_0'$ $a_1'b_{12}'$ $a_2'b_{11}'$ $a_3'b_{10}'$ $a_4'b_9'$ $a_5'b_8'$ $a_6'b_7'$ $a_7'b_6'$ $a_8'b_5'$ $a_9'b_4'$ $a_{10}'b_3'$ $a_{11}'b_2'$ $a_{12}'b_1'$

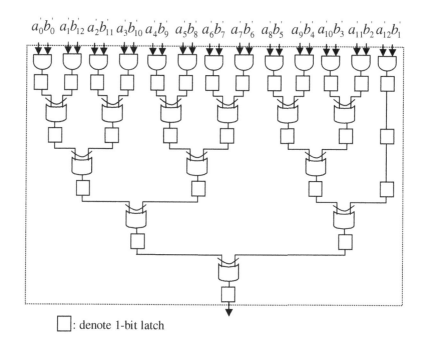

☐: denote 1-bit latch

Fig. 3. The pipeline XOR tree with m=4 and t=3

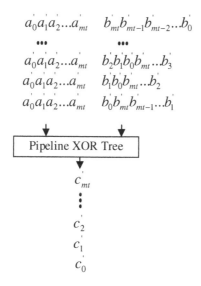

$a_0'a_1'a_2'...a_{mt}'$ $b_{mt}'b_{mt-1}'b_{mt-2}'...b_0'$

$\bullet\bullet\bullet$ $\bullet\bullet\bullet$

$a_0'a_1'a_2'...a_{mt}'$ $b_2'b_1'b_0'b_{mt}'...b_3'$

$a_0'a_1'a_2'...a_{mt}'$ $b_1'b_0'b_{mt}'...b_2'$

$a_0'a_1'a_2'...a_{mt}'$ $b_0'b_{mt}'b_{mt-1}'...b_1'$

| Pipeline XOR Tree |

c_{mt}'

$\vdots$

c_2'

c_1'

c_0'

Fig. 4. The proposed pipeline bit-parallel GNB multiplier

Table 1. Comparison results of various bit-parallel GNB multipliers with type t

Multipliers	Chuang et al. [35]	Azarderakhsh & Reyhani-Masoleh [36]	The proposed multiplier
#AND gates	$(mt+1)^2$	m^2	mt
#XOR gates	$(mt)^2+2mt$	$\leq \dfrac{t+4}{4}m(m-1)$	mt
#1-bit latches	0	2m	2mt
Critical path delay	$T_A + \left(\lceil \log_2(mt+1)\rceil+1\right)T_X$	$T_A + \left(\lceil \log_2 t\rceil+\lceil \log_2 m\rceil\right)T_X$	$T_A + \lceil \log_2 mt\rceil(T_X+T_L)$

Note: #: Number, T_A: propagation delay of 2-input AND gate, T_X: propagation delay of 2-input XOR gate, T_L: propagation delay of 1-bit Latch.

Table 2. Complexity comparisons for NIST suggested m values

Multipliers			Chuang et al. [35]	Azarderakhsh & Reyhani-Masoleh [36]	The proposed multiplier
m	t	No. of Gates			
163	4	#AND	426409	26569	652
		#XOR	426408	52812	652
		#1-bit Latch	0	326	1304
233	2	#AND	218089	54289	466
		#XOR	218088	81084	466
		#1-bit Latch	0	466	932
283	6	#AND	2886601	80089	1698
		#XOR	2886600	199515	1698
		#1-bit Latch	0	566	3396
409	4	#AND	2679769	167281	1636
		#XOR	2679768	333744	1636
		#1-bit Latch	0	818	3272
571	10	#AND	32615521	326041	5710
		#XOR	32615520	1139145	5710
		#1-bit Latch	0	1142	11420

5   Conclusions

The XOR tree in traditional bit-parallel GNB multiplier is a pure combinational circuit. Thus, huge number of XOR trees is required in existing bit-parallel GNB multipliers. This study presents a pipeline XOR tree. Only one XOR tree is employed in the proposed bit-parallel GNB multiplier. Computations for result bits can be calculated in pipeline method. The proposed pipeline bit-parallel GNB multiplier can save about 99% space complexity in both AND and XOR gates of XOR trees as compared to existing bit-parallel GNB multiplier.

References

1. Miller, V.S.: Use of Elliptic Curves in Cryptography. In: Williams, H.C. (ed.) CRYPTO 1985. LNCS, vol. 218, pp. 417–426. Springer, Heidelberg (1986)
2. Koblitz, N.: Elliptic curve cryptosystems. Mathematics of Computation 48, 203–209 (1987)
3. Savaş, E., Koç, Ç.K.: Finite field arithmetic for cryptography. IEEE Circuits and Systems Magazine 10(2), 40–56 (2010)
4. Bartee, T.C., Schneider, D.J.: Computation with finite fields. Information and Computing 6, 79–98 (1963)
5. Mastrovito, E.D.: VLSI architectures for multiplication over finite field GF(2^m), Applied Algebra, Algebraic Algorithms, and Error-Correcting Codes. In: Mora, T. (ed.) Proc. Sixth Int'l Conf., AAECC-6, Rome, pp. 297–309 (1988)
6. Koç, Ç.K., Sunar, B.: Low-complexity bit-parallel canonical and normal basis multipliers for a class of finite fields. IEEE Trans. Computers 47(3), 353–356 (1998)
7. Itoh, T., Tsujii, S.: Structure of parallel multipliers for a class of fields GF(2^m). Information and Computation 83, 21–40 (1989)
8. Lee, C.Y., Lu, E.H., Lee, J.Y.: Bit-parallel systolic multipliers for GF(2^m) fields defined by all-one and equally-spaced polynomials. IEEE Trans. Computers 50(5), 385–393 (2001)
9. Paar, C., Fleischmann, P., Roelse, P.: Efficient multiplier architectures for Galois Fields GF(2^{4n}). IEEE Trans. Computers 47(2), 162–170 (1998)
10. Wu, H.: Bit-parallel finite field multiplier and squarer using polynomial basis. IEEE Trans. Computers 51(7), 750–758 (2002)
11. Fan, H., Hasan, M.A.: A new approach to subquadratic space complexity parallel multipliers for extended binary fields. IEEE Trans. Computers 56(2), 224–233 (2007)
12. Huang, W.-T., Chang, C.H., Chiou, C.W., Tan, S.-Y.: Non-XOR Approach for Low-Cost Bit-Parallel Polynomial Basis Multiplier over $GF(2^m)$. IET Information Security 5(3), 152–162 (2011)
13. Chiou, C.W., Lin, J.M., Lee, C.-Y., Ma, C.-T.: Low complexity systolic Mastrovito multiplier over GF(2^m). European Journal of Scientific Research 65(4), 534–545 (2011)
14. Chiou, C.W., Lee, C.-Y., Yeh, Y.-C.: Multiplexer implementation of low-complexity polynomial basis multiplier in GF(2^m) using all one polynomial. Information Processing Letters 111(3.1), 1044–1047 (2011)
15. Wu, H., Hasan, M., Blake, A., New, I.F.: low-complexity bit-parallel finite field multipliers using weakly dual bases. IEEE Trans. Computers 47(11), 1223–1234 (1998)

16. Fenn, S.T.J., Benaissa, M., Taylor, D.: GF(2^m) multiplication and division over the dual basis. IEEE Trans. Computers 45(3), 319–327 (1996)
17. Wang, M., Blake, I.F.: Bit serial multiplication in finite fields. SIAM J. Disc. Math. 3(1), 140–148 (1990)
18. Berlekamp, E.R.: Bit-serial Reed-Solomon encoder. IEEE Trans. Inform. Theory IT-28, 869–874 (1982)
19. Wang, J.-H., Chang, H.W., Chiou, C.W., Liang, W.-Y.: Low-complexity design of bit-parallel dual basis multiplier over GF(2^m). IET Information Security 6(4), 324–328 (2012)
20. Hua, Y.Y., Lin, J.-M., Chiou, C.W., Lee, C.-Y., Liu, Y.H.: A novel digit-serial dual basis systolic Karatsuba Multiplier over GF(2^m). Journal of Computers 23(2), 80–94 (2012)
21. Lee, C.Y., Chiou, C.W.: Efficient design of low-complexity bit-parallel systolic Hankel multipliers to implement multiplication in normal and dual bases of GF(2^m). IEICE Transactions on Fundamentals of Electronics, Communications and Computer Science E88-A(11), 3169–3179 (2005)
22. Massey, J.L., Omura, J.K.: Computational method and apparatus for finite field arithmetic, U.S. Patent Number 4587627 (1986)
23. Wang, C.C., Truong, T.K., Shao, H.M., Deutsch, L.J., Omura, J.K., Reed, I.S.: VLSI architectures for computing multiplications and inverses in GF(2^m). IEEE Trans. Computers C-34(8), 709–717 (1985)
24. Reyhani-Masoleh, A.: Efficient algorithms and architectures for field multiplication using Gaussian normal bases. IEEE Trans. Computers 55(1), 34–47 (2006)
25. Chiou, C.W., Lee, C.Y.: Multiplexer-based double-exponentiation for normal basis of GF (2^m). Computers & Security 24(1), 83–86 (2005)
26. Agnew, G.B., Mullin, R.C., Onyszchuk, I.M., Vanstone, S.A.: An implementation for a fast public-key cryptosystem. Journal of Cryptology 3, 63–79 (1991)
27. Hasan, M.A., Wang, M.Z., Bhargava, V.K.: A modified Massey-Omura parallel multiplier for a class of finite fields. IEEE Trans. Computers 42(10), 1278–1280 (1993)
28. Kwon, S.: A low complexity and a low latency bit parallel systolic multiplier over GF(2^m) using an optimal normal basis of type II. In: Proc. of the 16th IEEE Symposium on Computer Arithmetic, Santiago de Compostela, Spain, pp. 196–202 (2003)
29. Fan, H., Hasan, M.A.: Subquadratic computational complexity schemes for extended binary field multiplication using optimal normal bases. IEEE Trans. Computers 56(10), 1435–1437 (2007)
30. Lidl, R., Niederreiter, H.: Introduction to Finite Fields and Their Applications. Cambridge Univ. Press, New York (1994)
31. Ash, D.W., Blake, I.F., Vanstone, S.A.: Low complexity normal bases. Discrete Applied Math. 25, 191–210 (1989)
32. Chiou, C.W., Chuang, T.-P., Lin, S.-S., Lee, C.-Y., Lin, J.-M., Yeh, Y.-C.: Palindromic-like representation for Gaussian normal basis multiplier over GF(2^m) with odd type-t. IET Information Security 6(4), 318–323 (2012)
33. Chiou, C.W., Chang, H.W., Liang, W.-Y., Lee, C.-Y., Lin, J.-M., Yeh, Y.-C.: Low-complexity Gaussian normal basis multiplier over *GF(2^m)*. IET Information Security 6(4), 310–317 (2012)
34. Lee, C.-Y., Chiou, C.W.: Scalable Gaussian normal basis multipliers over GF(2^m) using Hankel matrix-vector representation. Journal of Signal Processing Systems for Signal Image and Video Technology 69(2), 197–211 (2012)
35. Chuang, T.-P., Chiou, C.W., Lin, S.-S., Lee, C.-Y.: Fault-tolerant Gaussian normal basis multiplier over GF(2^m). IET Information Security 6(3), 157–170 (2012)

36. Azarderakhsh, R., Reyhani-Masoleh, A.: Low-complexity multiplier architectures for single and hybrid-double multiplications in Gaussian normal bases. IEEE Trans. Computers 62(4), 744–757 (2013)
37. ANSI X.962: Public key cryptography for the financial services industry: The Elliptic Curve Digital Signature Algorithm (ECDSA), Am. Nat'l Standards Inst. (1999)
38. FIPS 186-2: Digital Signature Standard (DSS), Federal Information Processing Standards Publication 186-2, Nat'l Inst. Of Standards and Technology (2000)
39. IEEE Standard 1363-2000: IEEE standard specifications for public-key cryptography (2000)
40. Vanstone, S.A.: Next generation security for wireless: Elliptic curve cryptography. Computers and Security 22(5), 412–415 (2003)

Symbolic Analysis Using Floating Pathological Elements

Hung-Yu Wang[1,*], Shung-Hyung Chang[2], Nan-Hui Chiang[1], and Quoc-Minh Nguyen[1]

[1] Department of Electronic Engineering, National Kaohsiung University of Applied Sciences,
Kaohsiung, Taiwan, R.O.C.
[2] Department of Microelectronics Engineering, National Kaohsiung Marine University,
Kaohsiung, Taiwan, R.O.C.
hywang@kuas.edu.tw

Abstract. The nullor-mirror pathological elements have been found useful in solving circuit analysis and design problems. They are further used to ideally represent various popular analog signal-processing properties that involve differential or multiple single-ended signals by utilizing the concept of floating mirror elements. For applying nodal analysis to the circuit containing such active devices with differential or multiple single-ended signals, we propose an efficient systematic analytical technique which directly performs symbolic analysis on the simpler RLC-nullor-floating mirror representations of circuits rather than their RLC-two-terminal pathological element-based counterparts. It releases the limitation of recently proposed symbolic analysis approaches and use simpler models which may be conductive to achieving high-performance symbolic nodal analysis. The feasibility and validity of the proposed method are demonstrated by practical circuit examples.

Keywords: Pathological element, RLC-nullor-floating mirror network, nodal analysis (NA), symbolic analysis.

1   Introduction

Tellegen first implicitly introduced a minimum-sized set of ideal elements in 1954 with which any linear and nonlinear driving-point impedance or transfer characteristic can be synthesized, and it is given the name of nullor by Carlin ten years later [1]-[2]. Since then, nullor has been accepted within the network theory community as a basic network element and it has been proven to be a very valuable network analysis, synthesis and design tool [3]-[5]. Recently, the mirror elements with grounded reference node have been extended to include the floating mirror elements [6]-[7]. With such extensions, a nullator and a norator can be represented in terms of a floating voltage mirror and a floating current mirror, respectively [8]. Moreover, the floating mirror elements are used to derive the pathological sections which ideally represent various popular analog signal-processing properties that involve differential

---

[*] Corresponding author.

J.-S. Pan et al. (eds.), *Genetic and Evolutionary Computing*,
Advances in Intelligent Systems and Computing 238,
DOI: 10.1007/978-3-319-01796-9_41, © Springer International Publishing Switzerland 2014

or multiple single-ended signals; like conversion between differential and single-ended voltages, differential voltage conveying and current replication. In virtue of the combination of various pathological sections, they are used to model many popular active devices, such as the differential voltage current conveyor (DVCC), the fully differential second generation current conveyor (FDCCII) and two-output CCII and ICCII family [7]-[9]. Since the better flexibility and simpler configuration of modeling active devices using the combination of nullor mirror elements, the pathological representations of many active devices have been reported in the literature [10]-[11].

To make good use of the simpler RLC-pathological element-based networks, a systematic analytical technique which performs direct analysis for nullor-mirror equivalent networks has been reported, and its improvement on increasing the efficiency of symbolic analysis was demonstrated [10, 12]. Furthermore, the parasitic resistors and capacitors of input-output terminals can be included in the model of an active device to perform symbolic analysis. However, the applied extent of the proposed approaches in [10] is limited because floating pathological element-based active devices cannot be included into the formulation process. In this paper, we intend to present a more comprehensive analytical technique for performing symbolic analysis on the RLC-nullor-floating mirror representation circuits. The proposed approach is applied to two practical circuit examples and some discussions about simplified pathological models of active devices are given.

2   Pathological Sections and Equivalents

The symbols and definitions of the nullor and mirror elements are shown in Table 1. The voltage mirror and current mirror in Tables 1(c) and (d) are lossless two-port network elements used to represent ideal voltage and current reversing property, respectively. Each of the voltage mirror and the current mirror symbols has a reference node, which is set to ground [6]. The symbols and definitions of the floating mirror elements are shown in Tables 1(e)-(f). It can be found that the mirror elements in Tables 1(c) and (d) can be regarded as the special cases of the floating mirror elements in Tables 1(e) and (f), respectively.

In the recent time, the floating mirror elements have been used to represent various popular analog signal-processing properties in concise forms. As shown in Tables 2(a) and (b), two floating voltage mirrors with a common reference node are used to represent the pathological differential voltage cell and differential voltage conveying cell. Also, two floating current mirrors with a common reference node can be used to express the pathological current replication cells as shown in Tables 2(c) and (d) [7]-[8]. From the current replication cell in Table 2(d), it can be found that if the I_x is the input current, the I_y and I_w can be recognized as the output currents of a pathological CM and a norator, respectively. Besides, the constructions in Tables 2(e) and (g) are equivalent to a nullator and a norator, respectively. The constructions in Tables 2(f) and (h) are respectively equivalent to a pathological VM and a CM.

Table 1. Symbols and definitions of nullor and mirror elements

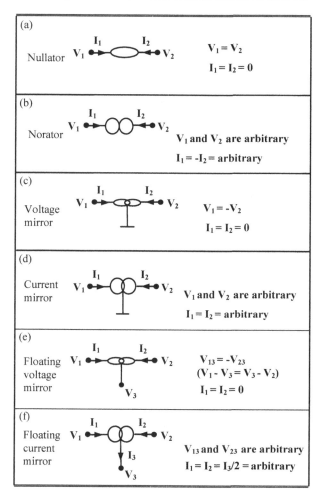

The above representation sections are concise comparing with the pathological representations using nullor and mirror elements with grounded reference nodes [13] because they comprise only two floating mirrors without the need of other elements such as resistors. With less internal nodes, the representations in Table 2 are helpful to enhancing the efficiency of symbolic circuit analysis. The sections were used to model some active devices with differential or multiple single-ended features, as shown in Fig. 1. In Fig, 1, it must be noted that the pathological representations of BOCCII and FDCCII are different from the ones in [8] by removing the dummy nullor elements. They can be used for symbolic nodal analysis using the proposed method in Section 3.

382   H.-Y. Wang et al.

Table 2. Some pathological sections and their characteristics

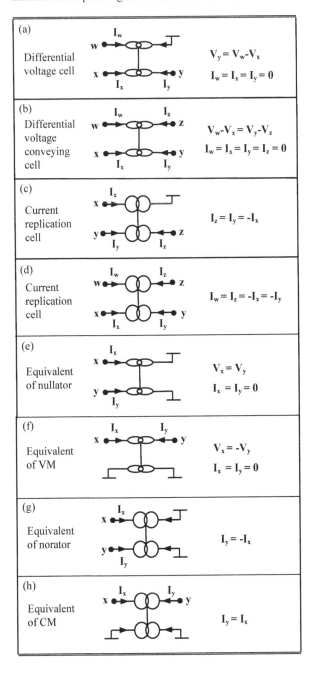

3   Symbolic NA of RLC-Nullor-Floating Mirror Network

The steps for symbolic NA of RLC-nullor-floating mirror networks are proposed below. They involve the symbolic nodal analysis of an arbitrary interconnection of RLC-nullor-floating mirror networks and independent current sources in an (N+1) node network (excluding the reference nodes between two mirror elements). The analytic process can be summarized as follows:

Step 1: Select a ground node and label all other nodes from 1 to N except the common reference nodes between two mirror elements (since we do not particularly wish to know the reference node voltages for the floating CMs and the reference node voltages for the floating VMs can be obtained from other node voltages). According to the element properties in Tables 1(b), (d), (f), and 2(c)-(e) denote the currents flow through every norator, current mirror and floating current mirror. It is known that no current flows through the nullators, voltage mirrors and floating voltage mirrors of the network. Also, remember that the sections in Tables 2(f), (g), (h) and (i) are equivalent to a nullator, a VM, a norator and a CM, respectively.

Step 2: Write the nodal admittance equations for every node except the reference nodes between two mirror elements. Express the nodal admittance equations in matrix form:

$$I = Y_{N \times N} V \qquad (1)$$

$I = \{I_1, I_2,....., I_N\}'$, where the ith component I_i is defined as the sum of the currents flowing into the ith node from the independent current sources, norators, current mirrors or floating current mirrors. $Y_{N \times N}$ is the passive nodal admittance matrix. Furthermore, V is the unknown column vector $\{V_1, V_2,..., V_N\}'$ of node voltages.

Step 3: For the floating VM-based sections with three or four ungrounded nodes, i.e., the differential voltage cell in Table 2(a) and the differential voltage conveying cell in Table 2(b), write their relations of nodal voltages and add these equations to (1). Hence the combined equations can be expressed in matrix form $I = BV$. It is clear that each cell adds an additional equation to the combined equations.

Step 4: For a nullator that is connected between the nodes p and q, for example, add the elements of column q to the elements of column p and delete column q of B. If q is the ground node, simply delete column q of B. The number of columns of the B matrix is thereby reduced by one. This operation is based on the voltages at the two terminals of a nullator with respect to the ground node being identical. Thus we can omit one unknown voltage variable in the V column vector.

Step 5: For a VM that is connected between the nodes r and s, for example, subtract the elements of column s from the elements of column r and delete column s of B. If s is the ground node, simply delete column r of B. The number of columns of the B matrix is thereby reduced by one. This operation is based on the voltage reversing property of a voltage mirror in Table 1(c).

Step 6: For a norator that is connected between nodes l and m, for example, add the equation in row m to the equation in row l and delete row m of the nodal equations. This involves adding I_m to I_l and the mth row of B to the lth row of B. If m is the ground node, simply delete row 1 of the nodal equations. The number of columns of the **B** matrix is thereby reduced by one. This operation is based on the current property of a norator in Table 1(b) and the number of equations being enough to solve the unknown independent node voltages after deleting the mentioned equations.

Step 7: For a CM that is connected between the nodes n and o, for example, subtract the equation in row o from the equation in row n and delete row o of the nodal equations. This involves subtracting I_o from I_n and the oth row of **B** from the nth row of **B**. If o is the ground node, simply delete row n of the nodal equations. This operation is based on the current reversing property of the terminals of a current mirror in Table 1(d) and the number of equations is enough to solve the unknown node voltages.

Step 8: For the pathological current replication cell with three ungrounded nodes x, y and z in Table 2(c), add the equation in row y to the equation in row x; subtract the equation in row z from the equation in row y and delete row z of the nodal equations. Similar manipulation process can be applied to the cells in Tables 2(d) and (e). This operation is based on the norator behavior between terminals x and y and the CM behavior between terminals y and z. One nodal equation is removed because the number of equations is enough to solve the unknown node voltages.

Step 9: The preceding steps 4-8 incur the reduction of the order of the system of equations derived in step 3. Solve the simplified combined equations for the unknown node voltages in V.

Following the above proposed procedure, it can be observed that each nullator (or VM) will incur the number of columns of the B matrix is reduced by one and each norator (or CM) will incur the number of rows of the B matrix is reduced by one. In Tables 2(a) and (b), it should be noted that each pathological section will result in the number of rows of the B matrix is increased by one and they cannot reduce the number of columns of the B. In addition, in Tables 2(c)-(d), each pathological section will result in the number of rows of the B matrix is reduced by one. Let us review the pathological models of active devices in Fig. 1. It can be found that the active devices models in Fig. 1 will incur the identical number of rows and columns of the B matrix. Hence the resulting matrix will still be square and hence solvable. So the pathological models in Fig. 1 can be used for the proposed symbolic NA approach.

4   Application Example

To demonstrate the feasibility of the proposed models and the symbolic NA method in Section 3, practical circuit example is illustrated. For comparison with previous techniques, we adopt the same circuit example as the second example in [13]. It is a DDCC+-based design in Fig. 2(a) of [14]. Fig. 2 shows its pathological representation with the equivalence of the inputted voltage source [15]. As shown in Fig. 2, the DDCC+ is modeled by using the pathological differential voltage conveying cell in Table 2(b) and a CM. Based on steps 1-3 of the proposed symbolic NA, we can write the (6×5) nodal admittance equations in matrix form $\mathbf{I} = \mathbf{BV}$ as

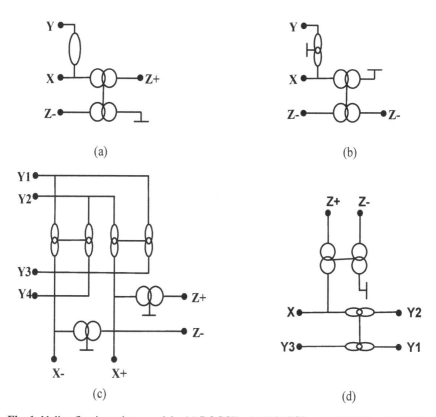

Fig. 1. Nullor-floating mirror models. (a) BOCCII   (b) DOICCII-- (c) FDCCII   (d) DDCC

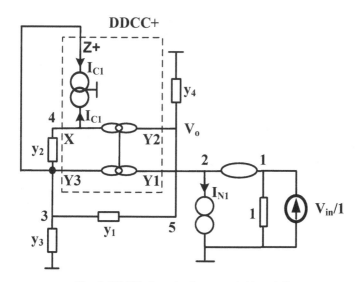

Fig. 2. DDCC+-based voltage-mode biquad filter

$$
\begin{bmatrix} V_{in}/1 \\ I_{N1} \\ I_{C1} \\ I_{C1} \\ 0 \\ 0 \end{bmatrix} = \begin{bmatrix} 1 & 0 & 0 & 0 & 0 \\ 0 & 0 & 0 & 0 & 0 \\ 0 & 0 & y_1+y_2+y_3 & -y_2 & -y_1 \\ 0 & 0 & -y_2 & y_2 & 0 \\ 0 & 0 & -y_1 & 0 & y_1+y_4 \\ 0 & -1 & -1 & 1 & 1 \end{bmatrix} \begin{bmatrix} V_1 \\ V_2 \\ V_3 \\ V_4 \\ V_5 \end{bmatrix} \tag{2}
$$

After applying the steps 4-8 of the proposed method leads to the derivation of the following (4×4) combined equations as (3). The output transfer function can be obtained and given in (4). The analyzed results are in accordance with the circuit functions in [14]. So the feasibility of the proposed method is verified. In additional, in this example the built symbolic matrix of dimension is (6×5) and it is reduced to (4×4) with 10 nonzero coefficients as in (3). Compared to the symbolic NA approach using nullor-mirror model of DDCC+ (as shown in Fig. 5 of [13]), the built symbolic matrix of dimension is (9×9) and it is reduced to (5×5) with 13 nonzero coefficients. So it reveals the improvement of this proposed approach.

$$
\begin{bmatrix} V_{in}/1 \\ 0 \\ 0 \\ 0 \end{bmatrix} = \begin{bmatrix} 1 & 0 & 0 & 0 \\ 0 & y_1+2y_2+y_3 & -2y_2 & -y_1 \\ 0 & -y_1 & 0 & y_1+y_4 \\ -1 & -1 & 1 & 1 \end{bmatrix} \begin{bmatrix} V_{1,2} \\ V_3 \\ V_4 \\ V_5 \end{bmatrix} \tag{3}
$$

$$
\frac{V_5}{V_{in}} = \frac{V_o}{V_{in}} = \frac{2y_1y_2}{2y_1y_2 + y_1y_4 + y_1y_3 + y_3y_4} \tag{4}
$$

5   Conclusion

In this paper, an efficient symbolic NA technique using of any pathological elements is proposed. Some pathological representations of active devices are modified by omitting the dummy pathological elements to perform symbolic NA. The simplified models with less internal nodes are conducive to enhancing the analysis efficiency. Practical application example is given which demonstrate the workability of the proposed symbolic NA approach.

Acknowledgments. This work has been supported by the National Science Council of the Republic of China (Grant Nos NSC 101-2221-E-151-074).

References

1. Tellegen, B.D.H.: La recherche pour una série complète d'éléments de circuit ideauxnonlinéaires. Rendiconti del Seminario Matematico e Fisico di Milano 25, 134–144 (1954)
2. Carlin, H.J.: Singular network elements. IEEE Trans. Circuit Theory CT-11, 67–72 (1964)
3. Schmid, H.: Approximating the universal active element. IEEE Trans. Circuits Syst. II 47(11), 1160–1169 (2000)
4. Carlosena, A., Moschytz, G.S.: Nullators and norators in voltage tocurrent mode transformations. Int. J. Circuit Theory Applicat. 21(4), 421–424 (1993)
5. Kumar, P., Senani, R.: Bibliography on nullors and their applications in circuit analysis, synthesis and design. Anal. Integr. Circuits Signal Process. 33(1), 65–76 (2002)
6. Soliman, A.M., Saad, R.A.: The voltage mirror-current mirror pairs as a universal element. Int. J. Circuit Theory Appl. 38(8), 787–795 (2010)
7. Saad, R.A., Soliman, A.M.: On the systematic synthesis of CCII-based floating simulators. Int. J. Circuit Theory Appl. 38(9), 935–967 (2010)
8. Saad, R.A., Soliman, A.M.: A new approach for using the pathological mirror elements in the ideal representation of active devices. Int. J. Circuit Theory Appl. 38(2), 148–178 (2010)
9. Soliman, A.M.: Pathological representation of the two-output CCII and ICCII family and application. Int. J. Circuit Theory Appl. 39(6), 589–606 (2011)
10. Sanchez-Lopez, C., Fernandez, F.V., Tlelo-Cuautle, E., Tan, S.X.D.: Pathological element-based active device models and their application to symbolic analysis. IEEE Trans. Circuits Syst. I: Reg. Papers 58(6), 1382–1395 (2011)
11. Tlelo-Cuautle, E., Sanchez-Lopez, C., Moro-Frias, D.: Symbolic analysis of (MO)(I) CCI(II)(III)-based analog circuits. Int. J. Circuit Theory Appl. 38(6), 649–659 (2010)
12. Wang, H.Y., Huang, W.C., Chiang, N.H.: Symbolic nodal analysis of circuits using pathological elements. IEEE Trans. Circuits Syst. II 57(11), 874–877 (2010)
13. Huang, W.C., Wang, H.Y., Cheng, P.S., Lin, Y.C.: Nullor equivalents of active devices for symbolic circuit analysis. Circuits Syst. Signal Process. 31, 865–875 (2012)
14. Ibrahim, M.A., Kuntman, H., Cicekoglu, O.: New second-order low-pass, high-pass and band-pass filters employing minimum number of active and passive elements. In: Proc. Int. Symp. Signal Circuits Syst., pp. 557–560 (2003)
15. Svoboda, J.A.: A linear active network analysis program suitable for a class project. IEEE Trans. Education E-27(1), 21–25 (1984)

Vehicle Driving Video Sharing and Search Framework Based on GPS Data

Chuan-Yen Chiang[1], Shyan-Ming Yuan[1], Shian-Bo Yang[1], Guo-Heng Luo[1],
and Yen-Lin Chen[2,*]

[1] Institute of Computer Science and Engineering, National Chiao Tung University,
Hsinchu, Taiwan
[2] Department of Computer Science and Information Engineering,
National Taipei University of Technology, Taipei, Taiwan
ylchen@ntut.edu.tw

Abstract. The driving dispute is a critical problem for drivers when the car accident is happened. The drivers usually install the car video recorder in their car to recode their driving images for many years ago. If a car accident is happened, the driver can provide a driving video file as an evidence to claim that they did not do any dangerous driving, and protect themself. However, in some situations, drivers may not in the car accident when driving the car, but they have the same requirement for a video record, because they want to use those video files to find out the crime in hit-and-run accident. Due to an evolution of social network, many people were post the required messages to find the driving videos recorded in some specific times and locations. In fact, such kind of messages can be beneficial for many people to solve the hit-and-run accident by using social networks and driving videos. The goal of this paper is to develop a framework which can provide a platform for users to upload their driving videos, and allow other users can search video by a given specific time, date and location from the frameworks' database. This framework can also provide the application on mobile devices for user to recorder their driving videos, and this application can upload their driving video into the frameworks' database automatically.

Keywords: driving recorder, video sharing, video searching, map matching algorithm, framework, GPS.

1   Introduction

Traffic accidents have become a major cause of death. Most traffic accidents are caused by driver carelessness on traffic conditions. Therefore, detecting on-road traffic conditions for assisting drivers is a promising approach to help drivers take safe driving precautions. Accordingly, many studies have developed valuable driver assistance techniques for detecting and recognizing on-road traffic objects, including

---

* Corresponding author.

J.-S. Pan et al. (eds.), *Genetic and Evolutionary Computing*,
Advances in Intelligent Systems and Computing 238,
DOI: 10.1007/978-3-319-01796-9_42, © Springer International Publishing Switzerland 2014

lane markings, vehicles, raindrops, and other obstacles. The objects are recognized from images of road environments outside the host car [1]. These driver assistance techniques are mostly developed based on camera-assisted systems, and can help drivers perceive possible dangers on the road or automatically control the apparatus of the vehicle (e.g., headlights and windshield wipers).

In the recent studies, most of them are focus on the issue of "how to prevent a traffic accident". However, if the accident is happened, how we can do to determine the attribution of the responsibility for this traffic accident. The drivers usually installed the driving recorder in their vehicle for protecting themselves, when the accident is happened. However, in most traffic accident those usually need more evidences to determine and recover the scene of the accident including road surveillance system, other driving videos and etc… to find out who is responsible for this accident. According to those situations, most drivers will post the message to the social network, such as Facebook to help for get more video files as evidences.

For now, the smartphone devices have become very widespread and powerful and the programmers can create many useful mobile applications, make smartphones become more useful. There are many driving video recorder application here in the smartphone, it can change your smartphone into the driving video recorder and store in the smartphone. This study proposes a vehicle driving video share and search framework based on GPS data. This framework can let people to share their car driving video and GPS track into database and search those driving video and GPS track data from database. The proposed framework includes the mobile application, video and GPS track database, users can ubiquitously upload and search the video records from the database. The mobile application is implemented for Android platform, it can help user to recorder their driving sense and GPS log when user active the program. After finished the recorder, it will upload the driving video file and GPS log file into database automatically when user allow the mobile program have a permission to do it. Otherwise the user did not allow the mobile program to upload file automatically, they also can use upload module to upload their video file and GPS log file into the database. Experimental results demonstrate that the proposed vehicle driving video share and search framework based on GPS data includes mobile application screenshot, user search and share screenshot.

2   Related Work

In some domain, a smartphone application can be used on analyze sport route, popular travel routes, hot place and traffic condition based on users' daily GPS tracker data [2, 3]. It also can be used on fleet management domain, a smartphone application can send out a GPS location data to the central management system. The central management system can track every member based on those GPS [4, 5]. However, most way to use those smartphone applications is to track, manager and analysis their daily information and discovers any useful data.

The GPS signal error has well known issue and there are many researchers provide some algorithm to solve this problem. In this study, we focus on develop the platform that allow user to share, search the driving video based on given latitude and longitude. However, the platform receive the geographical data from user may have

GPS error so we need to analyze and correct the data before stored it. The map-matching algorithm is proposed to solve GPS error based on road network. In previous research [6, 7, 8] their focus on provide more efficiency algorithm based on Kalman Filter and Hidden Markov Model to calibration latitude, longitude into road network. In other research [9] it uses probability of space and time probability to improve map-matching algorithm quality when the GPS signal is weak.

The proposed framework is to help user share their driving scene video file and GPS track data, and other people can search those video file based on given specific latitude and longitude data. To search those video file based on specific geographic data (latitude and longitude), so the data we stored must be a geographic data, PostGIS [10] is an open-source geographic database, and it provide useful method for spatial and geographic search and it also need road network dataset to provide those methods. The road network dataset is that contain the location of nodes and length of the links, the map-matching algorithm is use those dataset to analysis and correction each GPS signal, and corrected it into possible right position. The dataset that used in proposed system is from Institute of Transportation [11].

3   System Architecture

This study presents a video share, search platform and a mobile application to help users record their driver scene. Fig. 1 depicts the overview architecture of the proposed system.

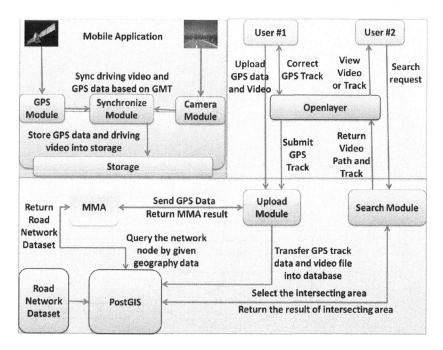

Fig. 1. System architecture of the proposed system

The overview of the proposed system can be divided into three major parts, Mobile Application, User Upload Module and User Search Module. The mobile application part shown in top-left of Fig. 1 combines camera, GPS sensor and synchronize module. The camera module intercepts the driving scene and store transfer into the synchronize module. The GPS sensor module captures GPS data in every second and passed transfer into the synchronize module. The synchronize module will sync each record frame and GPS signal based on GMT (Greenwich Mean Time) and then store into storage.

There are two user behaviors in the proposed system, upload and search. The top-right part of Fig. 1 shows the users' upload and search behaviors. When user #1 wants to share his/her driving video and GPS data through upload module, there are two steps to follow; the first step is submitting the GPS track data and the second step is uploading the video file. In the Submit GPS Track step, the Upload Module will use MMA (Map-Matching Algorithm) to relocate each GPS signal from GPS track data, and then show the result to the user #1 to confirm by the OpenLayer [12], which is an open-source JavaScript library which can display and render maps on web page. After confirming the GPS track data, the corrected GPS track data will be stored into PostGIS database, and then the system will notify the user to upload his/her video file to complete the upload procedure. When user #2 wants to search video file in specific time and location, first, the user should send specific location to the Search Module, and the search module will query the data from PostGIS based on given location, and then return the search result to the user. In the following section, we will describe the

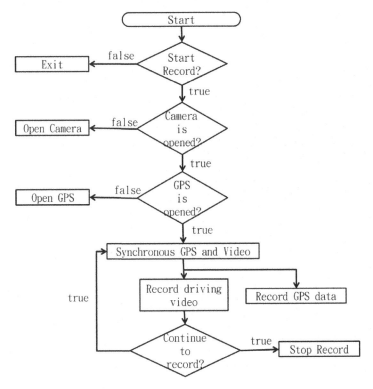

Fig. 2. The workflow of the mobile application

workflow of each proposed module includes mobile application, user upload and user share module in detail.

3.1   Mobile Application

As illustrated in the workflow of the proposed mobile application in Fig. 2, when the program is started, we first check the record state, if users want to record the driving scene; otherwise the program will not be started. When the user starts to record the driving video, then the program will check the camera state, to see if it is opened or not, and then ensure the GPS sensor has been opened. When both camera and GPS sensor are opened, and the user starts to record the driving video, the record procedure will be activate, store video file and GPS data into smartphones' memory. When record procedure is started, it will sync GPS data and driving video file based on GMT (Greenwich Mean Time). The mobile application also provides upload function to upload the stored driving video and GPS data into database.

3.2   User Upload Module

The workflow of the User Upload Module is shown in Fig. 3. It can be divided into five different parts: Client, OpenLayer [12], Server, MMA Model (Map-Matching Algorithm) and DB (Database). In the upload procedure, when Server receives the GPS track data from Client, it will transfer those data into MMA model to relocate

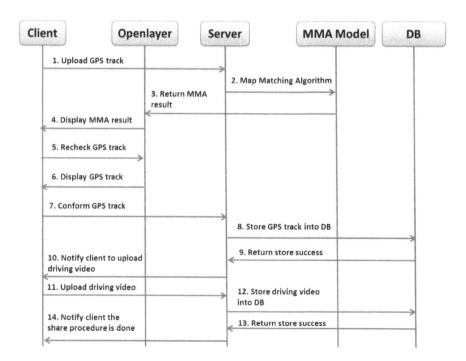

Fig. 3. Workflow of the user upload module

GPS signal position based on the road network dataset, and then return result and display relocated GPS signal track by OpenLayer on the website. The user has to confirm the relocated GPS result. After confirm the data, the server will notify user to submit driving video file related to the previous GPS track file. When the server received the GPS data and video file, it will store them into database and then show a successful message to the user.

3.3   User Search Function

The workflow of the user search module is show in Fig. 4. The web page will display the global map that can allow user to select the interest area. The selected area will be transferred into search module and it will query the video data and GPS track from database based on that selected area. And it will then return results list that contains video file and GPS track file into the web-server and display those results to user on web page. If the data on the list has been selected by user, the web-server will display the detail GPS tracks and show the driving video to user on web page.

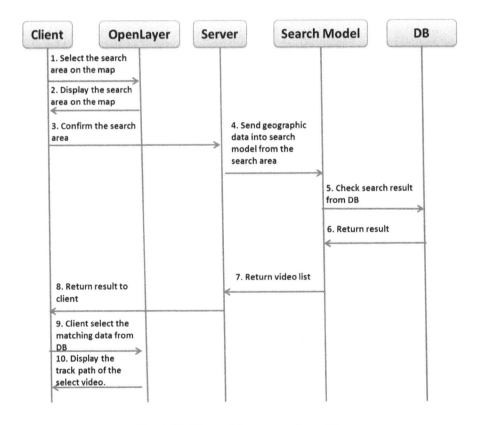

Fig. 4. Workflow of the user search module

4   Results of Proposed System

In the experimental results section, first we show the user interface of the proposed mobile application and then the user interface of the search module will be introduced, finally, the operate step of the proposed upload module will describe in detail.

The mobile application user interface (UI) is shown in Fig. 5. The Fig. 5(a) is the main UI of the proposed mobile application and Fig. 5(b) is a demonstration of notify checking GPS sensor and notify user to open it. When GPS sensor and camera sensor is opened as shown in Fig. 5(c) and it will sync GPS signal and video frame based on GMT every second. The Fig. 5(d) is a demonstration of record procedure and it will store GPS signal data and video file into the mobiles' storage.

The user interface of the upload module is shown in Fig. 6 and it includes four steps to help user to upload their GPS track file and video file. The first step as shown in the Fig. 6(a) is upload GPS track data section it notify user to upload their GPS track file and transfer track file to MMA, and then the MMA will return relocate GPS signal dataset and display those data on the map, and the screenshot of display GPS track data on the map is shown in Fig. 6(b). After confirmed the GPS track data by user, the server will notify user to upload the video file that is related to previous upload GPS track data and the screenshot of upload video web page is shown in Fig. 6(c). When user uploaded their driving video, the web-server will notify user the upload procedure is complete as a screenshot is shown in Fig. 6(d).

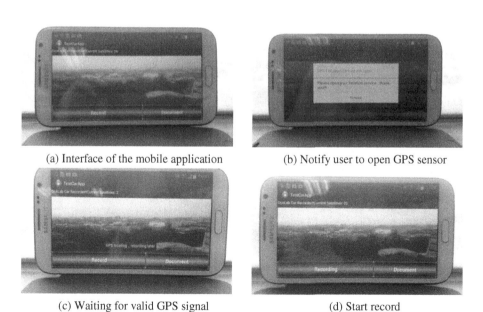

(a) Interface of the mobile application   (b) Notify user to open GPS sensor

(c) Waiting for valid GPS signal   (d) Start record

Fig. 5. The user interface of the mobile application

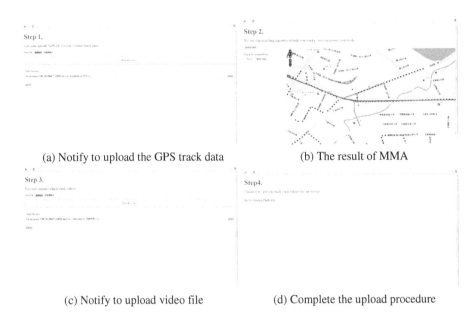

(a) Notify to upload the GPS track data   (b) The result of MMA

(c) Notify to upload video file   (d) Complete the upload procedure

Fig. 6. The user interface of the upload module

The user interface of the search module is shown in Fig. 7 and the Fig. 7(a) is a screenshot of select interest area on the map. When the area has been selected by user and then it will transfer into search module to query the video file and GPS track data based on select areas' geographic data. The search result of given area will show in the web page as a screenshot of Fig. 7(b) and the user can select the result to play video file and display GPS track file on the map.

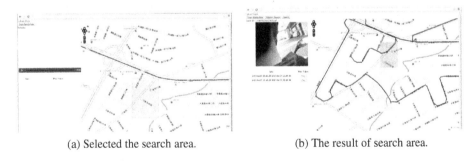

(a) Selected the search area.   (b) The result of search area.

Fig. 7. The user interface of the search module

5   Conclusion and Future Work

This study proposed the framework that contains mobile application, upload and search module. The mobile application can help user to record and store their driving sense in the mobile phone storage and those driving video can use to protect them if the car accident is happened. In the other hand the user can also share their driving video to other people through upload module that provide by proposed framework. The user can provide specific time and location to the search module of the proposed framework and it can search video file based on given data and then the user can select the result to watch the driving video and GPS track data on the map in web page. In the future, this framework not only use to search video which is used to solve the hit-and-run accident but also can use to find the most popular track which mean the popular track may cause traffic jam. In the other hand, the driving video can help user to know well the road environment that never been to.

Acknowledgment. This paper was supported by the Chunghwa Telecom Co., Ltd, under grant No. TL-102-G107.

References

1. Masaki, I. (ed.): Vision-based Vehicle Guidance. Springer, New York (1992)
2. Zheng, Y., Wang, L., Zhang, R., Xie, X., Ma, W.-Y.: GeoLife: Managing and Understanding Your Past Life over Maps. In: Proceeding of the Ninth International Conference on Mobile Data Management, pp. 211–212 (2008)
3. Lin, D.-H., Jeng, J.-T., Chuang, C.-C., Tao, C.-W.: Intelligent Video Car Recorder systems. In: Proceeding of International Conference on System Science and Engineering (ICSSE), pp. 7–12 (2012)
4. Chu, S.-T., Yeh, C.-C., Huang, C.-L.: A Cloud-Based Trajectory Index Scheme. In: Proceeding of International Conference on e-Business Engineering, pp. 602–607 (2009)
5. Almomani, I.M., Alkhalil, N.Y., Ahmad, E.M., Jodeh, R.M.: Ubiquitous GPS vehicle tracking and management system. In: Proceeding of Applied Electrical Engineering and Computing Technologies, pp. 1–6 (2011)
6. Yang, D., Cai, B., Yuan, Y.: An improved map-matching algorithm used in vehicle navigation system. In: Proceeding of Intelligent Transportation Systems, vol. 2, pp. 1246–1250 (2003)
7. Marchal, F., Hackney, J., Axhausen, K.W.: Efficient map-matching of large GPS data sets Tests on a speed monitoring experiment in Zurich. Journal of the Transportation Research Board 1935, 93–100 (2005)
8. Newson, P., Krumm, J.: Hidden Markov map matching through noise and sparseness. In: Proceeding of International Conference on Advances in Geographic Information Systems, pp. 336–343 (2009)
9. Lou, Y., Zhang, C., Zheng, Y., Xie, X., Wang, W., Huang, Y.: Map-Matching for Low-Sampling-Rate GPS Trajectories. In: Proceeding of International Conference on Advances in Geographic Information Systems, pp. 352–361 (2009)
10. PostGIS, http://postgis.net/
11. Institute of Transportation, http://www.iot.gov.tw/mp.asp?mp=1
12. OpenLayer, http://openlayers.org/

Personalized Cloud Storage System: A Combination of LDAP Distributed File System

Chen-Ting Hsu, Guo-Heng Luo, and Shyan-Ming Yuan

Institute of Computer Science and Engineering, National Chiao-Tung University, Taiwan
{flame861025,lasifu}@gmail.com, smyuan@cs.nctu.edu.tw

Abstract. "Cloud computing" gradually flourish, a wide range of distributed storage systems are increasingly diverse, Like of Gluster, Ceph, Lustre, as well as Hadoop, etc.. In this paper, we propose a personal cloud storage system Integrated with pNFS, it can be accessed in parallel for scalable performance. Besides, data backup and failover mechanism are designed. We expect that the function of the proposed system, not only can improve performance, but also can improve the user's privacy in order to provide better service quality in cloud storage.

Keywords: cloud storage, personalization, pNFS, LDAP.

1 Introduction

In recent years, the cloud technology industry is booming, Google is well-known representative of the cloud enterprise, the company provides many cloud application services. Users can access over the network at the various application services as well as information about individuals. Addition to providing computing services, cloud also need a powerful storage, therefore cloud storage technology development as future cloud technology is very important.

When it comes to cloud storage technology, can not fail to mention the Distributed File System, Distributed File System plays an important role in the large-scale application platform and cloud computing environments. It allows system administrators to simplify user access and manage files distributed across a network of. For users, files scattered across multiple servers like the same place on a network, users access files without knowing the actual location of the file.

Common distributed file system, such as Gluster File System[1], Hadoop Distributed File System[2], Ceph[3] and Lustre[4], etc. Like Gluster File System has a good performance and spatial scalability, but it must be the beginning it was decided the entire cluster system architecture, and can not be amplified space online. Ceph has no Single Point Of Failure (SPOF) shortcomings, because its algorithm is more complex, however, lead to the implementation complexity is high. Hadoop like Namenode single node failure problems, makes failure Namenode, can not recover automatically.

In response to the different distributed file system have different advantages and disadvantages, this paper is expected to create a personal cloud storage system that

J.-S. Pan et al. (eds.), *Genetic and Evolutionary Computing,*
Advances in Intelligent Systems and Computing 238,
DOI: 10.1007/978-3-319-01796-9_43, © Springer International Publishing Switzerland 2014

allows users to online storage and backup. In response to the different DFS have different advantages and disadvantages, this paper is expected to create a personal cloud storage system that allows users to online storage and backup. The system can, in accordance with the amount of disk space required by the user, as well as performance or backup requirements to service users. When the needs of the users increase or decrease (for example, increase the disk space and improve performance), the systems are also transparent to help the user to adjust the disk space and performance.

The system with other systems significant difference is this: with the personal use of units and parallel access. Require personal reason is because if they are companies, for example, to the department or person as the access unit, different people held files or messages will become no privacy, Any units or departments each other access to the contents of each file. Through personalized to separate the blocks of the user, and can be set according to the personal needs the services they need.

The parallel access part, is hoping to improve the speed of file access through the new NFS4.1 the protocol. Through separate files in different storage to read the file in parallel to enhance the speed. So that the system is able to obtain some advantage in the performance comparison with the previously mentioned DFS .And through the backup function allows the system to enhance better failover.

2   Related Work

In order to structure parallel access system. In this paper, we study the mature and stable NFS, that NFS the latest NFSv4.1 protocol provides new features that can be accessed in parallel.

PNFS is NFSv4.1 version of the protocol after the new function, the system architecture is shown in Fig. 1, pNFS Server plays the role of the Metadata Server, Client finished request and control operation with pNFS Server, subsequent could access directly through the parallel with the datanodes access data. This can break through the bottleneck of the previous NFS single metadata server, can significantly enhance the access performance [5-12].

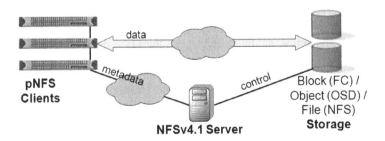

Fig. 1. Pnfs Archetecture

NFSv4.1 and pNFS by RFC5661 (NFSv4.1), RFC5662 (NFSv4.1 XDR), RFC5663 (pNFS Block / Volume Layout),

RFC5664 (Object-Based pNFS Operations) description, these standards have been 2010.01 release. At present, the pNFS commercial and open source products than less, and realize not mature. Linux kernel version 2.6.30 added pNFS support, the latest version is 2.6.38, pNFS is still in the experimental stage, including Server and Client.

And the other hands, the database is select the LDAP (Lightweight Directory Access Protocol), is an Internet protocol that email and other programs use to look up information from a server[13].

LDAP has some advantages, such as using a standard protocol (LDAP is a standard protocol proposed by the IETF). And easy to organize data management, and access to data access permissions (LDAP can classify the users, and provide different access rights). And coded security features to protect users' privacy, as well as provide fast and advanced search features, such as LDAP search keywords for each node, a simple method allows the user to be found in the the the desired information.

LDAP standardized and easy to use, you do not have to create an application or service-specific directory system, you can get benefits. Therefore hope that through a combination of LDAP database and the integration PNFS parallel access function, to develop a simple personal cloud storage system.

3   The Proposed Scheme

In this paper, we propose a personal cloud storage system.

The system has good spatial scalability can increase the storage space under the circumstances, and can automatically attached to a balance, and the data can be obtained well, there is no single point of failure, and then have the ease of management, the implementation of low complexity and support personal characteristics of the virtual file system.

System service process: a user first on demand registered with the server, and then by the server according to user required to allocate space and services. Then the system is responsible for deciding files location and storing the backup file policy, for example, the backup policy may be stored two or more storage by according to the needs . Server according to different user segments stored, in order to enhance the privacy features.

User information such as account passwords, and files assigned to nodes are stored in LDAP database. Part of the system for the single point of failure proposed different solutions approaches, such as the replacement of the backup node, and the slave of the metadata server to do a backup rescue.

The following section describes the architecture of the system contains a system consisting of three components (datanode, metadata, server, client), As well as detailed user operating procedures and the interactions between the server, The other part is the set of contingency strategies for the different single point of failure.

4   System Design

In this section, proposed system architecture and system operation flows will be described.

4.1   System Architecture

To provide personalized storage services, the system using the Cloud Node to manage information on an individual basis, once the user is assigned to a certain Cloud Node, its follow-up data access by the Primary Cloud node is responsible for service, and Backup Cloud Node is mainly responsible for the backup.

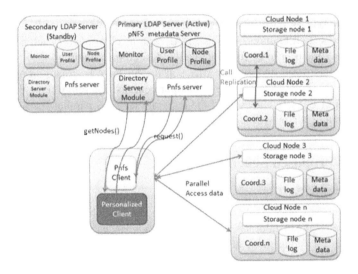

Fig. 2. System Architecture

The system will provide each user different storage space, that is, users could not see any other's data. The proposed system architecture is presented as Fig. 2. The system consists of three parts: LDAP Server, Cloud Node, and Client.

1. LDAP Server

In addition to the standard LDAP Server, it also contains the User profile Database and Self-developed Directory Server Module.

(a) User profile database.

User profile will contain the user can use Disk Space quota , Performance Policy, Backup Policy, And so on to describe the user's access and backup needs.

(b) *Node profile database*

Node profile database will be stored in each Cloud Node of information, including the ID, IP, Capacity, Free capacity, Allocated primary user number and so on.

(c) Directory Server Module

Directory Server Module will be responsible for deciding User Cloud Node by which to service and management, and is responsible for Cloud computing Node Balancing.

(d) monitor

Monitor is responsible for monitoring the the pNFS server send metadata to storage on the way if there is a problem occurs, and do troubleshooting.

(e) pNFS Server

Responsible for receiving files sent by the client, and do the stripe processing, after that the layout message back to the client, and then let the client can do with the storage parallel processing.

2. Cloud Node.

Cloud Node consists of Storage Node, Metadata and Coordinator.

(a) Storage Node

Storage Node will provide file access services, and through the action of the Coordinator for file backup.

(b) Metadata

The Metadata database will keep (User list) and the Cloud Node List (contains the Primary Nodes and Backup Nodes), as well as a list of files (File list).

(c) Coordinator

Client write files to the storage, file backup to the backup node, as well as the strain responsible for node failure handling.

3. Client

Client contains pNFS client and Personalized client.

(a) pNFS client:

The pNFS Server is responsible for the file submitted to the server side to do the stripe, and to obtain the file layout.

(b) Personalized client:

Personalized client is responsible for communicating with the Directory Server Module to get file location and personalized information, and then file parallel read or write.

4.2   User Registration

First, client will send registration message to primary LDAP server. The message contains information such as user account, password, capacity, performance, and backup requirement.

After primary LDAP server received the registration message, the system will perform several steps:

1. Primary LDAP server allocates cloud nodes for user and save information to user profile and node profile databases.
2. Primary LDAP server notifies cloud nodes to establish the volumes for the user with the backup information.
3. Storage nodes start the user volumes and coordinators starts backup services for newly created volumes.
4. Primary LDAP server returns success message and assigned cloud nodes' information to the client.

4.3   Write and Backup Operations

1. According to the assigned cloud nodes' information, the personalized client will automatically mount these nodes through PNFS Client. After requirement checking, any permitted I/O operations will be automatically passed to PNFS client for real I/O.
2. After client written or updated data to the storage node, the backup message will be sent to the corresponding coordinator as shown in Fig. 3.

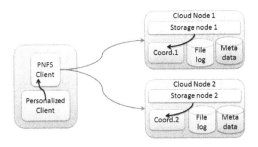

Fig. 3. Client writing or updating data to the storage nodes

3. The data will be copied to nodes 4 and 7. (assuming that the backup node for storage 1 are storage node 4 and storage node 7) And then after the completion of the backup, the complete message will be returned to coordinator 1 as shown in Fig. 4)

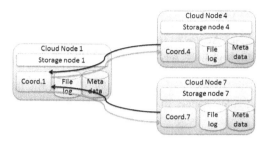

Fig. 4. Backing up data

4. After coordinator finished the backup operation, it will notify client that the write operation is complete as shown in Fig. 5.

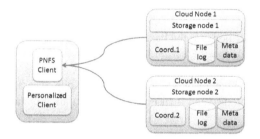

Fig. 5. Notifying a client with write completion message

4.4   Read Operations

1. Like write operations, the personalized client will automatically mount these nodes through PNFS Client.
2. Client perform parallel read from Node1 and   Node2 as shown in Fig. 6

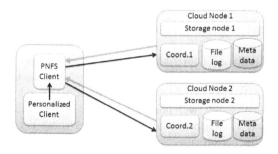

Fig. 6. Parallel read operations from Storage Node 1 and Node 2

4.5   Datanode Failover

Case 1 Client Could Not Access to the Point of Failure

Be able to reply at a preset time, and there is no client to access, not treated simply confirm their role failback for any change.

If not then continue to use the original role, to confirm the status of the file through the file log; role change, discard the original role and delete unnecessary information.

Case 2 Client Access to the Point of Failure

Server must for users to obtain new alternative node, and carried File Replication.

According to the failed node to distinguish between:

Major Cloud Node failure

From the backup list, select Node alternative of Major Node (and notify the MDS is assigned to the backup node), and then update the metadata backup list. The failure node recovery transformed into the backup node.

Backup Cloud Node failure

The backup when a Backup Node Fail, (Major Node message returned MDS to obtain new Backup Node), and then notify the other Node updated backup Metadata (List).

Whole list failure

The Client Node once again to MDS request for new Cloud Node List.

4.6   Metadata Server Failover

When the temporary failure of the primary Metadata Server, a backup of the Metadata Server then converted into a primary Metadata Server, such as after the primary Metadata Server restart, the transformation into a slave Metadata Server.

5   Experiment and Results

In this paper, we designed a preliminary experiment to test the performance of proposed system. Linux command dd[14] was employed as test tool. Write and read block size was set to 4K for sequential access. The tested systems are Gluster, HDFS, and the personal cloud system (with pNFS).

As shown in Table 1, Gluster (Distrib) write performance is about 21 to 25MB/s, Gluster (Replica) write performance is about 7 to 11MB / s, HDFS write performance is about 14 to 19MB/s, and write performance of proposed system is about 9 to 11MB/s.

Table 1. Command Dd Write Performance (Block Size=4k) Unit:Mb/S

File size	32 MB	64 MB	128 MB	256 MB	512 MB
Gluster (Distrib)	21.2	23.7	24.9	25.3	25.7
Gluster (Replica)	7.4	10.2	11.3	10.6	10.8
HDFS	14.4	17.7	16.8	19.3	17.5
Personal Cloud storage	9.4	9.6	10.3	9.2	10.1

As shown in Table 2, Gluster (Distrib) read performance is about 44 to 49MB /s , Gluster (Replica) read performance is about 45 to 49MB / s, HDFS read performance is about 38 to 43MB / s , and read performance of our system is about 19 to 22MB / s.

Table 2. Command Dd Read Performance (Block Size=4k) Unit:Mb/S

File size	32 MB	64 MB	128 MB	256 MB	512 MB
Gluster (Distrib)	48.4	45.5	47.2	51.8	44.1
Gluster (Replica)	46.7	45.6	46.5	48.3	46.3
HDFS	41.2	39.7	42.7	38.3	41.8
Personal Cloud storage	20.2	21.5	19.7	19.9	21.6

6   Conclusion

Cloud storage convenience brought us a great advantage, so that we can get the desired files on different devices, but not limited to the size of the installed capacity. However, the services provided and the performance of different distributed file system is not necessarily the same. To provide more convenient services and better performance is an important goal of the development of cloud storage.

In this paper, the proposed system focuses in Personalization. Provide personalized mechanism can provide higher for privacy and personalized service, for example, suppose a company uses a distributed storage system. But without personal user settings, it will lead to full disclosure of all data among various departments. One department can easily watch other department's data, thus, department privacy was not preserved.

On the other hand, departments or employees may have important files. They need better backup solution to ensure those files are safe. Personalized cloud storage can achieve needs of different users. The same situation can also be applied to the family, to provide personalized storage spaces, family members can easily save their file and privacies are preserved.

According to the above experimental data, although the deployment on VM impact the performance, which caused performance is not ideal. Parallelization leads to better performance in read performance significantly. In future, more experiments will be performed to find out the reasons for poor write performance. The system will be improved for better parallelization write performance.

Acknowledgement. This research was supported by Chunghwa Telecom Co., Ltd, under grant No. TL-102-G107.

References

1. GlusterFS, http://www.gluster.org/
2. Hadoop, http://hadoop.apache.org/
3. Sage, A., Weil Scott, A., Brandt Ethan, L., Miller Darrell, D.E.: Ceph: A Scalable, High-Performance Distributed File System. In: OSDI 2006: 7th USENIX Symposium on Operating Systems Design and Implementation (2006)
4. Lustre, http://www.lustre.org/
5. Hildebrand, D., Honeyman, P.: pNFS and Linux: Working Towards a Heterogeneous Future, CITI Technical Report 06-06
6. Hildebrand, D., Honeyman, P.: Direct-pNFS: Scalable, transparent, and versatile access to parallel file systems, CITI Technical Report 07-02
7. Yu, W., Drokin, O., Vetter, J.S.: Design, Implementation, and Evaluation of Transparent pNFS on Lustre. In: IEEE International Symposium on Parallel & Distributed Processing, IPDPS 2009 (2009)
8. Yu, W., Vetter, J.S.: Initial Characterization of Parallel NFS Implementations. In: Conference: Symposium on Parallel and Distributed Processing
9. PNFS, http://www.pnfs.com/
10. NFS 4.1 Protocol, http://tools.ietf.org/wg/nfsv4/
11. pNFS'_install_information, http://blog.csdn.net/liuben/article/details/6554866
12. Hildebrand, D., Eshel, M., Haskin, R.: Deploying pNFS across the WAN:First Steps in HPC Grid Computing
13. LDAP_Libraries, http://www.novell.com/developer/ndk/ldap_libraries_for_c.html
14. Linux_/_Unix_Command:_dd, http://linux.about.com/od/commands/l/blcmdl1_dd.htm

Author Index